Lecture Notes in Computer Science   1984

Edited by G. Goos, J. Hartmanis and J. van Leeuwen

T0216291

Springer

Berlin
Heidelberg
New York
Barcelona
Hong Kong
London
Milan
Paris
Tokyo

Joe Marks (Ed.)

Graph Drawing

8th International Symposium, GD 2000
Colonial Williamsburg, VA, USA, September 20-23, 2000
Proceedings

Springer

Series Editors

Gerhard Goos, Karlsruhe University, Germany
Juris Hartmanis, Cornell University, NY, USA
Jan van Leeuwen, Utrecht University, The Netherlands

Volume Editor

Joe Marks
Mitsubishi Electric Research Laboratories
201 Broadway, Cambridge, MA 02139, USA
E-mail: marks@merl.com

Cataloging-in-Publication Data applied for

Die Deutsche Bibliothek - CIP-Einheitsaufnahme

Graph drawing : 8th international symposium ; proceedings / GD 2000,
Colonial Williamsburg, VA, USA, September 20 - 23, 2000. Joe Marks
(ed.). - Berlin ; Heidelberg ; New York ; Barcelona ; Hong Kong ;
London ; Milan ; Paris ; Singapore ; Tokyo : Springer, 2001
    (Lecture notes in computer science ; Vol. 1984)
    ISBN 3-540-41554-8

CR Subject Classification (1991):G.2, I.3, F.2

ISSN 0302-9743
ISBN 3-540-41554-8 Springer-Verlag Berlin Heidelberg New York

Springer-Verlag Berlin Heidelberg New York
a member of BertelsmannSpringer Science+Business Media GmbH
© Springer-Verlag Berlin Heidelberg 2001
Printed in Germany

Typesetting: Camera-ready by author, data conversion by PTP Berlin, Stefan Sossna
SPIN:10781412   06/3142 - 5 4 3 2 1 0 – Printed on acid-free paper

Preface

This year's meeting marked the Eighth International Symposium on Graph Drawing. The organizing and program committees worked hard to make this year's symposium possible, and we were delighted that so many people came to Colonial Williamsburg, Virginia, for three days of the latest results in the field of graph drawing.

As in previous years, the review process was quite competitive. We accepted 30 out of 53 regular-length submissions, and 5 out of 15 short submissions, for a total acceptance ratio of 35 out of 68, or 51%. This year's program featured several new developments in the field. Four different approaches for handling very large graphs were presented in a session on force-directed layout. Two sessions were devoted to the latest advances in orthogonal graph drawing. And alongside the usual mix of theory and practice papers we had several contributions based on empirical studies of users and of systems.

Our invited talks were given by two speakers who were new to most members of the GD community, but who work in areas that are closely related to graph drawing. Professor Colin Ware of the University of New Hampshire told us how knowledge of human visual perception is useful for the design of effective data visualizations. And Professor David Jensen of the University of Massachusetts at Amherst talked about the process of knowledge discovery from graphs, a process that involves more than just graph drawing and visualization.

In addition to the program proper, we also had two additional events associated with the symposium. Uli Brandes organized a workshop on data-exchange formats for graph drawing, and Franz Brandenburg took charge of the annual graph-drawing contest. Reports on both of these events are included in the proceedings.

Finally, I would like to thank the members of the organizing and program committees for their hard work and dedication: their names are listed on the following pages. Special thanks goes to Kathy Ryall, the chair of the organizing committee, who chose the site and, with the aid of her able assistants, made sure that everything went smoothly throughout the symposium. I would also like to thank the sponsors of the graph-drawing contest: AT&T Research Laboratories, Daimler-Chrysler, and Tom Sawyer Software. And finally I would like to acknowledge the general and generous symposium sponsorship from MERL–Mitsubishi Electric Research Laboratories.

November 2000   Joe Marks

Organization

Program Committee

Therese Biedl, University of Waterloo
Peter Eades, University of Newcastle
Wendy Feng, Tom Sawyer Software
Ashim Garg, SUNY Buffalo
Michael Goodrich, Johns Hopkins
Michael Kaufmann, University of Tübingen
Jan Kratochvíl, Charles University
Giuseppe Liotta, University of Perugia
Joe Marks (chair), MERL
Stephen North, AT&T Research
Kathy Ryall, University of Virginia
Kozo Sugiyama, JAIST
Roberto Tamassia, Brown University
Robin Thomas, Georgia Institute of Technology
Dorothea Wagner, University of Konstanz
Stephen Wismath, University of Lethbridge

Organizing Committee

Renee Carabajal, MERL
Karen Dickie, MERL
Joe Marks, MERL
Janet O'Halloran, MERL
Kathy Ryall (chair), University of Virginia / MERL

Sponsoring Institutions

AT&T Research Laboratories
Daimler-Chrysler
MERL–Mitsubishi Electric Research Laboratories
Tom Sawyer Software

Steering Committee

Franz Brandenburg, University of Passau
Giuseppe Di Battista, University of Rome
Ioannis G. Tollis, University of Texas, Dallas
Jan Kratochvíl, Charles University
Joe Marks, MERL
Pierre Rosenstiehl, EHESS
Roberto Tamassia, Brown University
Takao Nishizeki, Tohoku University
Petra Mutzel, Technical University of Vienna

External Referees

François Bertault
Ulrik Brandes
Broňa Brejová
Sabine Cornelsen
Walter Didimo
Ugur Dogrusoz
Markus Eiglsperger
Irene Finocchi
Arne Frick
Seok-Hee Hong
Mike Houle
Stephen Kobourov
Wei Lai
Annegret Liebers
Anna Lubiw
Hugo A.D.Do Nascimento
Maurizio Patrignani
Maurizio Pizzonia
Aaron Quigley
Chris Riley
Tomáš Vinař
Thomas Willhalm
David Wood

Table of Contents

Invited Talk

Force-Directed Layout

k-Level Graph Layout

Orthogonal Drawing I

Orthogonal Drawing II

Theory II

Symmetry and Incremental Layout

Workshop and Contest

The Visual Representation of Information Structures

Colin Ware

Data Visualization Research Lab
Center for Coastal and Ocean Mapping
University of New Hampshire Durham, NH 03924.

Abstract. It is proposed that research into human perception can be applied in designing ways to represent structured information. This idea is illustrated with four case studies. (1) How can we design a graph so that paths can be discerned? Recent results in the perception of contours can be applied to make paths easier to perceive in directed graphs. (2) Should we be displaying graphs in 3D or 2D space? Research suggests that larger graphs can be understood if stereo and motion parallax depth cues are available. (3) How can heterogeneous information structures be best represented? Experiments show using structured 3D shape primitives make diagrams that are easier to discover and remember. (4) How can causal relationships be displayed? Michotte's work on the perception of causality suggests that causal relationships can be represented using simple animations. The general point of these examples is that data visualization can become a science based on the mapping of data structures to visual representations. Scientific methods can be applied both in the development of theory and testing the value of different representations.

1 Introduction

The human visual system provides by far the highest bandwidth channel into the brain (70% of all receptors, more than 40% of cortex devoted to vision). However, the visual machinery does not analyze all patterns equally well. This paper illustrates ways in which we may apply what is known about human perception to data representation.

2 Finding Paths in Directed Graphs

Recent work on the gestalt concept of "good continuity" provides us with a detailed description of the conditions under which continuous contours are perceived [2]. These results can be applied directly to the problem of emphasizing important paths in graphs. For example paths drawn so that edge curvature is minimized will be easier to be perceived. This is illustrated in Figure 1. In addition nodes should be placed in rough alignment and should consist of oriented elongated symbols.

J. Marks (Ed.): GD 2000, LNCS 1984, pp. 1-4, 2001.

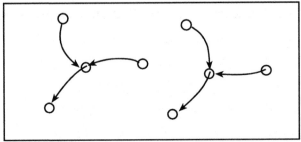

Fig. 1. Different paths are emphasized on the same graph.

3 Visualizing Graphs in 3D

There is a debate about whether information should be displayed in 3D or in 2D. A study by Ware and Franck examined the value of different depth cues in a task relating to tracing paths in graphs [8]. The results showed that over a wide range of graph sizes using stereopsis allowed about 60% more to be seen and using motion parallax allowed about 130% more to be seen. As expected, the perspective depth cue was not important for this particular task.

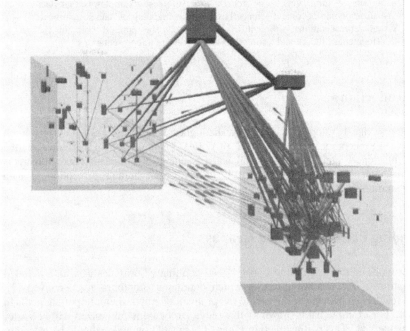

Fig. 2. This large graph represents the structure of a 6 million line program. (Image copyright Glenn Franck and Colin Ware. Reprinted with permission)

4 Using Geons to Represent Structured Information

Research by Marr [5] and by Biederman [1] suggests that complex objects are broken down by the visual system into 3d "geon" primitives. Irani and Ware have shown that diagrams constructed using these primitives are much easier to analyze and recognize in comparison with conventional box-line diagrams [3]. Figure 3. illustrates a "geon" diagram [3].

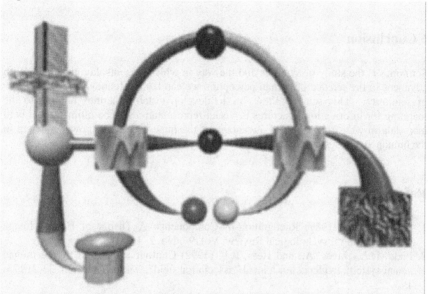

Fig. 3. Geons are 3D primitives of object perception. Diagrams constructed from geon elements may be both more accurately interpreted and recalled (image copyright Pourang Irani: reprinted with permission)

5 Simple Animation for Representing Causal Relations

The work of Michotte [4] suggests that given simple animations and the appropriate timing of events people can "directly" perceive causality in the same sense that they directly perceive a connection between two objects linked by a line. Ware et al [7] applied this phenomenon of causality perception to the representation of causal relationships in statistical graphs. One of the methods for representing causality was to have a ball move, striking a "node" in a graph apparently causing it to vibrate (Figure 4).

Fig. 4. Causality can be represented by a ball striking a node "causing" it to oscillate.

6 Conclusion

Semiotics is the study of symbols and the way in which they display meaning. With advances in the science of human perception we can lay the foundation for a science of semiotics. This science allows us to develop testable theories relating to the mapping for information structures to visual representations. The ultimate goal is to map data in ways that take full advantage of the huge processing power inherent in the human visual system [6].

References

1. Biederman, I. (1987) Recognition-by-Components: A Theory of Human Image Understanding, Psychological Review, Vol. 94, No. 2, 115-147.
2. Field, D.J., Hayes, A., and Hess, R.F. (1992) Contour integration by the human visual system: evidence for a local "association field", Vision Research, 33(2) 173-193.
3. Irani, P and Ware, C, (2000) Diagrams based on structured object perception. Advanced Visual Interfaces, AVI'2000, Palermo, Italy, May. Proceedings, 61-67.
4. Michotte, A. (1963) The perception of causality. Translated by T. Miles and E. Miles. Methuen.
5. Marr, D. (1982) Vision: A computational investigation into the human representation and processing of visual information. San Fransisco, CA: Freeman.
6. Ware, C. (1999) Information Visualization: Perception for Design. Morgan Kaufman. December.
7. Ware, C, Neufeld, E., and Bartram, L., (1999) Visualizing Causal Relations, IEEE Information Visualization. Proceeding Late Breaking Hot Topics, 39-42.
8. Ware, C. and Franck, G. (1996) Evaluating Stereo and Motion Cues for Visualizing Information Nets in Three Dimensions. ACM Transactions on Graphics.15(2) 121-139..

User Preference of Graph Layout Aesthetics: A UML Study

Helen C. Purchase, Jo-Anne Allder, and David Carrington

School of Computer Science and Electrical Engineering
The University of Queensland
Brisbane 4072, Australia
{hcp,joanne,davec}@csee.uq.edu.au

Abstract. The merit of automatic graph layout algorithms is typically judged on their computational efficiency and the extent to which they conform to aesthetic criteria (for example, minimising the number of crossings, maximising symmetry). Experiments investigating the worth of such algorithms from the point of view of human usability can take a number of different forms, depending on whether the graph has meaning in the real world, the nature of the usability measurement, and the effect being investigated (algorithms or aesthetics). Previous studies have investigated performance on abstract graphs with respect to both aesthetics and algorithms, finding support for reducing the number of crossings and bends, and increasing the display of symmetry.

This paper reports on preference experiments assessing the effect of individual aesthetics in the application domain of UML diagrams, resulting in a priority listing of aesthetics for this domain. The results reveal a difference in aesthetic priority from those of previous domain-independent experiments.

1 Introduction

The success of automatic graph layout algorithms which display relational data in a graphical form is typically measured by their computational efficiency and the extent to which they conform to aesthetic criteria (for example, minimising the number of crossings, maximising symmetry). In addition, designers of these algorithms often claim that by conforming to these aesthetic criteria, the resultant graph drawing helps the human reader to understand the information embodied in the graph. However, little research has been performed on the usability aspects of such algorithms: do they produce graph drawings that make the embodied information easy to use and understand? Is the computational effort expended on conforming to conventional aesthetic criteria justifiable with respect to better usability?

Usability studies investigating the merit of graph drawing algorithms from a human perspective can take several different forms, depending on the nature of the graph (syntactic or semantic), the nature of the usability measurement (preference or performance), and the nature of the effect being investigated (algorithms or aesthetics).

J. Marks (Ed.): GD 2000, LNCS 1984, pp. 5–18, 2001.

- *Syntactic* graph drawing experiments use a graph structure that has no meaning in the real world: it is merely an abstract collection of nodes with relationships between them. *Semantic* graph drawing experiments use a graph within a particular application domain: in this case, the graph has meaning in the real world (for example, a transport network, or a data-flow diagram).
- *Preference* experiments ask the subjects to state their preference of one drawing over another. *Performance* experiments require subjects to perform a particular task (or tasks) using a given graph, the data collected is the extent of the subjects' success in performing the task.
- Two possible effects on usability may be investigated in graph drawing experiments: the effect of individual *aesthetics* (e.g. reducing crossings, maximising symmetry) and the effect of the use of different *algorithms* (producing drawings conforming to different aesthetics to varying degrees).

Two previous studies have investigated syntactic performance. The first experiments considered the effect of individual aesthetics, and found support for reducing the number of crossings and bends, and increasing the display of symmetry. However, no support was found for maximising the minimum angle or increasing orthogonality (Purchase, 1997). The second experiment considered the effect of eight different algorithms, and revealed that it is difficult to say that one algorithm is 'better' than another in the context of syntactic understanding of the abstract graph structure (Purchase, 1998).

This paper reports on two semantic, preference experiments that investigated the effect of individual aesthetics. The application domains are the presentation of two types of UML diagrams: class diagrams and collaboration diagrams.

All these experiments are part of a larger project, the aim of which is to perform a thorough empirical investigation of the aesthetics underlying graph layout algorithms, and the algorithms themselves, in an attempt to influence the future design of graph layout algorithms through empirical human (rather then computational) experimentation.

2 Experimental Scope and Definition

2.1 The Application Domain: UML Diagrams

The Unified Modeling Language (UML) (Booch, Rumbaugh and Jacobson, 1998) was chosen as the semantic domain for these preference, aesthetics experiments. Many different methods and models have been proposed to capture a complete specification of requirements and a comprehensive design representation in a formal software engineering prcess. UML provides a mainly graphical notation to represent the artifacts of software systems. The notation is relatively new but it is rapidly being adopted as the accepted notation for object-oriented analysis and design. UML incorporates notations to describe systems at various levels of abstraction. UML diagrams can be used to model requirements, designs, implementations and tests. Since these diagrams are a means of communication between customers, developers and others involved in the software engineering process, it is critical that the diagrams present information clearly. Appropriate layout of these diagrams can assist in achieving this goal.

UML uses several different types of graph drawings that aim to describe a system to meet the users' needs at reasonable cost. Two UML diagram types, class and collaboration, were selected for the experiments.

Class diagrams describe the types of objects in the system, and the relationships between them. These relationships are either subtypes (representing inheritance) or associations (representing other types of relationship).

Collaboration diagrams show the interaction of objects and the sequence of events by numbering the events in the order in which they occur (along the arcs), referring to objects as nodes in the graph.

2.2 Aims

The aim of the experiment is to identify an ordered list of the aesthetic features preferred by subjects when embodied in UML class and collaboration diagrams. Such a list can indicate to interface designers of CASE tools the most suitable way to lay out their diagrams for the best response from users. It will also form the basis for more extensive experiments concentrating on performance of users in software engineering tasks.

The main focus of the study aimed to identify the 'subjectively pleasing' aesthetics in graph drawings. Evaluation of the graph drawings was done solely according to human, individual preference. No consideration was given to performance with respect to a task or correctness of interpretation.

2.3 Aesthetics Investigated

Using graphs from a semantic domain instead of an abstract graph structure introduces additional secondary notations that are particular to the formal semantic notation. Secondary notations are layout or graphical cues that tend not to be part of the formal notation (e.g. adjacency, clustering, white space) (Petre, 1995). Thus, while this experiment included some graph drawing aesthetics as advocated by designers of generic layout algorithms, it also included investigation of other layout features specifically related to the standard UML notation.

For each experiment (class and collaboration), a suitable subset of aesthetic features was identified. These were selected based on emphases in the literature, and as ones that could feasibly be applied to UML diagrams. Many of them were also considered in the prior experiments (Purchase 1997, 1998) and most could be used for both types of diagram.

Six aesthetics were evaluated for both class and collaboration diagrams:

- *minimize bends* (the total number of bends in polyline edges should be minimized (Tamassia, 1987))
- *minimize edge crossings* (the number of edge crossings in the drawing should be minimized (Reingold and Tilford, 1981))

- *orthogonality* (fix nodes and edges to an orthogonal grid (Tamassia, 1987; Papakostas and Tollis, 2000))
- *width of layout* (the physical width of the drawing should be minimised (Coleman and Stott Parker, 1996))
- *text direction* (all text labels should be horizontal, rather than a mixture of horizontal and vertical) (based on Petre, 1995)
- *font type* (all text fonts should be the same, rather than using different fonts for different types of labels) (based on Petre, 1995)

For UML class diagrams, two additional secondary notation features were investigated. Both versions of each notation have been found in published examples of UML notation (See Figure 1).

- *inheritance* (inheritance lines should be joined prior to reaching the superclass, rather than being represented as separate arcs)
- *directional indicators* (arcs should be labelled with two relationship labels and directional indicators, rather than one)

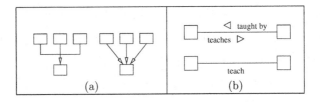

Fig. 1. The UML class diagram secondary notation features investigated, showing both alternatives: (a) inheritance, (b) directional indicators.

For UML collaboration diagrams, two additional secondary notation features were investigated. In both cases, the one option (long arrows adjacent to the arcs) is standard UML notation (See Figure 2).

- *adjacent arrows* (all arcs are undirected with an adjacent arrow indicating the direction of the message sent, rather than all arcs being directed)
- *arrow lengths* (the arrows adjacent to the arcs should be the same length as the arcs, rather than shorter than the arcs)

Usability measuring method. As these were our first experiments performed on layout aesthetics and secondary notations for UML diagrams, preference was chosen as the method of usability measurement. While increased preference does not necessarily correspond with improved performance, beginning with a preference study enables the most important layout features to be identified before a more substantial performance study is performed.

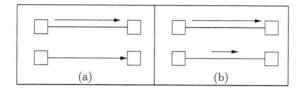

Fig. 2. The UML collaboration diagram secondary notation features investigated, showing both alternatives: (a) adjacent arrows, (b) arrow lengths.

By asking whether they prefer one UML drawing to another, subjects are likely to be anticipating the use of these drawings for a software engineering task: their responses are therefore likely to be related to their perceived usefulness of the diagrams.

Three experiments were performed:

Experiment 1: determined preferences for aesthetics embodied in UML class diagrams (70 subjects)

Experiment 2: determined preferences for aesthetics embodied in UML collaboration diagrams (90 subjects)

Experiment 3: a more focussed study which refined the results of the first two experiments, focussing on particular aesthetics that gave surprising results (6 subjects)

2.4 Methodology

A basic UML class diagram (depicting the relationships between students, lecturers, tutors and administrative staff, see Figure 3) and a basic UML collaboration diagram (depicting the procedure followed for organising honours students' seminars, see Figure 4) were created.

In both cases, the graph structures were complex enough to enable an appropriate and varied manipulation of the nodes and arcs within the diagram, but not so complex that the diagram would take a long time to understand.

The class diagram had 14 classes and 18 relationships, and the collaboration diagram had 12 objects and 17 messages.

Each diagram was drawn 16 times (twice for each aesthetic), with all the information within the drawing remaining constant: only the layout of the diagrams was altered. Each representation of the graph was drawn with attention to a specific graph-drawing aesthetic or secondary notation choice.

Graph drawings were grouped in pairs emphasising the contrast between the diagrams: one graph drawing in the pairwise comparison contained a higher presence of the aesthetic while the other graph drawing contained a lower presence. For example, one diagram was highly orthogonal while the paired diagram had minimal orthogonality. In the absence of computational metrics for measuring the presence of the UML and secondary notation features in the drawings, and due to the large number of aesthetics being investigated concurrently, the other

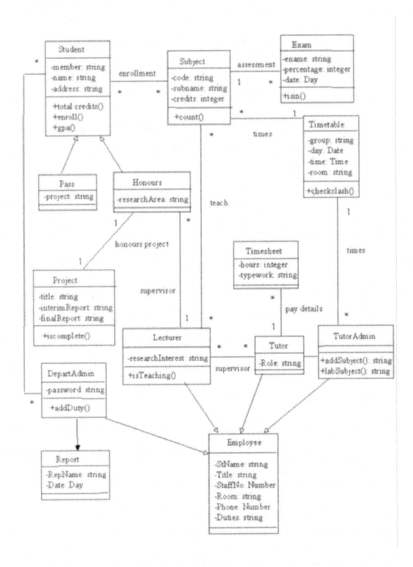

Fig. 3. A UML class diagram used in experiment 1.

aesthetics in each contrasting pair could not be controlled. We were aware that not controlling the other aesthetics could have resulted in confounding factors: for example, the diagram with lots of bends may have differed from its paired diagram in both the number of bends as well as in the extent of its perceived orthogonality. To prevent our overall conclusions being affected by potential con-

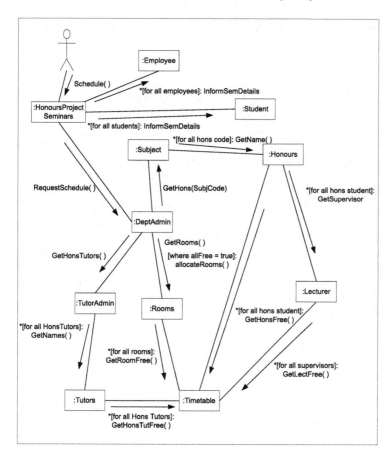

Fig. 4. A UML collaboration diagram used in experiment 2.

founding factors which may have biased the simple preference quantitative data, we collected additional qualitative and ranking data.

Subjects. Seventy student volunteers from the University of Queensland participated in the class diagram evaluation; ninety students participated in the collaboration diagram evaluation. The same experimental methodology and materials were used for both experiments.

All participants were third or fourth year Information Technology students who, although not generally proficient with UML, have experience with similar notations (e.g., Fusion, dataflow diagrams, entity relationship diagrams and Booch diagrams).

Prior to both experiments, pilot experiments with a small number of subjects were performed to check for problems in experimental materials and procedures.

Materials. Each subject was presented with an individual evaluation booklet designed to be completed without evaluator assistance and without a time limit. The booklet had the following structure:

- Instructions, the aim of the project, the task required, and an example.
- A questionnaire requesting information about the subject (prior knowledge and use of graph drawings, year of study etc.).
- A UML tutorial sheet, identifying the key points of the diagram under consideration (class or collaboration). This sheet was detached from the booklet so that subjects did not have to turn back to refer to it.
- Eight (facing) pairs of graphs drawings. For each pair, there was a space where the subjects had to indicate their preferred drawing, as well as write a brief explanation for their choice. The explanation was intended to identify any issues influencing the choice that had not been considered by the evaluator, and to provide qualitative data to support the quantitative data. The diagrams were presented in a random order in the booklet, in an attempt to counter any familiarity effect: after seeing several different representations of the same UML diagram, the subject may develop a deeper understanding of the information, and may therefore make different preference choices.
- A ranking sheet where subjects needed to rank the three diagrams that they most preferred (1-3), and the three that they least preferred (14-16). A separate large sheet, showing all 16 (reduced in size) diagrams on one page, was provided to assist in this ranking.

There was no time limit, and the subjects could go back and change any previous answers if they wished. Most subjects completed the task in about 20 minutes.

Data Analysis. The preference questions were analysed by calculating a percentage preference for each aesthetic, with the significance of the result computed using a standard binomial distribution. A result was considered significant (ie. not attributable to chance or random selections) if its probability was less than 0.05.

The written explanations for preferences were analysed by determining the percentage of subjects who stated that the targeted aesthetic comparison influenced their choice: this allowed us to identify whether there were any possible confounds (ie. other aesthetics unintentionally affecting the result) in the quantitative data obtained by the subjects' preferences.

The final ranking question was analysed to identify preferred diagrams by computing an overall weighted preference value for each diagram. A weight of 3 was given for a first choice, a weight of 2 for a second choice, a weight of 1 for a 3rd choice, a weight of -1 for a 14th choice, a weight of -2 for a 15th choice and a weight of -3 for a 16th choice.

aesthetic choice	% preference
fewer crosses	93%
fewer bends	91%
horizontal labels only	86%
joined inheritance arcs	76%
narrower	73%
more orthogonal	61%
no font variation	61%
directional indicators	60%

Fig. 5. Aesthetic preference results for UML class diagrams (all results significant).

3   Results

3.1   Results: UML Class Diagrams

The results for UML class diagrams are shown in Figure 5. By analysing the subjects' stated reasons for their preferences, the only class diagram aesthetic result that appeared to be affected by confounding factors was horizontal labels, where many subjects who preferred the horizontal labels drawing referred to direction of information flow. Both the crosses and bends aesthetic results were unaffected by other factors, while those subjects who did not like the orthogonal diagram did so because of the increased number of bends. Most of those subjects who preferred independent inheritance arcs, a wider diagram, variation in font or no directional indicators did so merely because of personal preference.

The results of the ranking question are shown in Figure 6, which presents the weighted overall ranking value for each of the 16 drawings.

3.2   Results: UML Collaboration Diagrams

The results for UML collaboration diagrams are shown in Figure 7. By analysing the subjects' stated reasons for their preferences, the only collaboration diagram result that appeared to be affected by confounding factors was longer arrows, where many subjects who preferred the longer arrows referred to diagram structure. Both the crosses and adjacent arrows aesthetic results were unaffected by other factors, while those subjects who did not like the orthogonal diagram did so because of increased bends. Most of those subjects who preferred a wider diagram or variation in font did so merely because of personal preference. The results for the use of horizontal labels only and bends were not significant, so no conclusions can be drawn for these two aesthetics as these results could be attributable to chance.

The results of the ranking question are shown in Figure 8, which presents the weighted overall ranking value for each of the 16 drawings.

class diagram drawing	weighted ranking value
joined inheritance arcs	54
narrower	53
directional indicators	42
fewer bends	41
horizontal text only	33
highly orthogonal	29
no directional indicators	21
wider	5
not orthogonal	3
no font variation	1
font variation	0
separate inheritance lines	-5
fewer crosses	-19
horizontal and vertical text	-39
many bends	-73
many crosses	-145

Fig. 6. Weighted ranking values for all sixteen class diagrams.

aesthetic choice	% preference
fewer crosses	91%
no adjacent arrows	90%
longer arrows	82%
no font variation	70%
more orthogonal	63%
narrower	57%
horizontal labels only	54% †
fewer bends	53% †

Fig. 7. Aesthetic preference results for UML collaboration diagrams († indicates a non-significant result).

4   Follow-Up Experiment

Experiment 3 was a follow-up experiment that focussed on the aesthetics common to both class and collaboration diagrams. Its aim was to investigate some of the unexpected results from experiments 1 and 2 that may have been due to confounding factors in the diagrams, and to create a final priority list of aesthetics. The subjects' comments had indicated that the direction of flow of information had sometimes influenced their preference. Direction of flow was an aesthetic feature that had not originally been targeted and which was introduced into experiment 3, with the following definition:

- *flow* (directed arcs should point in a consistent direction, preferably top-to-bottom and left-to-right) (Eades and Sugiyama, 1990)

class diagram drawing	weighted ranking value
highly orthogonal	101
longer arrows	77
no adjacent arrows	76
horizontal and vertical labels	30
horizontal labels only	27
shorter arrows	12
no font variation	3
fewer crosses	-2
font variation	-6
narrower	-7
not orthogonal	-10
wider	-11
fewer bends	-30
many bends	-31
adjacent arrows	-36
many crosses	-196

Fig. 8. Weighted ranking values for all sixteen collaboration diagrams.

The original UML diagrams were modified to correct any obvious confounding factors. For example, the original collaboration diagram with few bends had some crossing arcs that had been identified as influencing some subjects' preference for the diagram with more bends and no crossing arcs.

Six separate smaller experiments comprised this investigation. They were performed in intensive, focussed interviews where each subject was questioned about which aspects of a set of diagrams influenced their preferences. Six subjects took part, each being questioned about all six diagram sets, and providing extensive qualitative data with which to form a prioritised list of aesthetics.

The format of each of the six experiments differed according to the specific investigation of the experiment 1 and 2 results that were being considered. The general procedure was that subjects were shown two or three different diagrams, and were asked questions about their layout. Most of the diagrams used in experiment 3 were ones that had been used in experiment 1 or 2, although some were altered to target specific aesthetics.

The components of experiment 3. Two categories of unexpected quantitative results were investigated in these six smaller, focussed interview experiments. First, we investigated some surprising overall weighted ranking values. Second, we investigated some aesthetics for which the percentage preferences differed between the class and collaboration diagrams. In each case, we reviewed the diagrams that had been used for experiments 1 and 2 and tried to determine whether there were any confounding factors that may have led to these surprising results.

Overall weighed ranking values: In both the class and collaboration diagram experiments, the weighted ranking for the diagram with least crosses was negative. It was unlikely that this diagram was ranked according to its lack of crosses: careful inspection of both "least crosses" diagrams revealed that both appeared to be less orthogonal than the other fifteen diagrams in the corresponding set. The first two experiments of experiment 3 focussed on these diagrams, and supported our view that subjects' preference for the many crosses diagram over the few crosses diagram in experiments 1 and 2 may have been influenced by the lack of orthogonality and inconsistent flow direction in the "least crosses" diagram.

Difference in percentage preference between class and collaboration diagrams: Four differences in the initial preferences for the aesthetics were addressed: bends, font variations, text direction and width.

- *Bends:* The results for experiment 2 did not indicate a significant preference for the collaboration diagram with the least bends. Inspection of the pair of collaboration diagrams related to the bends aesthetic revealed that the diagram with more bends had a more consistent direction of flow, and the one with no bends had a single crossed arc. The third experiment of experiment 3 focussed on these diagrams, and supported our view that the direction of flow and the single cross influenced subjects' preference decisions.
- *Font variation:* Using consistent font was preferred more in the collaboration diagram than in the class diagram, and it was felt that this could have been because the collaboration diagram that used a variety of fonts included a more unusual font (cursive) than the corresponding class diagram (italic). This issue was addressed in the fourth experiment by producing identical diagrams with no font variation, and with more subtle font variation (bold). The results supported our view that the subjects' dislike of font variation is dependent on the type of fonts used.
- *Text direction:* The diagram that only used horizontal text was preferred to a much greater extent in class diagrams than in collaboration diagrams. On inspection of the collaboration diagram with both vertical and horizontal text, it appeared to have a more orthogonal shape. The fifth experiment confirmed that this orthogonality could have affected the subjects' preferences in the first experiment.
- *Width:* There was a stronger preference for narrow width in class diagrams than in collaboration diagrams. Although the qualitative data from experiments 1 and 2 suggested that preference for width may be an inexplicable subjective opinion, we also thought the amount of information associated with arcs in the collaboration diagram may have affected this preference decision. The sixth experiment confirmed this: while some subjects could not explain why they preferred the narrower or wider diagram, those that could explain their preference did so by referring to the amount of information in the collaboration diagram.

5 Conclusions

The follow-up experiments provided rich qualitative data which shed insight on the initial quantitative data. By analysing the extensive interview comments provided by the subjects, we concluded that the priority order of the aesthetics common to both diagram types is: arc crossings, orthogonality, information flow, arc bends, text direction, width of layout and font type. Of the UML-specific aesthetics, we concluded that joined inheritance arcs and directional indicators are preferred for class diagrams. For collaboration diagrams, no adjacent arrows are preferred (although this preference is incompatible with UML notation). This list provides a useful starting point for further studies on UML diagram layout aesthetics with respect to performance in a related task.

6 Discussion

The previous syntactic performance experiment found support for reducing the number of crossings and bends, and for increasing the orthogonality. Information flow and width were not considered.

The results of this semantic preference experiment confirm that the evidence is overwhelmingly in favour of reducing the number of arc crossings as the most important aesthetic to consider. While the results of the syntactic experiments did not highlight orthogonality as being important, in the domain of UML diagrams, this aesthetic moves up the priority list to second place: this is a clear signal that algorithms that are designed for abstract graph structures, with no consideration of their ultimate use, will not necessarily produce useful visualisations of semantic information.

This study is only the first step in assessing the usability of graph drawings produced by layout algorithms when used in application domains. Future work includes investigating aesthetics and algorithms with respect to measures of performance in UML related tasks, and extensions of this methodology to other domains. There is an increasing demand for software environments that can assist users in graph-based application tasks, for example, software engineering, social and transport network analysis, database design. Empirical research into the effectiveness of graph layout algorithms from a usability point of view can ensure that such software environments serve their users effectively.

Acknowledgements. We are grateful to the students in the School of Computer Science and Electrical Engineering at The University of Queensland who participated in these experiments. This research was partly funded by an Australian Research Council grant.

References

1. G. Booch, J. Rumbaugh, and I. Jacobson. *The Unified Modeling Language User Guide*. Addison-Wesley, 1998.

2. M.K. Coleman and D. Stott Parker. Aesthetics-based graph layout for human consumption. *Software – Practice and Experience*, 26(12):1415–1438, 1996.
3. P. Eades and K. Sugiyama. How to draw a directed graph. *Journal of Information Processing*, 13(4):424–437, 1990.
4. A. Papakostas and I.G. Tollis. Efficient orthogonal drawings of high degree graphs. *Algorithmica*, 26(1):100–125, 2000.
5. M. Petre. Why looking isn't always seeing: Readership skills and graphical programming. *CACM*, 38(6):33–44, June 1995.
6. H. C. Purchase. Which aesthetic has the greatest effect on human understanding? In G. Di Battista, editor, *Proceedings of Graph Drawing Symposium 1997*, pages 248–259. Springer-Verlag, Rome, Italy, 1997. LNCS, 1353.
7. H. C. Purchase. Performance of layout algorithms: Comprehension, not computation. *Journal of Visual Languages and Computing*, 9:647–657, 1998.
8. E. Reingold and J. Tilford. Tidier drawing of trees. *IEEE Transactions on Software Engineering*, SE-7(2):223–228, 1981.
9. R. Tamassia. On embedding a graph in the grid with the minimum number of bends. *SIAM J. Computing*, 16(3):421–444, 1987.

A User Study in Similarity Measures for Graph Drawing *

Stina Bridgeman and Roberto Tamassia

Center for Geometric Computing
Department of Computer Science
Brown University
Providence, Rhode Island 02912–1910
{ssb,rt}@cs.brown.edu

Abstract. The need for a similarity measure for comparing two drawings of graphs arises in problems such as interactive graph drawing
and the indexing or browsing of large sets of graphs. This paper builds
on our previous work [3] by defining some additional similarity measures,
refining some existing ones, and presenting the results of a user study
designed to evaluate the suitability of the measures.

1 Introduction

The question of how similar two drawings of graphs are arises in many situations.
In interactive graph drawing, the graph being visualized changes over time and
it is important to preserve the user's "mental map" [12] so she does not spend
a lot of time relearning the drawing after each update. Animation can provide
a smooth transition between drawings, but is of limited use if the drawings are
too different. In layout adjustment applications, an existing drawing is modified
to improve an aesthetic quality without destroying the user's mental map.

Another application is in indexing or browsing large sets of graphs. The
SMILE graph multidrawing system (Biedl et. al. [1]) produces many drawings
of a graph, and its graph browser arranges the drawings so that similar ones are
near each other to help the user navigate the system's responses. Similarities
between drawings can also be used as a basis for indexing and retrieval. In
character and handwriting recognition, a written character may be transformed
into a graph and compared to a database of characters to find the closest match.

Let M be a similarity measure defined so that measure's value is 0 when the
drawings are identical. In order to be useful, M should satisfy three properties:

Rotation: Given drawings D and D', $M(D, D'_\theta)$ should have the minimum
value for the angle a user would report as giving the best match, where D'_θ
is D' rotated by an angle of θ with respect to its original orientation.

---

* Research supported in part by the National Science Foundation under grants CCR-
9732327 and CDA-9703080, and by the U.S. Army Research Office under grant
DAAH04-96-1-0013.

J. Marks (Ed.): GD 2000, LNCS 1984, pp. 19–30, 2001.

Ordering: Given drawings D, D', and D'', $M(D, D') < M(D, D'')$ if and only if a user would say that D' is more like D than D'' is like D.

Magnitude: Given drawings D, D', and D'', $M(D, D') = \frac{1}{c}M(D, D'')$ if and only if a user would say that D' is c times more like D than D'' is like D.

This paper describes a user study intended to address rotation and ordering, and to explore a method for addressing magnitude. Data cannot be collected directly for the magnitude part because it is very difficult to assign numerical similarity values to pairs of drawings, much more so than judging ordering (Wickelgren [18]). As a result, other data (e.g., response times) presumed to be related to similarity must be collected instead. The assumption of the suitability of the data can be partially tested by determining if using the data to make ordering decisions between drawings is consistent with the user responses.

This study improves on our previous work [3] in several ways:

- **More Experimental Data:** A larger pool of users (103 in total) was used for determining the "correct" behavior for the measure.
- **Refined Ordering Part:** Users made only pairwise judgements between drawings rather than being asked to order a larger set.
- **Addressing of Magnitude Criterion:** Magnitude was not addressed in [3].
- **More Realistic Drawing Alignment:** The previous drawing alignment method allowed one drawing to be scaled arbitrarily small with respect to the other; the new method keeps the same scale factor for both drawings.
- **Refinement of Measures:** For those measures computed with pairs of points, pairs involving points from the same vertex are skipped.
- **New Measures:** Several new measures have been included.

We describe the experimental setup in Section 2, the measures evaluated in Section 3, the results in Section 4, and conclusions and directions for future work in Section 5.

2   Experimental Setup

This study focuses on similarity measures for orthogonal drawings of nearly the same graph. "Nearly the same graph" means that the graphs differ by only a small number of vertex and edge insertions and deletions. The focus on orthogonal drawings is motivated by the availability of an orthogonal drawing algorithm capable of producing many drawings of the same graph, and by the amount of work done on interactive orthogonal drawing (e.g., Biedl and Kaufmann [2], Fößmeier [8], Papakostas, Six, and Tollis [14], and Papakostas and Tollis [15]).

The graphs used were generated from a base set of 20 graphs with 30 vertices each, taken from an 11,582-graph test suite. [7] Each of 20 base graphs was drawn using Giotto [17]. Forty modified drawings were created by adding a degree 2 and a degree 4 vertex to separate copies of each base drawing. Each modified drawing is identical to its base drawing except for the new vertex and its adjacent edges, placed as a user might draw them in an editor. Finally, four new drawings were produced for each modified drawing using InteractiveGiotto [4]. The new drawings range from very similar to the base drawing to significantly different.

The experiment consisted of three parts, to address the three evaluation criteria. In all cases, the user was asked to respond as quickly as possible without sacrificing accuracy, and the user's response and time to answer were recorded. Each trial timed out after 30 seconds if the user did not respond.

Rotation Part. The rotation part directly addresses the rotation criterion. The user is presented with a screen as shown in Figure 1. The eight drawings on the right are different orientations of the same new (InteractiveGiotto) drawing; the one drawing on the left is the corresponding base drawing. The eight orientations are rotations by the four multiples of $\pi/2$, with and without an initial flip around the x-axis. For orthogonal drawings, only multiples of $\pi/2$ are meaningful since clearly rotation by any other angle is not the correct choice. The vertices are not labelled to emphasize the layout of the graph over the specifics of vertex names.

The user chooses the orientation that looks most like the base drawing. She may click the "can't decide" button if she cannot make a decision.

Ordering Part. The ordering part directly addresses the ordering criterion. In this part, the user is presented with a screen as shown in Figure 2. The two rightmost drawings are two different new drawings of the same modified drawing; the leftmost is the corresponding base drawing. The vertices are not labelled.

The user chooses which of the two rightmost drawings looks most like the base drawing. She may click on "can't decide" if she cannot make a decision.

Difference Part. The difference part addresses the magnitude criterion by gathering response times on a task, with the assumption that the user will complete the task more quickly if the drawings are more similar. Figure 3 shows the screen presented to the user. The drawing on the right is one of the InteractiveGiotto-produced drawings; the drawing on the left is the corresponding base drawing.

The user identifies the vertex in the right drawing missing from the left drawing. Vertices have random two-letter names, because the task is too difficult with unlabelled vertices; labels are often important in real-world situations. Corresponding vertices in drawings in a trial have the same name, but the names are different for separate trials to prevent the user from learning the answer.

A total of 103 students completed the three parts as part of a homework assignment in a second-semester CS course at Brown University. They had some familiarity with graphs through lectures and a programming project.

The students used an online system. A writeup was presented explaining how to use the system, and the directions were summarized each time it was run. Each of the three parts was split into four runs, so the students would not have to stay focused for too long without a break. The first run of each part was a practice run. In later runs, graphs were assigned to the student randomly so that 1/3 of the students worked with each graph. Within each run, the sequence of the individual trials and the ordering of the right-hand drawings in the rotation and

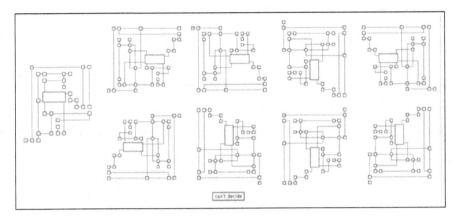

Fig. 1. The rotation part.

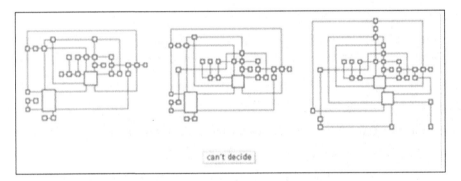

Fig. 2. The ordering part.

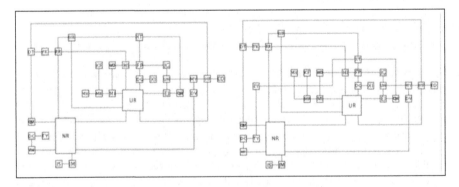

Fig. 3. The difference part.

ordering parts was chosen randomly. After the students completed all of parts, they answered a short questionnaire about their experiences. The questions were:

1. **(Ordering and Rotation)** What do you think makes two drawings of nearly the same graph look similar? Are there factors that influenced your decisions? Did you find yourself looking for certain elements of the drawing in order to make your choice?
2. **(Difference)** What factors helped you locate the extra vertex more quickly? Did you compare the overall look of the two drawings in order to aid your search, or did you just scan the second drawing?
3. **(All Parts)** As you consider your answers, think about what this means for a graph drawing algorithm that seeks to preserve the look of the drawing. What types of things would it have to take into account?

3   Measures Evaluated

All of the measures evaluated in this study are listed below. Most are the same as or similar to those described in [3]; the primary difference is that all have been scaled to have a value between 0 and 1, with 0 indicating identical drawings. The upper bound is often based on the worst-case scenario and may not be achievable by an actual drawing algorithm. Formal definitions and more extensive motivations for many of the measures can be found in [3]; only those measures which are new or have changed significantly are given more full treatment below.

First, we consider some preliminaries.

Corresponding Objects. The graphs in the drawings being compared are assumed to be the same; if not, only the common subgraphs are used. Thus, each vertex and edge of G has a representation in each of the drawings, and it is meaningful to talk about the *corresponding vertex or edge* in one drawing given a vertex or edge in the other drawing.

Point Set Selection. Most of the measures are defined in terms of point sets derived from the edges and vertices of the graph. Points can be selected in many ways; inspired by North [13], one point set contains the four corners of each vertex. A second point set, suggested by feedback from the study (section 4), contains only corner points near the edge of the drawing. Like vertices and edges, each point in one drawing has a corresponding point in the other drawing.

A change from the previous experiment [3] is the exclusion of pairs of points derived from the same vertex. This can have a great effect on measures which involve nearest neighbors, for example, because a point's nearest neighbor will often be another corner of the same vertex, which does not convey much information about how that vertex relates to other vertices in the drawing. This is not explicitly written in the definitions below for clarity of notation, but it should be assumed unless stated otherwise.

Drawing Alignment. For measures comparing coordinates between drawings, the value of the measure is very dependent on how well the drawings are aligned. Previously, drawings were aligned by simultaneously adjusting the scale and translation of one with respect to the other, which could reduce one drawing to a very small area if the drawings did not match well. The new alignment method treats the scale and translation factors separately. With orthogonal drawings, there is a natural underlying grid which can be used to equalize the scale. The translation factor is then chosen to minimize the distance squared between corresponding points. This alignment method is intended to better match how a person might align the drawings, since it does not seem likely that someone would mentally shrink or enlarge one drawing greatly with respect to the other.

Suitability for Ordering vs. Rotation and Ordering. Some measures do not depend on the relative rotation of one drawing with respect to the other. They are included even though they fail the rotation test because not all applications require determining the proper rotation for drawings. Also, a successful ordering-only measure could be combined with one which performs poorly on the ordering task but rotates well to obtain a measure which is good at both. Measures suitable for ordering only are marked [order only] below.

Notation. In the following, P and P' refer to point sets for drawings D and D', respectively, and $p' \in P'$ is the corresponding point for $p \in P$ (and vice versa). Let $d(p,q)$ be the Euclidean distance between points p and q.

The first group of measures measure the degree of matching between the point set as the maximum mismatch between points in one set and points in another.

Undirected Hausdorff Distance. The *undirected Hausdorff distance* is a standard metric for determining the quality of the match between two point sets. It does not take into account the fact that the point sets may be labelled.

Paired Hausdorff Distance. The *paired Hausdorff distance* is an adaptation of the undirected Hausdorff distance for labelled point sets, and is defined as the maximum distance between two corresponding points:

$$\mathrm{phaus}(P, P') = \max_{p \in P} d(p, p')$$

The position measures are motivated by the idea that the location of the points on the page is important, and points should not move too move far between drawings.

Average Distance. *Average distance* is the average distance a point moves between drawings.

Nearest Neighbor between. *Nearest neighbor between* is based on the idea that a point's original location should be closer to its new position than any other point's new position. The weighted version considers the number of points closer to the point's original location than the point's new location rather than simply whether or not the point is closest.

Relative position measures are based on the idea that the relative position of points should not change. "Relative position" includes both the distance between the points and the angles, though each measure is concerned with only one property.

Orthogonal Ordering. *Orthogonal ordering* measures the change in angle between pairs of points. In the constant-weighted version, all changes of angles are weighted equally; in the linear-weighted version changes in the north, south, east, west relationships are weighted more heavily than changes in angle which do not affect this relationship.

Ranking. The *ranking* measure considers the relative horizontal and vertical position of the point. (Ranking is a component of the similarity measure used in SMILE [1].) Let right(p) and above(p) be the number of points to the right of and above p, respectively.

$$\text{rank}(P, P') = \frac{1}{\text{UB}} \sum_{p \in P} \min\{ \, |\,\text{right}(p) - \text{right}(p')\,| + |\,\text{above}(p) - \text{above}(p')\,| \, , \, \text{UB} \, \}$$

where
$$\text{UB} = 1.5\,(|P| - 1)$$

Of note here is that the upper bound is taken as $1.5\,(|P|-1)$ instead of the actual worst-case value $2\,(|P| - 1)$ because it scales the measure more satisfactorily.

Average Relative Distance [order only]. The *average relative distance* is the average change in distance between pairs of points.

λ-Matrix [order only]. The λ-matrix model is used by Lyons, Meijer, and Rappaport [10] to evaluate cluster-busting algorithms. It is based on the concept of order type used by Goodman and Pollack [9], where two sets of points P and P' have the same order type if, for every triple of points (p,q,r), they are oriented counterclockwise if and only if (p',q',r') are also oriented counterclockwise.

The next group of measures are guided by the philosophy that each point's neighborhood should be the same in both drawings. They do not explicitly take into account either the point's absolute position or its position relative to other points.

Nearest Neighbor Within [order only]. For *nearest neighbor within*, a point's neighborhood is its nearest neighbor. The weighted version includes the number of points closer to the point than its nearest neighbor, whereas the unweighted version considers only whether or not the nearest neighbor remains the same.

ϵ-**Clustering [order only].** ϵ-*clustering* defines the neighborhood for each point to be its ϵ-cluster, the set of points within a distance ϵ, defined as the maximum distance between a point and its nearest neighbor.

Separation-Based Clustering [order only]. In the *separation-based cluste-ring* measure, points are grouped so that each point in a cluster is within some distance δ of another point in the cluster and at least distance δ from any point not in the cluster. The intuition is that the eye naturally groups things based on the surrounding whitespace.

Formally, for every point p in cluster C such that $|C| > 1$, there is a point $q \neq p \in C$ such that $d(p, q) < \delta$ and there is no point $r \notin C$ such that $d(p, r) < \delta$. If C is a single point, only the second condition holds.

Let clus(p) be the cluster to which point p belongs.

$$\text{sclus} = 1 - \frac{|S_I|}{|S_U|}$$

where

$$S_I = \{\, (p, q) \mid p, q \in P,\, \text{clus}(p) = \text{clus}(q) \text{ and } \text{clus}(p') = \text{clus}(q') \,\}$$
$$S_U = \{\, (p, q) \mid p, q \in P,\, \text{clus}(p) = \text{clus}(q) \text{ or } \text{clus}(p') = \text{clus}(q') \,\}$$

Edge measures are based on the graph's edges.

Shape. The *shape* measure treats the edges of the graph as sequences of north, south, east, and west segments. The sequences are compared using the edit distance. The normalized version of the measure uses the algorithm of Marzal and Vidal [11] to adjust for the length of the sequence.

4   Results

For rotation and ordering, each measure was evaluated according to what fraction of the time its choice (determined by its value for the drawings the user is choosing between) agreed with the user's response. If the measure had the same minimum value for two or more drawings, the trial was marked as a tie and was only considered correct if the user clicked the "can't decide" button or if the task timed out, even if one of the possibilities agreed with the user's choice.

Figure 4 shows the fraction of trials for which the rotation criterion was satisfied. Only unweighted nearest neighbor between had a significant number of cases where the user picked one of the drawings but the measure reported a tie. The column labelled "mode" shows the percentage of the trials for which the user's choice agreed with the most common choice (the mode) for that set of drawings. Since a single measure will always select the same rotation for the same pair of drawings, this value is the best a measure could expect to do.

It is disappointing that the correctness for the mode is below 50% since it means even the best measure will tend to rotate drawings incorrectly much of the

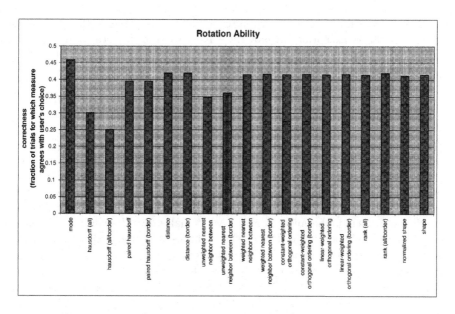

Fig. 4. Results for the rotation part. "border" indicates that the border point set was used; "all" indicates that pairs of points derived from the same vertex were included.

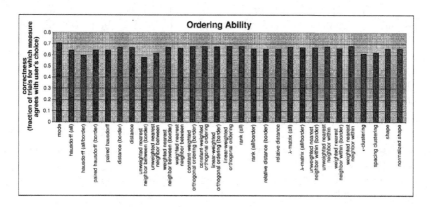

Fig. 5. Results for the ordering part. "border" indicates that the border point set was used; "all" indicates that pairs of points derived from the same vertex were included.

time. The results are better when the new drawing is more similar to the base drawing — the mode correctness averages 64% for the most similar drawings.

Figure 5 shows the fraction of trials for which the order criterion was satisfied. Tie results had only a slight effect on the unweighted nearest neighbor between, unweighted nearest neighbor within, and shape measures. The column labelled "mode" again shows how often the users agree with the most common opinion.

Most of the measures perform quite well when compared to the mode, and the most notable exceptions include those measures that performed most poorly on rotation. Also, as expected, the measures generally performed better when one of the drawings was clearly more like the base drawing than the other.

The goal in the difference part was to use the user's response times as an indicator of similarity, with the idea that a user can locate the new vertex faster if the drawings are more similar. To test the validity of this, the times on the difference part were used to order the pairs of drawings used in the ordering task. The results were very unsatisfactory, achieving only 45% correctness (compare to Figure 5, where even the worst measure reached nearly 58% correctness). As a result, the times on the difference task are not a good indicator of similarity and are not suitable for evaluating measures with respect to the magnitude criterion.

The responses to the final questionnaire yielded several interesting notes. As might be expected, the answers about what makes two drawings look similar included a sizable percentage (35%) who said preserving the position, size, number of large vertices was important and another large percentage (44%) who said they looked for distinctive clusters and patterns of vertices, such as chains, zigzags, and degree 1 vertices. More surprising was that 44% of the students said that borders and corners of the drawing are more important than the interior when looking for similarity. This is supported by research in cognitive science indicating that people often treat filled and outline shapes as equivalent, focusing primarily on the external contour (Wickelgren [18]). Many of these students mentioned the importance of "twiddly bits around the edges" — distinctive clusters and arrangements of vertices, made more obvious by being on the border. Related comments included that the orientation and aspect ratio of the bounding box should remain the same, and that the outline of the drawing should not change. Another sizable group (34%) commented that the "general shape" of the drawing is important.

On the difference part, several users said that the task was difficult and the system timed out frequently. The usefulness of the "big picture" view — looking at the overall shape of the drawing — was contested, with nearly equal numbers reporting that the overall look was useful in the task, and that it was confusing and misleading. About 16% of the users made limited use of the overall look, using it on a region-by-region basis to quickly eliminate blocks that remained the same and falling back on simply scanning the drawing or matching corresponding vertices and tracing edges when the regions were too different. Another 24% used vertex-by-vertex matching from the beginning. Similar-sized groups discovered and exploited shortcuts based on how the modified drawings was constructed (20%), and reported searching for vertices with extra edges rather than searching for new vertex directly (28%).

The most common answers about what a graph drawing algorithm should take into account to preserve the look of the drawing echoed those from the rotation/ordering question: maintaining vertex size and shape, the relative positions of vertices, the outline of the drawing, and clusters.

5   Conclusions and Future Work

The results from the rotation and ordering parts show that, for the most part, there is not a large difference in the performance of the tested measures. However, it is interesting to note that the worst-performing measures (undirected Hausdorff distance, unweighted nearest neighbor between, ϵ-clustering, and spacing clustering) give the least weight to absolute and relative point positions, suggesting that absolute and relative point positions are indeed important to similarity. It also suggests that point positions are less significant for ordering because the worst measures did not perform as badly with respect to the mode.

In the ordering task, the lack of difference between the full point set and the borders-only point set for the better measures seems to mesh well with the comments about the border being very important in the look of the drawing. More study is needed to determine if the results are because the borders really are more important, or if the degree of change in the borders is representative of the change in the whole drawing. This could be tested by comparing drawings where the border is largely unchanged but the interior is very different and those where the border is greatly changed but the interior is not. For the rotation task, only the lack of difference for the orthogonal ordering measures supports the students' comments, since most of the measures (except Hausdorff distance and orthogonal ordering) are already more sensitive to the borders of the drawing — rotation causes border points to move farther, giving them more weight.

The difficulty of the difference part suggests that the amount of difference between the drawings that is considered reasonable varies greatly with the task. When the user only needs to recognize the graph as familiar, the perimeter of the drawing and the position and shape of few key features are the most important. When trying to find a specific change, however, the drawings need to look very much alike or some other cues (change in color, more distinctive vertex names, etc.) are needed to highlight the change. The failure of using the times from the difference task to evaluate the magnitude criterion means that more study is needed to evaluate the measures in this way.

The responses on the questionnaire suggest several possible directions for future investigation. Large vertices are identified as being especially important, which could lead to measures which weight changes in the position and size of large vertices more heavily than other vertices.

Another major focus was clusters of vertices — both the presence of clusters in general, and the presence of specific shapes such as chains and zigzags. The relatively poor showing of the cluster-based measures indicates that they are not making use of clusters in the right way. The importance of specific shapes suggests an approach related to the drawing algorithms of Dengler, Friedell, and Marks [6] and Ryall, Marks, and Shieber [16]. These algorithms produce drawings employing effective perceptual organization by identifying Visual Organization Features (VOFs) used by human graphic designers. VOFs include horizontal and vertical alignment of vertices, particular shapes such as "T" shapes, and symmetrically placed groups of vertices. VOFs can also be used to identify features in an existing drawing that may be important because they adhere to a particular design principle. This is related to the work of Dengler and Cowan [5] on semantic attributes that humans attach to drawings based on the layout, such as that symmetrically placed nodes have common properties. A similarity measure might measure how well those structures are preserved, and an interactive graph drawing algorithm could focus on preserving the structures.

References

1. T. Biedl, J. Marks, K. Ryall, and S. Whitesides. Graph multidrawing: Finding nice drawings without defining nice. In *GD '98*, volume 1547 of *LNCS*, pages 347–355. Springer-Verlag, 1998.
2. T. C. Biedl and M. Kaufmann. Area-efficient static and incremental graph darwings. In *ESA '97*, volume 1284 of *LNCS*, pages 37–52. Springer-Verlag, 1997.
3. S. Bridgeman and R. Tamassia. Difference metrics for interactive orthogonal drawing. *J. Graph Alg. Appl.* to appear.
4. S. S. Bridgeman, J. Fanto, A. Garg, R. Tamassia, and L. Vismara. Interactive-Giotto: An algorithm for interactive orthogonal graph drawing. In *GD '97*, volume 1353 of *LNCS*, pages 303–308. Springer-Verlag, 1997.
5. E. Dengler and W. Cowan. Human perception of laid-out graphs. In *GD '98*, volume 1547 of *LNCS*, pages 441–443. Springer-Verlag, 1998.
6. E. Dengler, M. Friedell, and J. Marks. Constraint-driven diagram layout. In *Proc. IEEE Sympos. on Visual Languages*, pages 330–335, 1993.
7. G. Di Battista, A. Garg, G. Liotta, R. Tamassia, E. Tassinari, and F. Vargiu. An experimental comparison of four graph drawing algorithms. *Comput. Geom. Theory Appl.*, 7:303–325, 1997.
8. U. Fößmeier. Interactive orthogonal graph drawing: Algorithms and bounds. In *GD '97*, volume 1353 of *LNCS*, pages 111–123. Springer-Verlag, 1998.
9. J. E. Goodman and R. Pollack. Multidimensional sorting. *SIAM J. Comput.*, 12(3):484–507, Aug. 1983.
10. K. A. Lyons, H. Meijer, and D. Rappaport. Algorithms for cluster busting in anchored graph drawing. *J. Graph Algorithms Appl.*, 2(1):1–24, 1998.
11. A. Marzal and E. Vidal. Computation of normalized edit distance and applications. *IEEE Trans. on Pattern Analysis and Machine Intelligence*, 15(9):926–932, Sept. 1993.
12. K. Misue, P. Eades, W. Lai, and K. Sugiyama. Layout adjustment and the mental map. *J. Visual Lang. Comput.*, 6(2):183–210, 1995.
13. S. North. Incremental layout in DynaDAG. In *GD '95*, volume 1027 of *LNCS*, pages 409–418. Springer-Verlag, 1996.
14. A. Papakostas, J. M. Six, and I. G. Tollis. Experimental and theoretical results in interactive graph drawing. In *GD '96*, volume 1190 of *LNCS*, pages 371–386. Springer-Verlag, 1997.
15. A. Papakostas and I. G. Tollis. Interactive orthogonal graph drawing. *IEEE Trans. Comput.*, C-47(11):1297–1309, 1998.
16. K. Ryall, J. Marks, and S. Shieber. An interactive system for drawing graphs. In *GD '96*, volume 1190 of *LNCS*, pages 387–393. Springer-Verlag, 1997.
17. R. Tamassia, G. Di Battista, and C. Batini. Automatic graph drawing and readability of diagrams. *IEEE Trans. Syst. Man Cybern.*, SMC-18(1):61–79, 1988.
18. W. A. Wickelgren. *Cognitive Psychology*. Prentice-Hall, Inc., Englewood Cliffs, NJ, 1979.

Interactive Partitioning

System Demonstration, Short

Neal Lesh[1], Joe Marks[1], and Maurizio Patrignani[2]

[1] MERL — A Mitsubishi Electric Research Laboratory, Cambridge, MA 02139
{lesh,marks}@merl.com

[2] Dip. di Informatica e Automazione, Università di Roma Tre, Rome, Italy
patrigna@dia.uniroma3.it

Abstract. Partitioning is often used to support better graph drawing; in this paper, we describe an interactive system in which graph drawing is used to support better partitioning. In our system the user is presented with a drawing of a current network partitioning, and is responsible for choosing appropriate optimization procedures and for focusing their application on portions of the network. Our pilot experiments show that our network drawings succeed in conveying some of the information needed by the human operator to steer the computation effectively, and suggest that interactive, human-guided search may be a useful alternative to fully automatic methods for network and graph partitioning.

1 Introduction

Interactive, semi-automatic systems have been developed for several conventional optimization tasks, such as routing, scheduling, and layout; interactive systems have also been developed for unconventional optimization tasks such as geometric design and data analysis. One motivation to involve people in an optimization process is to combine a computer's fast processing speed with a human's superior ability in such areas as visual perception, learning from experience, and strategic assessment.

Bringing a human into the loop requires the development of operators with which a human user can steer or focus the optimization process, and visualizations that allow the user to apply these operators effectively. In recent work we have investigated a very tight integration of human and computer to solve optimization problems; we call our approach Human-Guided Simple Search (HuGSS), and we have applied it successfully to the problem of capacitated vehicle routing with time windows [3]. We are currently trying to apply the same general approach to the problem of k-way network partitioning, an important NP-hard problem that arises in VLSI design and elsewhere [2].

In this paper we demonstrate the visualization aspects of our interactive network-partitioning system. The networks that we consider are derived from chip circuits modeled as hypergraphs; these hypergraphs have up to $200,000$ nodes and as many hyperedges. We need to visualize those aspects of a network that are relevant to partitioning it optimally: the number and strength of

J. Marks (Ed.): GD 2000, LNCS 1984, pp. 31–36, 2001.

hyperedges in the cut set; the size and structure of the partition blocks; and ultimately the partition blocks and network nodes on which to focus the search for best results. Furthermore, the computational time of the whole visualization process should be appropriate for an interactive environment.

2   Problem Description

A *network* or *hypergraph* G is a pair (V, H), where V is a set of *nodes* and H is a set of nonempty subsets of V, called *hyperedges*. *Constrained k-way partitioning* is the NP-hard problem of partitioning the nodes of a network into k disjoint subsets, called *blocks*, so as to minimize the number of hyperedges spanning two or more blocks. The sizes of the blocks are allowed to vary around the value $\frac{n}{k}$, where $n = |V|$; a typical value for the allowed variance is 10%. In current state-of-the-art benchmark problems, $|V|$ and $|H|$ range from 10,000 to 200,000, and k is less than 10 [1]. A recent survey describes the many heuristics that have been devised for this problem [2].

Typical heuristics for difficult combinatorial optimization problems combine some form of gradient descent to find local minima with some strategy for escaping nonoptimal local minima and exploring the solution space. The HuGSS framework for interactive search [3] divides these two subtasks cleanly between human and computer: the computer is responsible only for finding local minima using a simple search method; using visualization and interaction techniques, the human user identifies promising regions of the search space for the computer to explore, and intervenes to help it escape nonoptimal local minima.

The HuGSS approach translates to the following search and focus operators available to the user in our network-partitioning system:

1. Manually edit the current partition by moving nodes between blocks.
2. Launch a refinement heuristic on the whole network, or on a focused subset of the blocks or nodes. Several heuristics are available in our current prototype, ranging from simple hill-climbing methods to more sophisticated techniques.
3. Navigate a history list of previous solutions and revert to an earlier one.

The key to our system is the effective selection and focusing of search heuristics in Step 2 by the human user. For this to be possible, the user must be able to identify fruitful areas of the network on which to concentrate the computer's search, and also be able to choose appropriate heuristics for the subproblems encountered in the focus areas. For example, he might try to identify groups of nodes that are loosely connected to nodes in their current block but strongly connected to nodes in other blocks. Including such nodes in a focused search will allow the computer to spend more effort looking at ways to rearrange just these nodes to achieve a better partition. And depending on the search focus and the nature of the current partition, some search heuristics might be better than others: for example, a tightly focused search might benefit from the application of a more thorough search heuristic than one could afford to use on the whole network.

3   Our Network Visualization

We designed our visualization to help the user make decisions about how to focus the search and to select search heuristics.

Because it is difficult to unambiguously draw a hypergraph, we convert the hypergraph into a graph with weighted edges. We replace each hyperedge of degree n with a clique of $\frac{n(n-1)}{2}$ edges whose weights are $\frac{2}{n(n-1)}$.

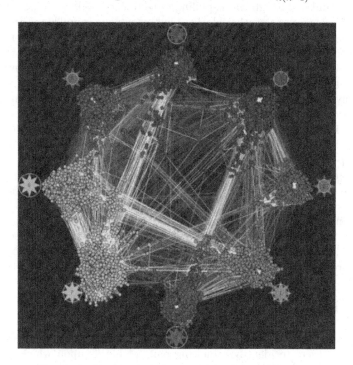

Fig. 1. A visualization of a partition of network ibm01.

We use a force-directed approach to determine the position of each node. Each edge in the induced graph is replaced by a spring that opposes stretching or compressing with a force proportional to its weight. Furthermore, a spring is attached between each node and the *hub* of its block. The hubs themselves are fixed and positioned uniformly around a circle. For reasons of efficiency, no other forces are considered, including repulsive forces between nodes.

A drawing produced with this algorithm is shown in Figure 1. Since each node is attracted towards the hub of its block, nodes in the same block tend to overlap near their hub, and only the ones that are strongly linked to other blocks stretch out and are easily distinguishable. These are typically the nodes that interest us the most, since they are those most likely to move usefully between blocks.

In addition to node position, other visual methods are used to convey information about the partition. Membership in a block is indicated by node color.

Edge weights are mapped onto intensity, so that edges of greater weight appear brighter. And at each hub an icon comprising a star-shaped polygon and circle indicates whether the block is near its minimum or maximum allowed size.

The visualization in Figure 1 is useful for comprehending the general structure of a partition and the relative sizes of the blocks. Figure 2 shows the result of a second force-directed system we designed specifically for visualizing pairs of blocks in isolation. We install a spring for each edge between nodes of the block pair and add a rightwards-pulling force to each node in proportion to the weight of its edges to the other blocks. Thus, each node is pulled towards the top or bottom of the screen, depending on its affinity for the top or bottom block and pulled towards the right depending on its affinity for the other blocks. The nodes in the left center of the screen are those that are strongly connected to both blocks and weakly connected to the other blocks (see Figure 2a); a focused search involving these nodes[1] might have several chances for reducing the cut set between the blocks by exchanging some of these nodes. A better scenario is shown in Figure 2d: it is clear that many nodes in the bottom block would prefer to be in the top block, but it is full—the circle surrounds the block icon in the top-left corner—and cannot accommodate them at the moment. However, if nodes can be moved out of the top block without incurring additional cut-set costs (we envision providing an operator that allows the user to request such an adjustment), the bottom-block nodes might then be included with some likely savings in cut-set cost. Finally, Figures 2b and 2c show less promising cases: there are few nodes loosely connected to the other blocks that are also strongly connected to both blocks in the pair; moving nodes between these blocks is less likely to reduce the cut-set cost.

4   Preliminary Experiments

We ran some initial experiments to measure the effectiveness of our visualization in enabling users to identify promising pairs of blocks upon which to focus a simple refinement algorithm. Although these "selection-only" experiments are only a crude test –we anticipate focusing searches at a finer granularity in the finished system–it served to validate that our visualizations contain useful cues.

In each experiment, a group of two or three users (the authors of the system) were presented with a partition of a large network. Using only the visualization capabilities of our system (i.e., they could not launch any optimization algorithms or move any nodes between blocks manually), their task was to select and rank the most promising pairs of blocks for subsequent focused searches.

For our experiments we obtained 8-way partitions of the ibm04,..., ibm08 networks from the benchmark set in [1] by running the Sanchis refinement algorithm ([5]) on the best solution produced by 200 runs of the hMeTis system [4], one of the most widely used network-partitioning systems. We computed a score

---

[1] A more useful search might focus on these nodes and their close neighbors in the hypergraph structure. Our interface allows the user to expand the current set of focused nodes to include all their neighbors.

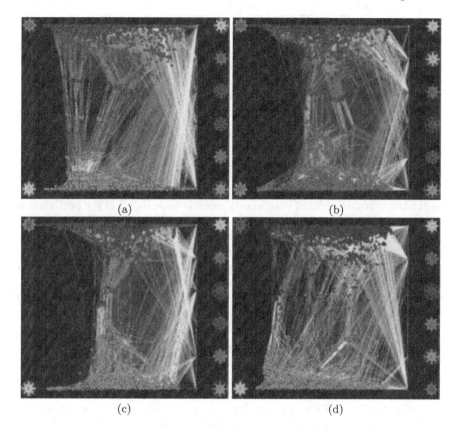

(a)   (b)

(c)   (d)

Fig. 2. Visualizing pairs of blocks

for each pair of blocks by running the hMeTis algorithm on just those blocks 20 times and then launching the Sanchis algorithm on the best partition found. To compute the score for each pair we repeated this process 10 times and averaged the results.

As shown by the solid line in Figure 3, our visualizations easily allow people to select the most promising pairs of blocks. For $n = 1$ to 14, the graph shows the average sum of the scores of the first n selected pairs divided by the total sum of the scores of all the pairs. For example, the scores of the first three pairs selected were, on average, about 50% of the sum of the scores of all 28 possible pairs. This shows that the user selections were far better than the expected value of random selections, as indicated by the dotted line in the chart.

However, the relatively simple heuristic of ranking the pairs of blocks by the total weight of the edges between them also produced excellent results, just slightly worse than that achieved by human selection. Although the difference is small (after three selections human selection is 8% better than the size heuristic on average), it is not insignificant: given the current state of the art and the com-

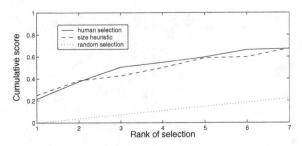

Fig. 3. Experimental results

mercial significance of VLSI design, very small improvements in the performance of partitioning systems are both hard to achieve and very useful. Furthermore, when we correctly selected block pairs out of order relative to the size heuristic, we usually did so based on our observation of complex patterns in the visualizations, as indicated by the discussion of Figure 2. We conjecture that future work will allow us to take greater advantage of these observations by employing search and focus operators that are more sophisticated than simply applying the same refinement algorithm to all the nodes in a block pair.

5   Conclusions and Future Work

Traditionally graph drawing has been used to support analysis-oriented tasks involving databases, software engineering, and computer and communication networks. In this paper we demonstrate the potential use of graph drawing to support an optimization-oriented task, network partitioning. Although at an early stage in our research, we have implemented a fully working system and established that graph drawing can supply useful information about large network partitions to a human user. In future work we plan to refine our set of search and focus operators and devise experiments to see if human users can use our system to improve state-of-the-art solutions to network-partition problems.

References

1. C. J. Alpert. The ISPD98 circuit benchmark suite. In *Proc. of the Intl. Symposium of Physical Design (ISPD'98)*, pages 80–85, 1998.
2. C. J. Alpert and A. B. Kahng. Recent directions in netlist partitioning: A survey. *Integration: The VLSI Journal*, 19:1–81, 1995.
3. D. Anderson, E. Anderson, N. Lesh, J. Marks, B. Mirtich, D. Ratajczak, and K. Ryall. Human-guided simple search. To Appear in Proc. of AAAI 2000. Also http://www.merl.com/reports/TR2000-16/index.html.
4. G. Karypis and V. Kumar. Multilevel *k*-way hypergraph partitioning. In *Proc. of the 36th Design Automation Conference*, pages 343–348, 1999.
5. L. A. Sanchis. Multiple-way network partitioning. *IEEE Trans. on Comp.*, 38:62–81, 1989.

An Experimental Comparison of Orthogonal Compaction Algorithms⋆

(Extended Abstract)

Gunnar W. Klau[1], Karsten Klein[2], and Petra Mutzel[3]

[1] Technische Universität Wien, Austria, gunnar@ads.tuwien.ac.at
[2] MPI für Informatik, Saarbrücken, Germany, karsten@mpi-sb.mpg.de
[3] Technische Universität Wien, Austria, mutzel@ads.tuwien.ac.at

Abstract. We present an experimental study in which we compare the state-of-the-art methods for compacting orthogonal graph layouts. Given the shape of a planar orthogonal drawing, the task is to place the vertices and the bends on grid points so that the total area or the total edge length is minimised. We compare four constructive heuristics based on rectangular dissection and on turn-regularity, also in combination with two improvement heuristics based on longest paths and network flows, and an exact method which is able to compute provable optimal drawings of minimum total edge length.

We provide a performance evaluation in terms of quality and running time. The test data consists of two test-suites already used in previous experimental research. In order to get hard instances, we randomly generated an additional set of planar graphs.

1 Introduction

Orthogonal graph drawing is getting increasing attention from industry because of its numerous applications, *e.g.*, in database design, software engineering and many more. For many of these applications, the *topology-shape-metrics approach* leads to the best results. Here, a first phase (planarisation) determines the topology of the drawing. The aim is to generate a drawing with a small number of edge crossings, *e.g.*, by computing a large planar subgraph and carefully reinserting the removed edges. Then, the crossings are replaced by artificial vertices resulting in a planar (also called planarised) graph. A second phase determines the shape of the final layout. This phase is often restricted to planar orthogonal drawings (orthogonalisation). A widely accepted optimisation criterion is to minimise the number of bends without changing the topology. Finally, the third phase (compaction) deals with computing the metrics of the layout. Here, the optimisation problem is to compute a layout of minimum area or with minimum total edge length.

⋆ This work is partially supported by the Bundesministerium für Bildung, Wissenschaft, Forschung und Technologie (No. 03–MU7MP1–4).

J. Marks (Ed.): GD 2000, LNCS 1984, pp. 37–51, 2001.
© Springer-Verlag Berlin Heidelberg 2001

Although Batini *et al.* [1,2] suggested the topology-shape-metrics approach already in 1984, it took some years until the approach was getting popular. The reason for this is twofold: On the one hand, expertise in planarity algorithms, combinatorial embeddings, planar graph drawing algorithms and combinatorial optimisation is necessary in order to deal with the upcoming combinatorial graph problems—most of them are *NP*-hard, including the compaction problem we are dealing with in this paper. On the other hand, it is time consuming to implement the approach, since many different kinds of algorithms are needed.

Recently, a lot of research has been done to solve many of the upcoming problems. Major improvements have been made concerning various aspects of the planarisation phase, *e.g.*, [3,4,5], the orthogonalisation phase, *e.g.*, [6,7,8,9], and the compaction phase [10,11]. Moreover, today there exist some software libraries containing the topology-shape-metrics approach [12,13,14,15]. Often it is implemented in a modular form, so that it is easy to experiment with.

This enables experimental comparisons between various algorithms in order to understand their influence on the final drawing. Already in [16], the impact of choosing two different algorithms for the orthogonalisation phase has been analysed and compared to two different orthogonal drawing methods. The experimental results showed that the topology-shape-metrics approach is superior to other orthogonal methods. Experimental comparisons for various hierarchical drawing methods appeared in [17,18] and for force-directed methods in [19].

In [11], we have suggested an exact method for compacting orthogonal drawings. Our implementation is able to solve instances of up to 1,000 vertices to provable optimality in short computation time. Independently, Bridgeman *et al.* came up with a new heuristics for the compaction problem in [10]. We were interested how the behaviour of the various compaction heuristics are related to the optimal solutions. Moreover, we wanted to find the best strategies among the heuristic methods. These questions led to the present computational study.

We compare the following constructive methods for orthogonal compaction: the original dissection method introduced in [20] based on longest paths, the same method based on network flows, two methods based on turn-regularity described in [10], and the exact method suggested in [11]. An orthogonal drawing can be improved by using iteratively one-dimensional compaction methods based on longest paths or network flows. We have tested all possible combinations of construction and improvement heuristics.

Section 2 introduces the orthogonal compaction problem formally. The compaction methods under evaluation are described in Section 3. We present computational experiments on various benchmark sets of graphs in Section 4. The paper closes with our conclusions in Section 5.

2   The Compaction Problem

In this section, we present a precise formulation of the compaction problem we are considering in this paper. We assume familiarity with planarity and basic graph theory.

Let $G = (V, E)$ be a 4-planar graph, *i.e.*, a planar graph whose maximum vertex degree does not exceed four. We associate with each undirected edge $(v, w) \in E$ two directed *half-edges* (v, w) and (w, v). A *planar representation* P for G describes the topology of a drawing for G in the plane by specifying the set of faces F as lists of counter-clockwise ordered half-edges and one explicit external face f_0.

An *orthogonal representation* H for G is an extension of P and describes, in addition to the topology, the *shape* of a drawing for G by specifying the bends in the edges and angles inside the faces. It is an equivalence class of planar orthogonal drawings. Two orthogonal drawings belong to the same class if one can be obtained from the other by rotating the drawing and modifying the lengths of the horizontal and vertical edge segments without changing the angles formed by them.

We call an orthogonal representation *simple* if its number of bends is zero. In the following, we always assume that an orthogonal representation is simple by treating bends as artificial vertices. A simple orthogonal representation H is defined by giving for each half-edge in P the angle it forms with its cyclic successor in the same face. A *planar orthogonal grid drawing* for a 4-planar graph maps vertices and bends to distinct grid points and edge segments to horizontal or vertical non-crossing non-empty line segments in the grid which connect the images of their endpoints. A drawing for a simple orthogonal representation H is a planar orthogonal drawing for the corresponding 4-planar graph which respects the shape coded in H. As with representations, we call orthogonal drawings simple if they do not contain bends.

Problem (Two-dimensional compaction problem in orthogonal graph drawing). *Given a simple orthogonal representation H for a 4-planar graph, find a simple planar orthogonal drawing for H of minimum total edge length.*

Patrignani shows in [21] that the compaction problem—and the related problems which ask for minimum area and minimum maximum edge length—are NP-complete.

3   Orthogonal Compaction Algorithms

In this section we briefly introduce the compaction algorithms under evaluation. We first describe the constructive techniques which produce a drawing for a given orthogonal representation H. The key idea behind these methods is to transform H into an auxiliary representation H' by introducing artificial edges and vertices and to find a drawing for H' in polynomial time. Removing the artificial vertices and edges in the auxiliary drawing leads to a drawing for H. Unfortunately, in the general case, these drawings can be far away from an optimal solution. We therefore present techniques which operate directly on drawings in order to improve the total edge length and area. Finally, we summarise an approach based on an integer linear program (*ILP*) to compute an optimal drawing for H.

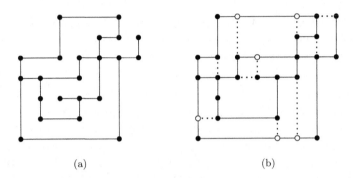

(a)   (b)

Fig. 1. The dissection method. (a) Original representation H, (b) Transformed representation H'. Dashed lines and empty vertices represent artificial edges and vertices

3.1 Constructive Heuristics

Tamassia mentions the first and still most common method to produce an auxiliary representation H' which is easier to deal with in his ground-breaking paper on bend-minimisation [20]. He introduces the *dissection method* which consists of decomposing each internal face of the given simple orthogonal representation H into a set of faces each of which has rectangular shape by introducing artificial vertices and edges. Figure 1 illustrates this method with an example—please note that the method works at the level of the representation; the coordinates have not yet been assigned. This process can be done in $O(n)$ time where n denotes the number of vertices in H. In the resulting orthogonal representation H', all interior faces have rectangular shape. Of course, the artificial vertices and edges impose additional constraints on the geometry which may lead to suboptimal total edge length and area in the resulting drawing.

Bridgeman *et al.* present in [10] another, more sophisticated, approach to produce a polynomial-time compactable auxiliary representation H'. Using the concept of *turn-regularity*, they manage to introduce a significantly lower number of artificial vertices and edges. Let f be a face in H. With each occurrence of a vertex v on the boundary of f, one or two *corners* are associated with v, depending on the angle internal to f between the edges preceding and following v. For each ordered pair of corners (c_i, c_j) associated with vertices of f, let $\rho(c_i, c_j)$ be the difference of the left and right turns along the boundary of f between c_i (included) and c_j (excluded). The value $\rho(c_i, c_j)$ defines the net angle between the edges preceding the vertices associated with c_i and c_j. Two corners at angles of at least 270 degrees are called *kitty corners* if $\rho(c_i, c_j) = 2$ or $\rho(c_j, c_i) = 2$. A face of an orthogonal representation is *turn-regular* if it has no kitty corners. Observe that rectangular faces are turn-regular. A representation is called *turn-regular*, if all its faces are turn-regular. We describe two methods from [10] to transform H into a turn-regular representation H'.

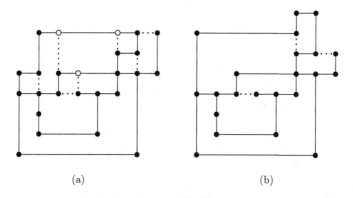

(a)   (b)

Fig. 2. The two turn-regularity-based dissection methods

The first heuristics uses the rectangular dissection method described above, but only for faces in H which are not turn-regular (see Fig. 2(a)). The second heuristics recursively adds an artificial edge between each pair of kitty corners until the face has been decomposed into smaller turn-regular, but not necessarily rectangular, faces. The direction of the inserted edge (vertical or horizontal) is chosen randomly. See Fig. 2(b).

Two methods can be used to compute drawings for an orthogonal representation H' in which all internal faces have rectangular shape; variants of these methods generate a drawing if H' is general turn-regular. The methods, longest path-based compaction and flow-based compaction, are one-dimensional methods, *i.e.*, they assign horizontal and vertical coordinates separately—we therefore restrict our description to the computation of x-coordinates, the same techniques can be used for the y-coordinates.

We construct a directed graph D_x as follows: Each maximally connected vertical path in H' corresponds to a node in D_x—for a vertex v in H' we denote by $\text{vert}(v)$ the unique node in D_x it belongs to. For each horizontal edge $e = (v, w)$ in H' we assume that it is directed from left to right and insert an arc $(\text{vert}(v), \text{vert}(w))$ into D_x. Figure 3(a) shows this construction for the orthogonal representation from Fig. 1(b). Generally, we will refer to the pair of graphs in which nodes correspond to maximal horizontal and vertical paths and arcs to distance relations between these paths as *constraint graphs*.

Topologically sorting the nodes in D_x by computing longest paths and setting the x-coordinate of each vertex in H' to the topological number of $\text{vert}(v)$ leads to a feasible assignment of x-coordinates; y-coordinates result from a similar computation in the directed graph D_y. This approach, illustrated in Fig. 3(a), yields a drawing for H' with minimum width, height and area, but in general not minimum edge length. The running time is $O(n)$.

A more elaborate method which also minimises the total edge length for H' is the flow-based compaction. It assigns topological numbers for the nodes in D_x

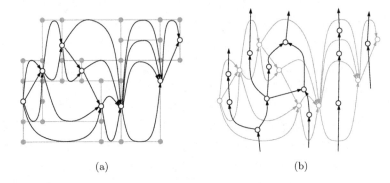

(a)   (b)

Fig. 3. (a) The graph D_x in the longest path-based compaction method. (b) The dual network D_x^* in the flow-based compaction method. Unconnected arcs are linked with an undisplayed node corresponding to the external face

by computing a minimum cost flow in its dual graph D_x^*. Flow through an arc a in D_x^* determines the length of the edge corresponding to the primal arc in D_x, thus a has a lower bound of one, infinite capacity and unit cost. See Fig. 3(b) for an example of the method which has a running time of $O(n^{7/4} \log n)$. If H' is turn-regular but not all faces have rectangular shape, the construction of the flow network is more complicated. In this case additional arcs which correspond to so-called *saturating* edges connecting switches in two oriented copies of H' have to be included in the flow network. For details see [10].

3.2   Improvement Heuristics

The constructive heuristics have a serious drawback: Though removing the artificial objects in the drawing for H' leads to a feasible drawing for H it is in general not the best one. Fortunately, variants of the longest path-based and flow-based compaction methods can be used to operate directly on the layout (this has also been used in VLSI-design; see, *e.g.*, [22]).

Unlike above, where the arcs in D_x correspond to edges in H', we insert arcs based on the visibility properties in the layout. If a maximally connected vertical path "sees" another one to its right, we insert an arc between the corresponding nodes in D_x. This will preserve the one-dimensional relative positions. Topologically sorting D_x or computing a flow in D_x^* results in new x-coordinates. Alternating the direction of the compaction and performing another step results in an iterative process. However, at each step the decisions are purely local, and compaction in one direction may prevent greater progress in the other direction. Furthermore, the layout may be blocked in both dimensions, but still be far away from an optimal solution.

3.3   Optimal Compaction

In [11], Klau and Mutzel present an ILP-based approach to solve instances of the two-dimensional compaction problem to optimality. It is based on a characterisation of the set of feasible solutions in terms of paths in the pair of constraint graphs. Given a pair of constraint graphs in which only the relative positions known from the shape of H are present (*shape graphs*), the compaction problem can be seen as optimising over the set of certain extensions of these graphs. The quality that such an extension must comply is based on geometric properties and establishes a direct link between the two graphs D_x and D_y. The new combinatorial task can be naturally formulated as an ILP which can be solved using a branch-and-cut algorithm. If there is only one possible extension of the given shape graphs, the authors show that their algorithm runs in polynomial time. Even if the algorithm does not find an optimal solution, it will report a feasible solution with a quality guarantee.

4   Computational Experiments

In this section we present a selection of our computational results for the comparison of compaction techniques. We concentrate on the—in our opinion—most interesting and usable experiments. We provide the full data at [23].

Experimental Settings. Each of the compaction strategies introduced in Sect. 3 is available as a module inside the AGD library (see [13]). Many heuristics rely on flow computations: In all modules we use the LEDA-function for this task. Our implementation of the ILP-based method uses CPLEX, an ABACUS-version is available. Here, we set a time limit of 15 minutes CPU time and return the upper and lower bound.

We test the implementations of the constructive heuristics from Sect. 3.1 both stand-alone and in combination with implementations of the two improvement heuristics from Sect. 3.2. Additionally, we test the implementation of the ILP-based algorithm. We run the implementations of the 13 resulting compaction techniques on a Sun Enterprise 450 with 1.1 GB main memory and two 400 MHz-CPUs and refer to them using the following scheme: We call the implementations of the constructive heuristics LP, FL, T1 and T2, corresponding to the rectangular dissection method with longest path compaction and with flow compaction and the two variants of turn-regularity-based dissection with flow compaction. If we apply an improvement heuristics, we append either LP or FL. We call the implementation of the ILP-based approach OPT. Figure 4 shows the output of some of the methods for an example.

For our experiments, we use three different groups of graphs which can roughly be divided in easy instances, practical instances and hard instances. Graphs which are relatively easy to compact and which have already been used in [10] are 4-planar biconnected graphs. We use a set of 500 graphs with 10 to 100 vertices. The set of more than 11,000 practical instances has been introduced in [16] and has since then become a widely used test-suite in experimental graph drawing.

Additionally, we generated 540 hard instances for the compaction problem with a graph generator in the LEDA library: To generate a planar graph with n vertices, the generator chooses n segments whose endpoints have random coordinates of the form x/K, where K is the smallest power of two greater or equal to n, and x is a random integer in $[0, \ldots, K-1]$. It then constructs the arrangement defined by the segments and keeps the n nodes with the smallest x-coordinates. Finally, it adds edges to make the graph connected. We call this test-suite quasi-trees because large subgraphs of the resulting graphs are trees. In [11], quasi-trees have shown to be hard instances of the compaction problem; they have many fundamentally different drawings which makes the compaction task difficult. We transform each test graph G into an instance of the two-dimensional compaction problem in the following way:

1. We compute a planarised graph G' using the planarisation method in the AGD library. Note that the number of vertices in G' equals the number of vertices in G plus the number of crossings. We then compute a planar embedding of G'.
2. We run the transformation phase of the Giotto algorithm [2]. The phase creates an auxiliary 4-planar graph G'_4 by replacing vertices of degree greater than four by artificial faces.
3. We use a variant of Tamassia's bend minimising algorithm in which the underlying network differs from the original one in [20]: a minimum cost flow corresponds to a bend-minimum shape in which, due to a second optimisation goal, the number of 180-degree angles between edges is maximum. This avoids unnecessary staircase-like structures in the shape and makes the following compaction task easier. We replace bends by artificial vertices in the resulting orthogonal representation and get a simple orthogonal representation H'_4 which we use as the input for the compaction algorithms.

We choose this strategy, because it is among the orthogonal methods which yield the best layouts in practice. Note that the number of nodes in H'_4 is higher than the number of nodes in G. Especially for the non-planar practical graphs, where not only the bends but also the crossings count as vertices, this leads to an uneven distribution of graph sizes. Because the number of crossings has a strong influence on the behaviour of the compaction algorithms, we provide the number of crossings and details on the distribution of the input data at [23].

Total Edge Length. First we consider the set of 4-planar graphs. We divide the instances in subgroups according to their sizes with steps of 20 vertices. For each group we compute the average total edge length first for the four constructive heuristics (see Fig. 5(a)), for the four methods with longest path-based improvement (undisplayed) and for the methods with flow-based improvement (Fig. 5(b)). We display the results relative to the optimum edge length which is provided by OPT. Longest path-based post-compaction does not lead to significant improvements in terms of total edge length. This behaviour could also be observed with the practical graphs and the quasi-trees, we therefore will not display the data for improvement with the longest path method (see also Fig. 4(e)).

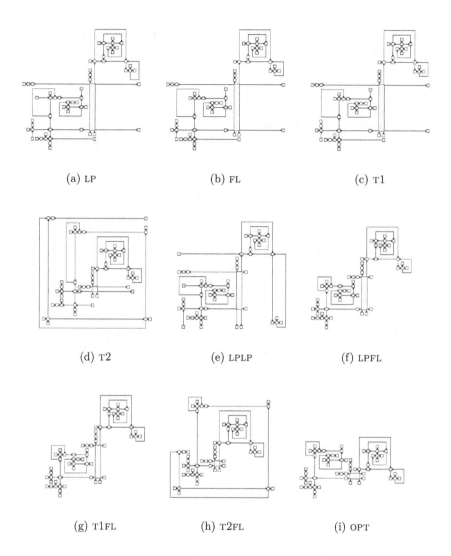

(a) LP   (b) FL   (c) T1

(d) T2   (e) LPLP   (f) LPFL

(g) T1FL   (h) T2FL   (i) OPT

Fig. 4. quasiTree.60.5.lgr compacted by different methods

Obviously the biconnected graphs are easy to compact: Already the worst constructive method in terms of edge length, LP, achieves quite good results. Improving the drawings with the flow method often results in an optimal drawing. The plot also indicates that the methods LPFL, T1FL and FLFL result in very similar values—which is also true for the other test-suites.

We partition the more than 11,000 graphs corresponding to practical data in the same way as the biconnected graphs. Figure 6 shows the resulting total edge length for different methods relative to the optimum value computed by OPT. On the one hand, the practical graphs behave like the biconnected graphs: The heuristics are close to the optimum value and almost reach it when using improvement with flow. On the other hand, bigger instances are easier to compact, whereas the biconnected graphs indicate the opposite. This is due to the high number of crossings as produced by the prior planarisation step. The higher this number, the simpler the shapes of the faces. For the bigger graphs, we often observe that almost all faces have rectangular shape; this explains the good performance of all methods for big planarised graphs.

A different view of the improvement with the flow method shows its influence on the quality more drastically. In Fig. 7, we grouped the graphs according to their size using a step size of 50 vertices and showed for each constructive heuristics its value before and after the improvement step. Again, it can be observed that the big planarised graphs are almost optimally compacted by the heuristics. Additionally, the plot illustrates that the choice of the constructive heuristics in the first step does not have a big impact of the final quality: what matters is the improvement with the flow method.

The most challenging instances for the compaction problem are the quasi-trees. We first consider the smaller instances where OPT could find the optimal solution or at least a very good bound within the time limit. These are 75 graphs with 40 to 85 vertices in the original graph, resulting in 44 to 121 vertices in the simple orthogonal representation. Figure 8 illustrates the quality of the heuristics with respect to the optimum value. Again, the best heuristics perform very well: Even for these hard instances, three of them (T1FL, T2FL and FLFL) always stay below 1.1 times the optimum value, alternating at the top position among the

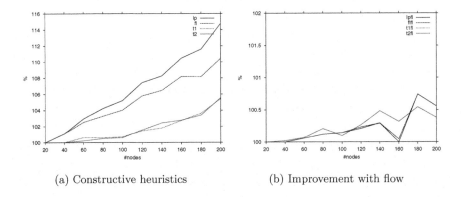

(a) Constructive heuristics   (b) Improvement with flow

Fig. 5. 4-planar biconnected graphs: total edge length relative to optimal value

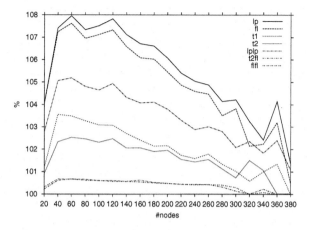

Fig. 6. Practical graphs: total edge length relative to optimal value

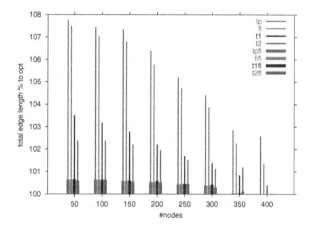

Fig. 7. Practical graphs: impact of the flow method

heuristic methods. Another observation is that the quality of the methods is relatively independent of the graph size.

We also look at the trend when the sizes of the quasi-trees grow and investigate the quality of the methods for 490 bigger instances in the range of 100 to 2,500 original vertices. Here, we choose T2FL as the comparison method, see Fig. 9. Again, T1FL, FLFL and LPFL are very close together and manage in some cases to beat the comparison method. In the plot, we display LPFL and the four constructive heuristics whose quality decreases as the instance sizes grow. It can be seen that the methods based on rectangular dissection perform similarly, as a stand-alone heuristics, T2 is the best. But using flow compaction as an improvement almost nullifies this advantage.

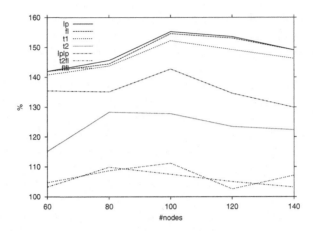

Fig. 8. Small quasi-trees: total edge length relative to optimal value

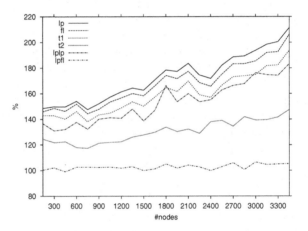

Fig. 9. Big quasi-trees: total edge length relative to T2FL

Running Time. All implementations run very fast on the biconnected and practical instances, where the constructive heuristics stay below .5 seconds and the improvement heuristics stay below .8 seconds on all instances. Even OPT which computes a provable optimal solution stays below four seconds on all instances. In the case of the hard compaction instances, most methods stay inside a five second time limit for graphs up to 1,000 vertices. On the large instances, the running time increases to up to 75 seconds for the flow improvement heuristics. Generally, there is a typical pattern for the performance of the methods which can be observed in all graph sets: Heuristics LP beats every other implementation in terms of running time. For the flow-based methods, FL has an advantage over T1 and T2, which are slowest among the constructive methods. This order remains the same when applying an improvement heuristics.

Other Criteria. In [10], Bridgeman *et al.* report a ratio of turn-regular faces of about 95% in their biconnected graphs suite. We have the same results with exception of the small quasi-trees, where only about 81% of the faces are regular.

We count the number of edges inserted during the dissection heuristics for LP, T1 and T2. For the biconnected and practical instances, the number of inserted edges for T1 is only about one tenth of the number for LP, and for T2, only about 3 percent of that number has to be inserted to achieve regular faces. In the case of harder instances, however, the number of edges to be inserted for T1 raises to roughly half the number of edges for LP and also the number for T2 raises relative to LP. Even for a small number of inserted edges, the choice which edge to insert in heuristics T2 can have a strong impact on the compaction result.

The maximal edge lengths and the area in the computed drawings behave similar to the total edge lengths for the given graph sets. For the biconnected and practical graphs, the average maximal edge lengths per subgroup lie close together. Among the constructive methods, T1 and T2 have a slight advantage over the other methods. As for the total edge length, the flow improvement has only a small impact on the results. The results for the quasi-trees, however, show a bigger variation. Heuristics T2 performs best among the constructive methods, but combination with the flow improvement can drastically reduce the lengths for all constructive methods. Here, too, the flow improvement dominates the quality of the result, the differences in the results of the initial methods vanish. The area and perimeter results show the same behaviour: In case of hard instances, flow improvement can help to enhance the quality and to close the gap created by the initial methods, otherwise the constructive heuristics lie close together and show only a slight improvement when flow is applied on their initial layout.

5   Conclusions

We have evaluated the state-of-the-art compaction techniques for orthogonal graph drawing in terms of quality and run time performance. We propose to divide the heuristic methods into constructive heuristics and improvement heuristics which yields many different compaction techniques. We have compared the results of the heuristics both against the optimal drawings and against each other. In our experiments, we have used three different test-suites to test the algorithms: easy, practical and hard instances and come to the following main conclusions:

Heuristics perform very well on most instances of the compaction problem. Especially for the data from the easy and practical instances we could observe an excellent behaviour, both in terms of quality and running time. We want to emphasise, however, that in some cases it is desirable to get an optimal drawing, *e.g.*, when quality is more important than running time. Moreover, the implementation of the ILP-based algorithm is essential to compare the heuristics.

The choice of the constructive heuristics does not matter as long as flow compaction is used as post-processing. Although the results for the constructive methods differ significantly, the differences vanish when using flow-based compaction in an improvement step. We therefore propose to use a simple constructive method, *e.g.*, rectangular dissection and longest path-based coordinate assignment, followed by a flow-based post-processing step, which yields also the best running time among the good heuristics. The implementational effort for this variant is low as compared to the relatively complex turn-regularity-based techniques.

Graphs with many crossings are easy to compact.
In the topology-shape-metrics approach, crossings are treated as artificial vertices. The more crossings a drawing contains, the simpler are the shapes of its faces. For the big planarised graphs in the practical test-suite we observe that in many cases almost all faces have rectangular shape. Such graphs provide easy instances of the compaction problem.

Experimental studies help to improve the algorithms.
When we had the idea of doing a comparison between the different compaction techniques, we had in mind to simply provide the missing methods as modules inside our graph drawing library and run everything on a large number of graphs. This turned out to be more difficult than expected but at the same time very useful. During the more than 500,000 compaction runs we still found a considerable number of errors in our implementations which we considered stable in the beginning. This helped us to understand the two-dimensional compaction problem in graph drawing more deeply and we are now convinced that testing algorithms on a really large number of graphs—including pathological instances—helps a lot to provide good implementations.

References

1. C. Batini, E. Nardelli, and R. Tamassia. A layout algorithm for data-flow diagrams. *IEEE Trans. Soft. Eng.*, SE-12(4):538–546, 1986.
2. R. Tamassia, G. Di Battista, and C. Batini. Automatic graph drawing and readability of diagrams. *IEEE Trans. Syst. Man Cybern.*, SMC-18(1):61–79, 1988.
3. M. Jünger and P. Mutzel. Maximum planar subgraphs and nice embeddings: Practical layout tools. *Algorithmica, Special Issue on Graph Drawing*, 16(1):33–59, 1996.
4. P. Mutzel and R. Weiskircher. Optimizing over all combinatorial embeddings of a planar graph. In G. P. Cornuéjols, R. E. Burkard, and G. J. Woeginger, editors, *Integer Programming and Combinatorial Optimization (IPCO '99)*, volume 1610 of *LNCS*, pages 361–376. Springer-Verlag, 1999.
5. C. Gutwenger, P. Mutzel, and R. Weiskircher. Inserting an edge into a planar graph. Technical report, Technische Universität Wien, 2000. Submitted for publication.
6. U. Fößmeier and M. Kaufmann. Drawing high degree graphs with low bend numbers. In F. J. Brandenburg, editor, *Graph Drawing (Proc. GD '95)*, volume 1027 of *LNCS*, pages 254–266. Springer–Verlag, 1996.
7. P. Bertolazzi, G. Di Battista, and W. Didimo. Computing orthogonal drawings with the minimum number of bends. In *Proc. 5th Workshop Algorithms Data Struct. (WADS '97)*, volume 1272 of *LNCS*, pages 331–344, 1997.

8. M. Eiglsperger, U. Fößmeier, and M. Kaufmann. Orthogonal graph drawing with constraints. In *Proc. 11th Symposium on Discrete Algorithms (SODA '00)*. ACM-SIAM, 2000.
9. G. W. Klau and P. Mutzel. Quasi-orthogonal drawing of planar graphs. Technical Report MPI-I-98-1-013, Max–Planck–Institut für Informatik, Saarbrücken, 1998.
10. S. Bridgeman, G. Di Battista, W. Didimo, G. Liotta, R. Tamassia, and L. Vismara. Turn-regularity and optimal area drawings for orthogonal representations. *Computational Geometry Theory and Applications (CGTA)*. To appear.
11. G. W. Klau and P. Mutzel. Optimal compaction of orthogonal grid drawings. In G. P. Cornuéjols, R. E. Burkard, and G. J. Woeginger, editors, *Integer Programming and Combinatorial Optimization (IPCO '99)*, volume 1610 of *LNCS*, pages 304–319. Springer–Verlag, 1999.
12. C. Gutwenger, M. Jünger, G. W. Klau, and P. Mutzel. Graph drawing algorithm engineering with AGD. Technical report, Technische Universität Wien, 2000.
13. AGD. *AGD User Manual*. Max-Planck-Institut Saarbrücken, Universität Halle, Universität Köln, 1999. http://www.mpi-sb.mpg.de/AGD.
14. Graph drawing toolkit: An object-oriented library for handling and drawing graphs. http://www.dia.uniroma3.it/~gdt.
15. N. Gelfand and R. Tamassia. Algorithmic patterns for orthogonal graph drawing. In S. Whitesides, editor, *Graph Drawing (Proc. GD '98)*, volume 1547 of *Lecture Notes in Computer Science*, pages 138–152. Springer-Verlag, 1998.
16. G. Di Battista, A. Garg, G. Liotta, R. Tamassia, E. Tassinari, and F. Vargiu. An experimental comparison of four graph drawing algorithms. *Computational Geometry: Theory and Applications*, 7:303–316, 1997.
17. G. Di Battista, A. Garg, and G. Liotta. An experimental comparison of three graph drawing algorithms. In *Proceedings of the 11th Annual Symposium on Computational Geometry (SoCG'95)*, pages 306–315, 1995.
18. M. Jünger and P. Mutzel. Exact and heuristic algorithms for 2-layer straightline crossing minimization. In F. J. Brandenburg, editor, *Proceedings of the 3rd International Symposium on Graph Drawing (GD'95)*, volume 1027 of *LNCS*, pages 337–348. Springer–Verlag, 1995.
19. F. J. Brandenburg, M. Himsolt, and C. Rohrer. An experimental comparison of force-directed and randomized graph drawing algorithms. In F. J. Brandenburg, editor, *Proceedings of the 3rd International Symposium on Graph Drawing (GD '95)*, volume 1027 of *LNCS*, pages 76–87. Springer–Verlag, 1996.
20. R. Tamassia. On embedding a graph in the grid with the minimum number of bends. *SIAM J. Comput.*, 16(3):421–444, 1987.
21. M. Patrignani. On the complexity of orthogonal compaction. In F. Dehne, A. Gupta, J.-R. Sack, and R. Tamassia, editors, *Proc. 6th International Workshop on Algorithms and Data Structures (WADS '99)*, volume 1663 of *LNCS*, pages 56–61. Springer–Verlag, 1999.
22. T. Lengauer. *Combinatorial Algorithms for Integrated Circuit Layout*. John Wiley & Sons, New York, 1990.
23. G. W. Klau, K. Klein, and P. Mutzel. An experimental comparison of orthogonal compaction algorithms. Technical Report TR-186-1-00-03, Technische Universität Wien, 2000. Online version at
    http://www.ads.tuwien.ac.at/publications/TR/TR-186-1-00-03.

GraphXML — An XML-Based Graph Description Format

Ivan Herman and M. Scott Marshall

Centre for Mathematics and Computer Sciences (CWI)
P.O. Box 94079, 1090 GB Amsterdam, The Netherlands
Email: {I.Herman, M.S.Marshall}@cwi.nl

Abstract. GraphXML is a graph description language in XML that can be used as an interchange format for graph drawing and visualization packages. The generality and rich features of XML make it possible to define an interchange format that not only supports the pure, mathematical description of a graph, but also the needs of information visualization applications that use graph–based data structures.

1 Introduction

Most graph drawing systems could benefit from a textual description language to input and, possibly, to output graphs in a human readable form. Ideally, such a description language could also serve as a standardized format for data exchange between systems, enabling information exchange. Defining such a description language is not, by itself, a particularly difficult task: after all, a graph is simply a collection of nodes and edges. Several formats have been proposed in the past such as GML[1] and WebDot's DOT format[2], and the formats used by Rigi[3], LEDA[4], and GDS[5]. None of these are universally supported and they are usually bound to specific systems.

An important development of recent years, which also influences the choice of interchange formats, is the synergy between graph drawing techniques and information visualization. Information visualization has become a well-known research area, with important industrial and scientific applications. Graph drawing techniques play a prominent role in information visualization because the data structures to be visualized can often be described as graphs. The demands of information visualization pose new challenges with respect to graph description formats. It is necessary to include features in a graph description language that are not directly relevant to pure graph drawing. For example, the graph description should be able to include application dependent data[1], either embedded in the description itself or externally referenced (e.g., network statistics for a Web visualizer, genetic data for consensus trees as used in evolutionary research, database references for the result of a database search, etc.). Furthermore, it should be possible to describe composite structures such as nested

---

[1] Although GML, for example, is a capable description language for graph drawing purposes and includes provisions for extension, the mechanism for associating external data with a graph element is not well defined.

J. Marks (Ed.): GD 2000, LNCS 1984, pp. 52-62, 2001.

hierarchies and clusters. Another less obvious feature is to support the description of the evolution of graphs in time for applications such as animation.

As part of a larger project in graph visualization, we needed a graph description language. Although, initially, we looked for an existing standard, like the ones cited above, none could fully support the needs of information visualization. Consequently, we decided to develop our own format with particular attention to the requirements of information visualization. The result is GraphXML, a graph description language based on XML. Description of this language is the subject of this paper. We hope that this format will enter widespread use. In our view, this would be beneficial for the graph visualization community.

2 Why XML?

XML is a language specification developed by the World Wide Web Consortium that has received significant attention in the last few years[2]. The future evolution of the Web, both in the traditional Internet domain and in mobile communications, is based on XML. There are several reasons for choosing XML as the basis for a graph description format:

- XML defines clear syntactic rules for specifying a "language" for a particular application. An application-specific language is defined in a file called a Document Type Definition (DTD). The end–user also has the option of adding extensions to the language specified in the DTD.

- XML is used by many different applications to define data formats including those for databases, chemical compound definitions, e-commerce, mobile devices, and schematic graphics. A graph interchange format based on XML has a greater chance of being accepted by other application communities.

- There are a number of XML–based specifications that are being defined by communities, both within and outside the World Wide Web Consortium. GraphXML can take advantage of some of these existing standards. See Section 0 for examples.

- Software tools are emerging, which are either based on XML or work with XML. For example, there are a number of both commercial and public domain XML editors. These tools can be very helpful in managing graph description files that are based on GraphXML.

- Several XML parser tools are freely available in Java, C, and C++. A full-featured parser that provides error management and syntax checking can easily be generated using these tools. The main task is to define the semantic interpretation specific to the application (in our case, GraphXML).

---

[2] There are hundreds of books on XML, so rather than pick a specific reference we refer to the original specification[6], which is available on-line.

In what follows, an overview of the main features of GraphXML is given. A more complete description of GraphXML, including the exact specification, is available in [7].

3   Graph Structures in GraphXML

3.1 A Simple Example

The following code segment shows the simplest possible use of GraphXML. It describes a graph with two nodes and a simple edge:

```
1    <?xml version="1.0"?>
2    <!DOCTYPE GraphXML SYSTEM "file:GraphXML.dtd">
3    <GraphXML>
4      <graph>
5        <node name="first"/>
6        <node name="second"/>
7        <edge source="first" target="second"/>
8      </graph>
9    </GraphXML>
```

This example shows the basic style of a graph description in GraphXML. It resembles the way HTML documents are written, albeit using different tags. This example contains all the elements that are necessary to describe a purely mathematical graph.

The first line is required in all XML files. The second line specifies that this is an XML application based on the GraphXML DTD contained in the file GraphXML.dtd[1]. Finally, the third and the last lines enclose the real content of the files. The real content begins with line number 4, which defines a full graph. We delineate graph definitions with the <graph> tag so that a file can contain several graph definitions. The body of the graph description is straightforward: two nodes and a connecting edge are defined.

Attributes can also be defined for each of the elements: these are key–value pairs. The GraphXML DTD defines the set of allowable attributes for each element and a validating parser will to check those. It is partly through those attributes that additional information about nodes, edges, or graphs can be conveyed to the application. For example, the <graph> tag can use keys such as version, vendor, preferred-Layout, isPlanar, isDirected, isAcyclic, or isForest.

3.2   Application-Dependent Data

Application data can be added to different levels of the graph description through a series of additional elements. These elements are meant to represent the different types of application or domain-dependent data that can be associated with a graph, a node, or an edge. GraphXML defines the following application data:

---

[1] This line specifies that the DTD is on the local file system but could be replaced by a URL specification. In this way, it is possible to make the DTD publicly available on the World Wide Web.

- **Labels** (`<label>` tag): whereas the `name` attribute is used for unique identification, the label can contain any kind of text and does not have to be unique. Applications can use these to label nodes and edges.

- **Data** (`<data>` tag): domain-dependent data represented by a node, edge, or even the full graph[1].

- **Data references** (`<dataref>` tag): data that is referenced externally rather than embedded in the graph description file.

The format of external references (within the `<dataref>` tag) follows the specification of a separate document of the World Wide Web consortium, called XLink[8]. For our purposes, the virtually identical URL format used in HTML will suffice for external references.

The following example describes the same graph as in the previous section, except that the first node has associated data[2].

```
1        <node name="first">
2         <label>Project Home page</label>
3          <data> CWI Information Visualization project</data>
4          <dataref>
5           <ref xlink:role="Lead"   xlink:href="http://…/~ivan"/>
6           <refxlink:role="Descr"xlink:href="http://…/InfoVisu"/>
7          </dataref>
8         </node>
9        <node name="second"/>
10       <edge source="first" target="second"/>
```

Note the use of the `xlink:role` attribute in the example (see lines 0 and 0), which can be used to describe what the exact role of the link is. This can provide a useful indication to the application.

3.3 Hierarchical Graphs

All the examples mentioned until now refer to a single graph. However, information visualization applications are often used on very large graphs, and one of the ways of handling the size problem is to define the information in terms of a hierarchy of graphs, or clusters, instead of one single graph. What this means is that the nodes of one graph can refer to other graphs, these can refer to yet another graph, and so on. Powerful techniques exist to hierarchically cluster graphs and to visualize the hierarchies (see the survey of Herman *et. al.*[9]). GraphXML offers a way to describe such graph hierarchies. Consider the following example:

```
1        <graph id="L-1">
2          <node name="first"/>
3          <node name="second"/>
```

---

[1] Although XML describes everything in terms of strings, it has its own formalism, called entities and notations, which can be used to include binary data, too. However, using external references, through the `<dataref>` tag, may be more appropriate for this.

[2] From now on, we are omitting the header part from the examples to save space.

```
4              <edge source="first" target="second"/>
5          </graph>
6          <graph id="L-2">
7            <node name="third"/>
8            <node name="fourth"/>
9            <edge source="third " target="fourth"/>
10         </graph>
11         <graph id="levelTwo">
12           <node isMetanode="true" name="cluster1" xlink:href="#L-1"/>
13           <node isMetanode="true" name="cluster2" xlink:href="#L-2"/>
14           <edge source="cluster1" target="cluster2"/>
15         </graph>
```

The example shows the tools introduced in GraphXML to describe graph hierarchies. A GraphXML document can contain several graph descriptions. Each graph description can use a (unique) identifier, using the id key. A "meta" node in a higher level in the hierarchy uses this identifier to "link" to another graph, using the xlink:href attribute key (see line numbers 0 and 0). The isMetanode attribute is used to unambiguously identify a node that refers to another graph. Using these definitions, this example describes a two level graph with two nodes and one connecting edge, where each node represents another graph. The full format of the xlink:href value is: URL#identifier where the identifier refers to a graph identifier *within* the document referred to by the URL. If the target is in the same document as the source, the URL part can be left out (as in the example).

This simple adaptation of HTML introduces an powerful feature to GraphXML. It is possible to define a hierarchy of graphs, consisting of graphs that that are located in another file, or possibly in another Internet location than the hierarchy description itself. Applications that make use of this capability can create their own hierarchical or clustered views of public datasets.[1]

3.4 Dynamic Graphs

If a graph visualization system is used interactively, the system may be asked to store the history of the user's actions in some form of journaling. What this means is that an interchange format should be able to describe not only the initial graph, but also any editing steps that have changed the structure or the attributes of the graph. This is the reason for the use of <edit> tags in GraphXML.

The edit sections in a GraphXML document are syntactically similar to graph specification, except for the use of the <edit> tag instead of <graph>. Furthermore, the <edit> tag has a required attribute key action, whose value can be remove or replace. Here is an example:

---

[1] Note that WebDot's DOT format[2] includes facilities to describe clusters, but the format is limited to subgraphs defined in the same file.

```
1          <graph version="1.0" vendor="cwi" id="theGraph">
2            <node name="first">
3              <label>A label to display for this node</label>
4              <dataref> <ref xlink:href="BigIcon.gif"/> </dataref>
5            </node>
6            <node name="second"/>
7            <edge name="thisEdge" source="first" target="second">
8          <dataref> <ref xlink:href="ExternalData.bmp"/> </dataref>
9            </edge>
10         </graph>
11         <edit action="replace" xlink:href="#theGraph">
12           <node name="first">
13           <label>Another label</label>
14           <dataref> <ref xlink:href="anotherImage.gif"/> </dataref>
15           </node>
16           <edge name="thisEdge" source="first" target="second"/>
17         </edit>
```

The semantics of the editing element (line 0) is based on identifying the element in the edit block and in the original graph. This controls what is being edited. The value of the `action` attribute determines what happens: the corresponding element will either be removed or replaced by the content in the `<edit>`. The semantics of matching elements is more involved (see the full description[7] for details).

The result of carrying out the editing action in the above example can be represented by the following GraphXML description:

```
1          <graph version="1.0" vendor="cwi" id="theGraph">
2            <node name="first">
3              <label> Another label </label>
4              <dataref> <link xlink:href="anotherImage.gif"/> </dataref>
5            </node>
6            <node name="second"/>
7            <edge name="thisEdge" source="first" target="second"/>
8          </graph>
```

Note the disappearance of the data references in the edge (line 0 in the previous example). This is because the editing action has replaced those with their "empty" counterpart from within the editing element (line 0 in the previous example).

If, in the same example, the `action` attribute were set to `remove`, the result would be as follows:

```
1   <graph version="1.0" vendor="cwi" id="theGraph">
2     <node name="first"> </node>
3     <node name="second"/>
4     <edge name="thisEdge" source="first" target="second">
5     </edge>
6   </graph>
```

The `xlink:href` attribute used by the edit element has the same syntax and semantics as described for hierarchical graph descriptions. In other words, an edit element can refer to a graph in another file or even another Internet location.

As an additional tool for editing, GraphXML also defines the `<edit-bundle>` tag, which is simply a sequence of edit tags:

```
1 <edit-bundle>
2   <edit …> … </edit>
3   <edit …> … </edit>
4 </edit-bundle>
```

This simple grouping of editing elements can be useful if the user wants to animate the result of editing, but doesn't want to display each individual editing step. Using this bundling mechanism, the granularity of animation can be controlled by the creator of the GraphXML file.

4   Storing Geometry

Section 0 describes only structural elements: nodes, edges, and hierarchies. Visualization systems have to layout the graph before presenting it to the user. GraphXML tags for this purpose are described in the next section.

4.1   Positions and Size

The position of a node can be described by adding the `<position>` tag as a child to `<node>`:

```
1     <position x="0.0" y="0.0"/>
```

The size of the node can also be described with the `<size>` tag:

```
1     <size width="3.0" height="5.0"/>
```

This tag can be especially important for layout algorithms that take the node size into account when laying out the graph.

The `<size>` tag can also be used as a direct child of `<graph>`. It then denotes the size, or bounding box, of the full graph. Applications can benefit from such information because it allows them to allocate the necessary area on the screen and coordinate transformations in advance.

4.2   Edge Geometry

Edges differ from nodes insofar as a sequence of coordinates may be necessary. This is achieved through the `<path>` tag:

```
1 <path type="polyline">
2     <position x="0.0" y="0.0"/>
3     <position x="0.1" y="0.0"/>
4     <position x="0.1" y="0.1"/>
5 </path>
```

The `<path>` tag contains a sequence of control points. The `type` attribute can take the value of `polyline`, `arc`, or `spline`, depending on whether the edge is to be drawn as a polyline or a spline curve. In the case of a spline, the positions indicate the spline control points.

4.3 Geometry for Hierarchical Graphs

The geometry definition described earlier is insufficient for the description of hierarchical graphs: the same graph might be included in more than one place in the higher-level graph and the geometry must be adapted to the metanode's position. The solution is to use the `<transform>` tag, which is a child of `<node>`. This element describes the transformation to be applied to each coordinate value in the referenced graph. For example, the following code fragment:

```
1   <node name="SecondOrder" isMetanode="true" xlink:href="#basic">
2       <transform matrix="1.0 0.0 0.5 0.0 1.0 0.5"/>
3   </node>
```

translates all the referred nodes and points to the (0.5,0.5) point. The `<transform>` element contains 6 numbers to describe a 2×3 matrix. See [10] for details on how these transformation matrices are used in computer graphics.[1]

5 Visual Properties

In addition to layout, the appearance of a graph is determined by visual properties, such as line width, colour of the components, icons replacing nodes, etc. In GraphXML, the user can control these properties through the `<style>` tag. A style can include the tags `<line>` or `<fill>`. In the case of a node, the line tag controls the border of the symbol drawn for the node, whereas the fill tag controls the interior. For example, the block:

```
1  <node name="first">
2   <style>
3     <line linestyle="dashed" linewidth="2" colour="red"/>
4     <fill fillstyle="solid" colour="blue"/>
5   </style>
6  </node>
7  <edge source="first" target="second">
8     <style>
9       <line linestyle="solid" linewidth="1" colour="cyan"/>
10      <fill fillstyle="none"/>
11    </style>
12 </edge>
```

defines a node symbol to have a red dashed boundary drawn with a line width of 2, and filled with solid blue[2]. The edge should be drawn in cyan without filling the interiors (in case the edge is drawn as a polygon).

The fill element can also refer to an image file instead of specifying the colour and the fill style. This instructs the application to use the image as an icon to display the node. For example, line 0 could be replaced by:

```
<fill xlink:href="http://www.some.site/imagefile.gif"/>
```

---

[1] All positions, as well as the transformations, can also be extended to 3D.

[2] Note that this specification does not specify the exact glyph to be drawn by the application. This is either left to the implementer of the visualization system, or specified via a more sophisticated control tag, called `<implementation>`. See [7] for further details.

The mechanism described so far would lead to repeated visual control tags, greatly increasing the size of the graph file. It might also become cumbersome to adapt a graph file to a new environment with other visual characteristics. To solve these problems, an inheritance mechanism for visual properties is available. The goal is to provide a general control mechanism that allows for easy adaptation. A style element can be added on the graph level, to control the overall appearance of the graph:

```
1  <graph>
2   <style>
3    <line tag="node" linestyle="dashed" linewidth="2" colour="red"/>
4    <fill tag="node" fillstyle="solid" colour="blue"/>
5    <line tag="edge" linestyle="solid" linewidth="1" colour="cyan"/>
6    <fill tag="edge" fillstyle="none"/>
7   </style>
8   ...
```

This results in the same visual effect as before, except that the visual properties are valid for *all* nodes and edges in the graph (note the use of the tag attribute in the line and fill elements to differentiate between nodes and edges).

Nodes and edges can also use the class attribute to categorize elements with common visual properties. Using this additional identification, finer control over the visual attributes can be achieved by applying a specific visual attribute to a class of nodes or edges. For example, in following block of code:

```
1  <graph>
2   <style>
3    <line tag="node" linestyle="dashed" linewidth="2" colour="red"/>
4    <fill tag="node" fillstyle="solid" colour="blue"/>
5    <fill tag="node" colour="green" class="special"/>
6   </style>
7   ...
8   <node name="first"/>
9   <node name="second" class="special"/>
10  ...
11  <node name="nth" class="special"/>
12  ...
```

the node first will be displayed the same way as before; however the nodes second and nth will become green instead of red. This is because line 0 specifies that "all nodes of class 'special' should be filled in green".

Metanodes require a special style control facility to affect the style of all included elements. The reader should refer to [7] for details.

Using the entity mechanism of XML, it is possible to include an XML file within another. This is particularly handy when controlling styles: it is possible to collect all the style elements into a separate file and include it in a graph specification. If a change is made to the style file, it will automatically affect the visual properties of all the graph description files that reference it.

6 User Extensions

Although the specification of GraphXML includes rich facilities for the association of data with nodes and edges, it is not possible to predict all the possible attributes an application might want to add to an element. For example, if the application is for web visualization, the user might want to associate a MIME type with a node. In other words, the application would need to have its own, `<mime>` tag for each node, in addition to the tags defined by the GraphXML DTD.

Such extensions are possible using GraphXML. This is how an extension for a mime tag can be added to a graph:

```
1  <!DOCTYPE GraphXML SYSTEM "file:GraphXML.dtd" [
2  <!ENTITY % nodeExtensions "|mime">
3  <!ELEMENT mime EMPTY>
4  <!ATTLIST mime
5     type         CDATA #REQUIRED
6     application CDATA #IMPLIED
7  >
8  ]>
9  <GraphXML>
10    <graph>
11      <node name="first">
12        <label>Project Home page</label>
13        <dataref> <ref xlink:href="Description.pdf"/> </dataref>
14        <mime type="application/pdf" application="Adobe Acrobat"/>
15      </node>
16      ...
```

The file contains what is called an "internal DTD" (lines 0–0), which extends the elements that can be included in a node definition. The definition states that a `<mime>` element can be added as the child of a node, that this is an element that contains only attributes (i.e. no sub–elements), and that the attributes can be `type` and `application` (the first being compulsory, the second optional).

The XML syntax is a bit cryptic. However, the end–user does not necessarily have to include such internal DTD's into all GraphXML files. Instead, using standard XML mechanisms, it is possible to define a *separate* DTD containing the application-specific extensions (and a reference to the `GraphXML.dtd`, of course). Using such extension, the header part becomes simply:

```
1  <!DOCTYPE GraphXML SYSTEM "file: WebVisualizer.dtd ">
2  <GraphXML>
3    <graph>
4      ...
```

In other words, the intricacies of the XML DTD syntax remain invisible to most users. Furthermore, because the extension is made through a standard XML mechanism, the basic GraphXML parser remains unchanged. Only the application–dependent part has to be adapted for the new extensions. Herman and Marshall [7] describes application-specific DTD's in more detail.

7   Future Developments

As stated REFearlier, one of the advantages of using XML is that the future evolution of XML-based specifications can be used to develop new tools for GraphXML. Here are just two examples:

- An upcoming specification is the RDF Schemas[11], which will replace DTD's in future. Schemas also include various data types with possible constraints on the values. Schema based parsers will be able to make on-the-fly checks on the input data (e.g., coordinates should be numbers), relieving the GraphXML parser and applications from having to perform these checks themselves.

- W3C is currently developing a standard for graphics on the Web, called SVG[12]. It will be possible to define simple tools to transform GraphXML specifications into SVG[1]. This means that the result of graph drawing systems can be published on the Web in the form of vector based graphics, rather than screen dumps, yielding better quality and smaller bandwidth requirements.

8 Public Availability

The full description of GraphXML, as well as the GraphXML DTD, is publicly available at the URL: http://www.cwi.nl/InfoVisu/GraphXML. The parser is also available as a collection of Java 1.2 classes, which can be embedded into an application. The implementation uses publicly available XML parsers in Java. See the full description at the above URL for details.

References

1. M. Himsolt, *GML - Graph Modelling Language,* http://infosun.fmi.uni-passau.de/Graphlet/GML/, 1997.
2. S. C. North, *Dot Abstract Graph Description Format,* http://www.research.att.com/~north/cgi-bin/webdot.cgi/dot.txt.
3. K. Wong, *Rigi Users' Manual,* http://www.rigi.csc.uvic.ca/rigi/manual/user.html, 1996.
4. S. Thiel, *LEDA Graph Input/Output Format,* http://www.mpi-sb.mpg.de/LEDA/information/graph_io_format/graph_io_format.html, 1999.
5. *GDS Supported File Formats,* http://loki.cs.brown.edu:8081/geomNet/gds/formats.shtml, 2000.
6. "Extensible Markup Language (XML) 1.0", World Wide Web Consortium, (eds. T. Bray, J. Paoli, C.M. Sperberg–McQueen), Recommendation February 1998, http://www.w3.org/TR/REC-xml.
7. I. Herman and M. S. Marshall, "GraphXML — an XML Based Graph Interchange Format", Centre for Mathematics and Computer Sciences, Amsterdam INS-R0009, 2000, http://www.cwi.nl/InfoVisu/GraphXML/GraphXML.pdf.
8. "XML Linking Language (XLink)", World Wide Web Consortium, (eds. S. DeRose, D. Orchard, B. Trafford), Working Draft July 1999, http://www.w3.org/TR/WD-xlink.
9. I. Herman, M. S. Marshall, and G. Melançon, "Graph Visualisation and Navigation in Information Visualisation: A Survey", *IEEE Transactions on Visualization and Computer Graphics*, vol. 6, pp. 24-43, 2000.
10. J. D. Foley, A. v. Dam, and S. K. Feiner, *Computer Graphics: Principles and Practice* Reading, Addison–Wesley, 1990.
11. "Resource Description Framework (RDF) Schema Specification 1.0", World Wide Web Consortium, (eds. D. Brivkley and R.V. Guha), Candidate Recommendation March 2000, http://www.w3.org/TR/2000/CR-rdf-schema-20000327/.
12. "Scalable Vector Graphics (SVG) 1.0 Specification", World Wide Web Consortium, (eds. D. Ferraiolo et al.), Working Draft March 2000, http://www.w3.org/Graphics/SVG/Overview.htm8.

---

[1] Standard XML-based tools can perform such a transformation. Also, a graph vizualization application that reads GraphXML and saves to SVG format can be found at http://www.cwi.nl/InfoVisu/GVF/ .

On Polar Visibility Representations of Graphs

Joan P. Hutchinson*

Department of Mathematics and Computer Science,
Macalester College,
St. Paul, MN 55105, USA
hutchinson@macalester.edu

Abstract. We introduce polar visibility graphs, graphs whose vertices can be represented by arcs of concentric circles with adjacency determined by radial visibility including visibility through the origin. These graphs are more general than the well-studied bar-visibility graphs and are characterized here, when arcs are proper subsets of circles, as the graphs that embed on the plane with all but at most one cut-vertex on a common face or on the projective plane with all cut-vertices on a common face. We also characterize the graphs representable using full circles and arcs.

1 Introduction

Visibility graphs are now a well-established area of graph drawing [10]. Much has been written about their importance and application; however, they continue to pique the imagination of mathematicians with their intrinsic appeal and intriguing questions [2]. There has been a natural progression from bar-visibility graphs (BVGs) [11, 17] to rectangles [1, 3, 4, 6], from bars with visibilities in the plane to those on the sphere and cylinder [12, 13], on a (flat) torus [8], or on the Möbius band [5]. These rectilinear representations are natural ones for most applications; however, we turn instead to the realm of polar representations with arcs of circles and radial visibility. In many ways circular representations and related polar coordinates are equally natural and in some contexts more applicable than rectilinear ones. With this change of perspective we can and do represent a class of graphs, larger than with bars in the plane, though ultimately constrained by the real projective plane. Thus with a planar representation of arcs of circles nonplanar graphs are drawn in a natural way, resulting in diagrams often reminiscent of time-exposed shots of the North Star and surrounding stars.

We introduce the layout of graphs as polar visibility graphs (PVGs) using arcs of concentric circles (arcs that are proper subsets of a circle) with radial visibility, including visibility through the origin, the center of all the concentric circles. These graphs, though arising naturally from visibility in the plane, corres-

---

* Research supported in part by NSA Grant #MDA904-99-1-0069.

J. Marks (Ed.): GD 2000, LNCS 1984, pp. 63–76, 2001.

pond to graphs embedded on the (real) projective plane, the nonorientable surface of Euler characteristic 1. PVGs are characterized as the planar graphs that can be drawn in the plane with all but at most one cut-vertex on a common face plus the graphs that can be embedded on the projective plane with all cut-vertices on a common face. We also consider the variation in which full circles are allowed along with arcs, and characterize the graphs so representable (CVGs) in terms of their block-cutpoint tree.

2 Background

Just as visibility wider than along a line is required for BVGs, we ask that radial visibility in PVGs be available through a nondegenerate cone. Define a (nondegenerate) *cone* in the plane to be a 4-sided region of positive area with two opposite sides being arcs of circles, centered at the origin, and the other two sides, possibly intersecting, being radial line segments on lines through the origin. Thus, both $\{r, q: 1 \pounds r \pounds 2, 0 \pounds q \pounds p \ 6\}$ and also

$$9\{r, q: 0 \pounds r \pounds 1, 0 \pounds q \pounds \frac{p}{6} \text{ or } p \pounds q \pounds \frac{7p}{6} == 9\{r, q: -1 \pounds r \pounds 1, 0 \pounds q \pounds \frac{p}{6} =$$

are considered to be cones, respectively, not containing and containing the origin. Given a set of arcs, all centered at the origin, two of these arcs a_1 and a_2 are said to be *radially visible* if there is a cone that intersects only these two arcs and whose two circular ends are subsets of the two arcs; the same definition holds for visibility between an arc and a circle and between two circles. A graph is called a *polar visibility graph* if its vertices can be represented by arcs, including endpoints, of circles centered at the origin, having pairwise disjoint relative interiors, so that two vertices are adjacent if and only if the corresponding arcs are radially visible; see Figure 1a below for a PVG representation of K_6. If this model is used, but without visibility through the origin, the graphs arising are one of the cylindrical types characterized in [13]. Note that for a 2-connected graph (a graph without cut-vertices) there is no loss in taking arcs as proper subsets of circles

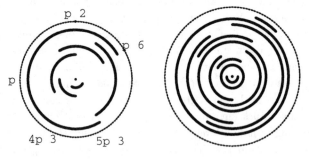

Fig. 1a. K_6 **1b.** A circular visibility layout

since a full circle can be cut down to a smaller arc, leaving the same visibilities. Arcs in a PVG layout spanning more than half its circle will provide interesting variations, full circles even more. We use graph theoretic terminology as in [15], topological notions as in [9], and algorithmic ideas following the BVG presentation in [10].

Similarly a graph is called a *circular visibility graph* if its vertices can be represented by arcs and circles with radial visibility between arcs and circles determining edges as for PVGs. When possible we prefer, but do not require, arcs over circles; that is, in a layout we will decrease a circle to become a proper arc if no additional visibilities are introduced. We shall see that some planar and projective planar graphs with cut-vertices on an arbitrary number of faces are CVGs, but not PVGs, but that these faces must be nested appropriately. Figure 1b shows such a planar CVG. In that layout the inner circle contains one arc; if instead, it contained four mutually visible arcs, encircling the origin and forming a K_5, the example becomes a nonplanar CVG.

Note that in a PVG or CVG layout of a graph G, we may draw each arc and circle on a distinct circle, and we may take these circles to have radii 1, 2, ..., n where $n =$ ''VHG·'. This naturally leads to another layout of the graph in a disk of radius n+1 and centered at the origin by inverting each circle and arc through the circle of radius Hn+ 1L 2. That is, each point with polar coordinates Hr qI, $0 < r < n + 1$, is mapped by the inversion to the point Hn+ 1 − r, qI. This inversion preserves circles, arcs, and the angles defining these arcs. If the original layout was L, we denote this inverted layout by *IHL*.

Recall that the (real) projective plane can be obtained by taking a circular disk and identifying opposite (or antipodal) points. Thus if we identify opposite points of the circle of radius $n + 1$, we create a projective plane. Two arcs in *IHL* (or an arc and a circle or two circles) that were previously radially visible in a cone, not containing the origin, are still radially visible, and a pair visible in a cone through the origin are now visible in a "generalized cone" that crosses the boundary of the projective plane, reemerging on the other side. The coordinates of such a generalized cone are given by 8H$r$$q$L, r^x £ r £ $n + 1$ or − Hn+ 1L£ r £ −s^x, q_1 £ q £ q_2< where r^x, s*, $q_1 < q_2$ are constants, 0 £ r^x, $s^x < n + 1$. In addition, the interior of no two of these new cones intersect. Fig. 2a shows the inverted layout of K_6 on the projective plane with dashed lines indicating a conical area of visibility, and in 2b we see an embedding of K_6 created by shrinking each arc to a vertex. The first proposition is then clear since each inverted arc and circle on the projective plane can be replaced by a single vertex. Then the visibility cones can each be shrunk and transformed to a set of nonintersecting edges on the projective plane.

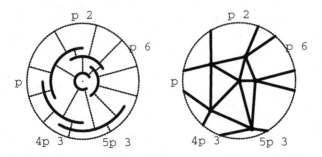

Fig. 2a. $I\!H\!K_6 I$ and $I\!H\!K_6 L'$ on the projective plane
2b. For $G = K_6$, $I\!H\!L_G L = H\!H\!L\!I\!I_G$

Proposition 2.1. A PVG or a CVG embeds on the projective plane.

Recall that a graph G is said to *embed* on a surface S if it can be drawn there without any edge crossings, and that each maximal connected component of $S \setminus 8V\!H\!G\!L\!E\!H\!G\!L$ is called a *face* of the embedding (we do not require that the faces be simply connected).

Theorem 2.2. A graph G is a PVG if and only if either a) G has an embedding in the plane with all but at most one cut-vertex on a common face, or else b) G has an embedding on the projective plane with all cut-vertices on a common face.

Note that condition (a) allows for the representation of planar graphs that are not BVGs; for example $K_{2,3}$ with three additional vertices of degree 1 appended, one each to a vertex of degree two, is a PVG. Similarly $K_4 + 4e$ (K_4 plus a pendant vertex and edge at each vertex) is a PVG (see Fig. 3); these are the smallest graphs that are not BVGs. Condition (b) also allows for more planar graphs; for example, two vertices joined by three internally disjoint paths of length three (i.e., three edges each) plus six vertices of degree 1, each adjacent to a different vertex of degree two, satisfies (b), but not (a).

Every graph G can be decomposed into its blocks and their connecting cut-vertices (a *block* is either an edge or a 2-connected subgraph; see [15]), and these connections determine a tree, called the *block-cutpoint tree* of the graph, $BC\!H\!G$. This tree has a vertex for each block and for each cut-vertex of G, and two vertices of BCHG are adjacent if and only if they correspond to an incident cut-vertex and block. We call a block *planar* if it represents a planar graph.

Theorem 2.3. A graph G is a CVG if and only if $BC\!H\!G\!I$ consists of a path $P = H\!q, e_2, \ldots, e_{2k+1} I$, $k \neq 0$, with e_{2i} representing a planar block, $i = 1, \ldots, k$, so that

1a) e_1 is also incident with one additional (nonempty) block representing a (2-

connected) projective planar graph, or

1b) e_1 is also incident with one or more (nonempty) planar blocks, and

2) e_{2k+1} is also incident with an arbitrary tree structure T so that T $\&_{2k+1} <$ represents a planar graph that can be drawn in the plane with all cut-vertices, except possibly for that representing e_{2k+1}, on a common face.

When $k = 0$, these conditions reduce to those of Theorem 2.2. On the other hand, it may be that each cut-vertex of G, represented by $e_1, e_3, ..., e_{2k+1}$, lies on a different face, as in Figure 1b. This example is the first of an infinite family of CVGs with an increasing number of cut-vertices, all on different faces; the family is obtained by nesting repeatedly the same pattern of arcs and circles. Most of the details of the proofs of Theorems 2.2 and 2.3 are included below.

As described in [10], planar layouts and the block-cutpoint tree of a graph can be determined in linear time. Projective planar graphs can also be recognized and embedded in linear time [7]. It can quickly be determined whether all cut-vertices of a graph lie on a common simple cycle and, if so, whether there is an embedding in either surface in which this cycle bounds a face. The proofs of Theorems 2.2 and 2.3, together with standard BVG algorithms, lead to a $O\text{HN}\mathcal{E}^2\text{L} = O\text{HN}^3\text{L}$ time algorithm for laying out a PVG G with N vertices and E edges, given an embedding of G in the projective plane as a rotation scheme (defined below), as in [7, 9].

3 Main Results on PVGs

We develop theory that will also allow extension to CVGs. We focus on simple graphs and their characterizations as in Theorems 2.2 and 2.3. Thus we say that two arcs are radially visible if there is at least one maximal cone providing mutual visibility; however, we can also obtain more precise results by keeping track of multiple and even self-visibility between arcs and circles.

First we need more precise topological and geometric definitions. Consider a PVG or CVG layout L of a graph G and its inverse layout on the projective plane, $I\text{HL}$. We let   (respectively, $I\text{HL}'$) denote the visibility depiction obtained by shrinking each maximal visibility cone of L (resp., $I\text{HL}$) to a distinct line segment by reducing its angles $b_1 \leq q \leq b_2$ to some constant $q = b$, $b_1 < b < b_2$; strict inequality ensures distinct visibility segments. For $G = K_6$, $I\text{HL}'$ is shown in Fig. 2a.

Also let L_G (resp., HHL_C) denote the graph obtained from L^* (resp., $I\text{HL}'$) by shrinking each arc to a vertex, consisting of one point, and transforming each visibility line segment to an edge that intersects no other edge except possibly at the origin (resp., an edge that intersects no other edge on the projective plane). If   or $I\text{HL}'$ contains a circle, it is replaced by a point as vertex. Thus HHLL_C is a graph embedded on the projective plane, see Fig. 2b. Note that

$L^{\times}$, $IHLL'$, IHL_GI, and $HHLL_l$ have visibility segments and edges for each distinct, maximal visibility cone so that multiple edges and loops may be present in these depictions; however, a pair of multiple edges will not form an embedded digon with empty interior.

Note that the complement of the arcs, circles, and lines of $L^{\times}$ divide up the plane into faces; similarly $IHLL'$ divides up the projective plane. One face of $L^{\times}$ is the exterior face, possibly containing the origin; this exterior face is the one in which most cut-vertices of a PVG and their blocks can be placed. We say that an arc or circle of a layout L lies on the exterior face if it lies on the exterior face of $L^{\times}$.

We use the following combinatorial description of an embedded graph and of a PVG or CVG layout. If a graph is embedded on any surface, then for each vertex there is naturally defined a cyclic rotation of its neighbors, given by the order, say clockwise, of its edges in the embedding; such a collection of rotations, one for each vertex, is called a *rotation scheme*. (See, for example, [16] where it is shown that an embedding is equivalent to a rotation scheme.) Such a description is generally used for algorithms on embedded graphs [9]. Similarly, given a PVG layout L in the plane and its inverse layout IHL in the projective plane, one can define the *arc-rotation scheme* to be the set of cyclic rotations of neighbors about each arc of its visibilities to other arcs; note that the rotations at the arcs of L and of $IHLI$ are inverses of each other. We say that an embedding of a PVG graph G in the plane or on the projective plane and its polar visibility layout L are *equivalent* if the arc-rotation scheme of IHL, when translated into a set of vertex-neighbor cycles, yields the rotation scheme of the embedded graph; see Figs. 1, 2. Given a circle in a CVG layout L or IHL, the neighbors divide into two cyclic rotations of the inner and outer visibilities, called the *circle-rotation scheme*. Then a drawing of a CVG and its layout L are equivalent if the arc/circle-rotation schemes of IHL agree with those of the embedded graph.

It is not hard to see the following, by bending or straightening corresponding BVG and PVG layouts.

Proposition 3.1. A connected graph has a PVG layout with no visibilities through the origin if and only if the graph is a BVG.

A PVG layout with no visibilities through the origin contains arcs in sectors, alternating about the origin, so that some can be reflected through the origin, leaving the layout in two quadrants, and this can be straightened to form a BVG.

Of course there are planar graphs with layouts as PVGs including visibilities through the origin and with cut-vertices represented on the exterior face. Note that whenever there are visibilities through the origin in a layout L, then the equivalent graph $HHLL_G$ is embedded on the projective plane. It turns out that in some PVG layouts there is a (sneaky) hiding place for a cut-vertex and its connecting blocks, but the resulting graphs turn out to be planar. In a PVG layout we call an arc a^* a *long arc* if its angular span is greater than p. Suppose $a^* = 8H^*$, qL, $0 \leq q \leq p + x$, for some $0 < x < p$. Then the cone defined by $CHaL = 8H$, qL, $-r^* \leq r \leq r^*$, $0 \leq q \leq x$ is an area in which interior arcs can see the arc a^* and possibly no others; see Fig. 3.

In preparation for CVG layouts, we require special PVG layouts. Let a^* be a long-arc at radius 1, spanning $q_1 \leq q \leq q_1 + p + x$ for some $x > 0$. Arcs a^* and b^* are called a *long-arc pair at the origin* if they are mutually visible, together they span 2p, and if b^* lies at radius $r^* > 1$, no arcs intersect the *long-arc cone* $8Hr$, qL: $0 \leq r < r^*$, $q_1 + p + x < q < q_1 + 2p$. (For example, when $r^* = 2$, no arcs can meet the designated cone.) Similarly if a^* is a long-arc at the outermost radius $n = "VHGL$, spanning $q_2 \leq q \leq q_2 + p + y$ for some $y > 0$, then a^* and b^* are a *long-arc pair at infinity* if they are mutually visible, together span 2p, and if b^* lies at radius $r^* < n$, no arcs intersect the long-arc cone $8HrqL$: $r^* < r < n$, $q_2 + p + y < q < q_2 + 2p$; see Figs. 1a, 3. Notice that in long-arc pairs the long arc at radius 1 or at radius n could be extended to form a full circle without changing visibilities.

Fig. 3. $K_4 + 4e$.

Here are the building-block results needed for the PVG characterization.

Proposition 3.2. Let G be laid out as a PVG L including a long arc a^* that represents a cut-vertex x^*, not lying on the exterior face, and let B be a block of G incident with x^* and whose representation lies within CHu^*I in L. Then G is a planar graph and can be drawn in the plane with one face including all vertices whose arcs lie on the exterior face of L.

The proof consists of observing that in IHI and in $HHILL_C$ the representation of a block B incident with x^* lies within a (noncontractible) sector of the projective plane that divides the space into two contractible (planar) regions.

The next result is the necessary topological argument needed to character-ize PVGs; it is a contraction proof similar to that of [14] and [8]. This result is carried out for multigraphs, those embedded with no digon face with empty interior except for two special faces. For these graphs we achieve layouts with a one-to-one correspondence between distinct, maximal visibility cones and edges of G. If x is a vertex of a PVG, we let a_x denote its arc in the layout, and conversely x_a is the vertex corresponding to an arc a.

Proposition 3.3. (i) Let G be a loopless 2-connected plane multigraph, let F be a face in the embedding, and let c be a vertex of G. Suppose G has at most two digon faces, possibly F and, when c does not lie on F, possibly one incident with c. Then $G^{\varsigma} = G$ plus a loop at c has a PVG layout L^{ς} in which all vertices of F are represented on the exterior face of L^{ς} and Ha_c, a_dI is a long-arc pair at the origin for some neighbor d of c. In addition, G^{ς} has an embedding on the projective plane that is equivalent to L^{ς}.

(ii) Let G be a loopless 2-connected plane multigraph with v_1 and v_2 designated, distinct vertices and with no digon face. Then $G^{\varsigma} = G$ plus a loop at v_2 has a PVG layout L^{ς} with v_i represented by arc a_i, $i = 1, 2$, with Ha_1, b_1I a long-arc pair at infinity, and with Ha_2 b_2I a long-arc pair at the origin, where for $i = 1$ and 2, arc b_i corresponds to some neighbor of v_i. Also G^{ς} has an embedding on the projective plane that is equivalent to L^{ς}.

(iii) Let G be a loopless 2-connected multigraph with a 2-cell embedding on the projective plane, with F a face in the embedding, and with no digon face except possibly for F. Then G has a PVG layout L that is equivalent to the embedding of G with exterior face corresponding to F.

(iv) Let G be a loopless 2-connected multigraph with a 2-cell embedding on the projective plane, with no digon face, and with v_1 a designated vertex. Then G has a PVG layout L, equivalent to the embedding of G, in which v_1 is represented by arc a_1 with Ha_1, b_1I a long-arc pair at infinity and with arc b_1 corresponding to some neighbor of v_1.

Sketch of proof of (i). For most cases (when c does not lie on F), the proof is by induction on n.

We can always find a nonloop, nonmultiple edge $e = Hx$, yI of G so that G with e contracted, $G\ e$, is 2-connected, loopless, embedded on the plane, and F is still bounded by at least two edges. $G\ e$ satisfies the inductive hypothesis and so has PVG layout L_e, equivalent to an embedding of $G\ e$ plus a loop on the projective plane. If the contraction combines vertices x and y into new vertex x^*, let a^* be its representation in L_e, at say radius r. Because the embeddings of G/e and L_e are equivalent, the lines of visibility to arcs representing vertices adjacent to x in G are consecutive in the rotation of visibility lines about a^* in L_e. Then, when a^* is not one of the special long arcs at the origin, it can be replaced by two arcs a_x and a_y at radii $r - 0.5$ and $r + 0.5$ (or vice versa), representing vertices x

and y of G, so that their visibilities give all edges incident with x and y and preserve the arc-rotations at x and y in G; see Fig 4. This alteration gives the desired PVG layout L for G. The argument is similar, though a bit more intricate, when a^* is part of the long-arc pair at the origin. The proofs for (ii–iv) are analogous.

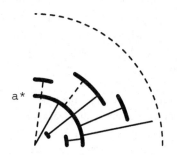

Fig. 4. An arc and its neighbors

We then obtain the following.

Proposition 3.4. If G has a PVG layout L, then the embedding HH$L L_l$ of G on the projective plane has cut-vertices on at most two faces. If the embedding has cut-vertices on two faces, then on one face there is only one cut-vertex, represented in L by a long arc.

Corollary 3.5. If G has a PVG layout L with a long arc a^*, representing a cut-vertex x^* and not lying on the exterior face, then G has a planar embedding with all cut-vertices except for x^* lying on a common face.

Theorem 3.6. A simple planar graph G has a PVG representation if it has a planar embedding with all but at most one cut-vertex on a common face.

Sketch of the proof. Assume that G is not a BVG and so can be drawn in the plane with cut-vertices lying on the exterior face F_1 and an additional cut-vertex c lying on $F_2 \,_{,\!,} F_1$. Consider the block-cutpoint tree BCHG of G; c may lie on several blocks, but at least one, call it B_0, contains a cut-vertex $c^c \,_{,\!,} c$ lying on F_1. Both of the faces F_i are bounded by a facial walk W_i, and each W_i contains a unique simple subcycle C_i, lying in B_0 and containing c^c and c, respectively. If G has cut vertices $c_1, \ldots, c_i$ lying on F_1, we label the blocks other than B_0 incident with $c_1, \ldots, c_i$ $B_1, B_2, \ldots, B_j$, and the blocks $D_1, \ldots, D_k$ incident with c. Then we prove by induction on j that there is a PVG layout L of G with F_1 represented by the exterior face of L, with Hq, a_dI a long-arc pair at the origin for some neighbor d of c, and with the blocks incident with c represented within CHa_cI.

Theorem 3.7. If a simple graph has an embedding on the projective plane with all cut-vertices on a common face, then it is a PVG.

Proof. Let G have an embedding on the projective plane P with all cut-vertices on a common face F. We prove by induction on $n = $ "$VHGT$ that G has a PVG layout with arcs representing cut-vertices on the exterior face and with its embedding equivalent to that of G. When $n < 5$, the graph has a BVG layout and so a PVG one by Prop. 3.1; each such graph containing a cycle also has a 2-cell embedding on the projective plane and an equivalent PVG layout.

If G has no cut-vertex, then we apply Prop. 3.3(iii) for graphs on the projective plane to get the PVG layout of G.

If G has a cut-vertex, we consider the block-cutpoint tree T of G, and, if possible, let c be a cut-vertex incident with a leaf of T with that leaf-block planar and embedded in a contractible region of P; call this block B. Deleting the vertices and edges of $B \setminus 8c$ leaves G^{c} on the projective plane with face F now a face F^{c}, containing all remaining cut-vertices. By induction G^{c} has a PVG layout L^{c} that is equivalent to G^{c} and with exterior face representing F^{c}. Then there is a BVG layout of B with the bar representing c bottommost and extending the width of the layout, and by Prop. 3.1 B has a corresponding PVG layout L_B. Then a_c in L_B can be inserted as a subarc of a_c on the exterior face of L^{c} so that L_B together with L^{c} gives the desired layout of G.

Otherwise every leaf-block B is embedded in a noncontractible region of P and contains a noncontractible cycle in its embedding. If blocks B and B^{c} are two such leaves, they must intersect at a cut-vertex c since every pair of noncontractible cycles on P intersects. If there are additional blocks, there are additional leaves which must also all meet at c so that T is a star $K_{1,i}$ with the non-leaf vertex of T representing c, the only cut-vertex of G, and each block is embedded in a wedge of P, all wedges meeting at, say, the origin. Such a graph is planar with one cut-vertex c and so by Theorem 3.6 is a PVG.

Proof of Theorem 2.2 for simple graphs. By Theorems 3.6 and 3.7 the graphs described are PVGs. Conversely if L is a layout of a PVG G, then G has an embedding on the projective plane by Prop. 2.1 with embedding HHLLL_t. If L has no visibility through the origin, then by Prop. 3.1 G is a BVG and so embeds in the plane with all cut-vertices on a common face. Otherwise, if L contains a long arc, satisfying the conditions of Cor. 3.5, then G embeds in the plane with all but one cut-vertex on a common face. Otherwise G embeds in the projective plane with all cut-vertices on a common face by Prop 3.4.

4 Results on CVGs

As the example in Fig. 1b and its extensions demonstrate, cut-vertices on many faces can be achieved using circles in layouts. We characterize CVGs in this section, as given in Theorem 2.3.

Suppose G has a layout L with circles $c_1, c_2, ..., c_k$ at radii $r_1 < r_2 < \quad < r_k$ and with no circle replaceable by an arc so that the same visibilities are achieved. The circles c_i divide up the plane into annular regions and one projective planar region; note that neither the interior of c_1, denoted intHc₁I, nor the exterior of c_k, extHₖI, is empty in L since neither circle can be replaced by an arc. Then the corresponding vertices $v_1, v_2, ..., v_k$ of G are cut-vertices, and G is the union of graphs whose layouts lie in the annular regions plus the innermost region: $G = G_1 \quad G_2 \quad ... \quad G_k \quad G_{k+1}$ where G_1 is the subgraph whose layout in L lies on c_1 intHq I, G_{k+1} lies on c_{k+1} extHc_{k+1} I, and for $i = 2, ..., k$, G_i lies on the annulus given by $c_{i-1} \quad c_i$ intHqL extHq_{i-1} L. Thus $G_2, ..., G_{k+1}$ are each planar. In addition for $i = 2, ..., k$ G_i is 2-connected since each block of G_i contains some vertices adjacent to v_{i-1} and some to v_i. Thus the block-cutpoint tree for G, BCHG, contains a path of $2k - 1$ vertices, representing consecutively $v_1, G_2, v_2, ..., G_k, v_k$. What sorts of graphs are possible for G_1 and for G_{k+1}, and what additional tree structure in BCHG is possible at the two ends of this path?

Consider G_1, laid out on c_1 intHc₁I, with c_1 opened up to become an arc a_1 so that this is a PVG layout of G_1. If G_1 is planar, by Prop. 3.4 and its proof, G_1 can have at most one additional cut-vertex, not on the exterior face but represented by a long arc a^* at radius 1. If there is no long arc a^* besides a_1, then v_1 may be attached to an arbitrary positive number of planar blocks. If there is a long arc a^* ″ a_1, then each block represented between a^* and a_1 sees these two arcs and so there is only one block lying in this annular region. Inside and attached to a^* may be any number i_a ǂ 0 of 2-connected, planar graphs, but in any case, BCHGI has attached to the path-end v_1 either $i_1 > 0$ leaves or else one additional block vertex b, representing part or all of G_1, then a vertex for a^* that is also adjacent to $i_a > 0$ vertices of degree one. (Thus the latter case corresponds to having v_3 represented by c_1 and v_1 by a^*.) If G_1 is not planar, by Prop. 3.4 and Cor. 3.5 it is 2-connected so that the path of BCHG is extended at v_1 by one additional vertex representing G_1.

The layout for the planar graph G_{k+1} lies in the infinite region, c_k extHq I. In this layout of G_{k+1} the circle c_k can be opened up to a long arc with empty interior to form a PVG layout; by Prop. 3.4 G_{k+1} has all its cut-vertices on a common face, the exterior face, and so can have arbitrarily many cut-vertices with arbitrarily many connected blocks, provided all cut-vertices lie on the infinite face. Thus attached to v_k in BCHG is any tree representing a planar graph with all cut-vertices, except possibly for v_k, on a common face. These remarks prove the necessity of Theorem 2.3.

Lemma 4.1. Let L be a layout of a PVG G with n vertices and with a long-arc pair at infinity or at the origin (or both). Then L can be laid out as a CVG with a circle on the exterior face at radius n or a circle about the origin at radius 1 (or both).

As noted in Section 3, a long arc at radius 1 or at n can be extended to a full circle, changing no visibilities.

Proof of the sufficiency of Thm. 2.3. Suppose G has BCH$\mathcal{G}$ satisfying (1a) and (2) so that BCH$\mathcal{G}$ is Hb_0, e_1, e_2, ..., e_{2k+1}, TI where for $i = 1$, ..., k, each e_{2i-1} represents a cut-vertex v_i of G, each e_{2i} represents a 2-connected planar graph, b_0 is a 2-connected projective planar graph, and T represents a plane graph with all cut-vertices on a face F. Such a graph embeds on the projective plane; in the layout each cut-vertex v_i will be represented by a circle c_i.

By Prop. 3.3(iv) the projective planar subgraph of G corresponding to b_0 has a PVG layout L_0^c with the arc a_1 representing v_1 in a long-arc pair at infinity with some neighbor of v_1. By Lemma 4.1 L_0^c can be changed to the CVG L_0 so that a_1 becomes a circle surrounding L_0. By Prop. 3.3(ii) the planar subgraph of G corresponding to e_2 can be represented as a PVG L_1^c with a_1, representing v_1, part of a long-arc pair at the origin and with a_2, representing v_2, part of a long-arc pair at infinity. By Lemma 4.1 L_1^c can be changed to the CVG L_1 so that a_1 and a_2 each become circles inside and surrounding L_1 respectively. Then L_1 is joined with L_0 by identifying the two copies of the circle a_1, placing L_1 wholly outside of L_0. This process of expansion can be repeated for e_4, ..., e_{2k}. Finally by Prop. 3.3(i) T can be laid out as a PVG with v_k represented by a_k, part of a long-arc pair at the origin. Again by Lemma 4.1 a_k can be extended to a full circle inside of T's layout and can be identified with the circle representing a_k on the exterior of the layout previously constructed. In this way G is laid out.

If BCH$\mathcal{G}$ satisfies (1b) and (2), it can be laid out similarly, only differing within c_1.

Since v_1 is incident with one or more planar blocks, we can lay these out in radial segments within c_1. Each planar block can be represented as a BVG with v_1 represented top-most and a neighbor bottom-most, then as a PVG via Prop. 3.1, and then inserted with v_1's arc as a subarc of c_1 within a distinct wedge of, say, $0 \pounds q \pounds p$, giving the desired visibilities. Thus in all cases the graph can be laid out as a CVG.

5 Concluding Thoughts

It is clear that more complex graphs can be achieved in the polar visibility model by allowing visibility through the origin and diagonally across the boundary of a disc with antipodal points identified; call such a layout a doubly polar visibility layout and the resulting graphs doubly polar visibility graphs (DPVGs). These naturally lead to graphs that embed on the Klein bottle, the nonorientable surface of Euler characteristic 0. Analogous proofs to those given on the projective plane give the following results.

Proposition 5.1. a) A DPVG embeds on the Klein bottle.
b) If G has a layout L as a DPVG with no long arcs, then G contains no cut-vertex.
c) If a 2-connected graph G has an embedding on the Klein bottle, then G is a DPVG and has an equivalent doubly polar visibility layout.

It seems that a DPVG that is neither a BVG nor a PVG can have at most two cut-vertices, represented by a long arc about the origin and at infinity.

Acknowledgements The author wishes to thank Alexandru Burst, Alice Dean, Michael McGeachie, Bojan Mohar, William Owens, and Stan Wagon for useful conversations and insightful examples concerning this work.

References

1. Bose, P., Dean, A., Hutchinson, J., Shermer, T.: On rectangle visibility graphs. In: North, S. (ed.): Proc. Graph Drawing '96. Lecture Notes in Computer Science, Vol. 1190. Springer, Berlin (1997) 25–44
2. Chang, Y-W., Jacobson, M. S., Lehel, J., West, D. B.: The visibility number of a graph (preprint)
3. Dean, A., Hutchinson, J.: Rectangle-visibility representations of bipartite graphs. Discrete Applied Math. 75 (1997) 9–25
4. _____, Rectangle-visibility layouts of unions and products of trees, J. Graph Algorithms and Applications 2 (1998) 1–21
5. Dean, A.: Bar-visibility graphs on the Möbius Band. In: Marks, J. (ed.): Proc. Graph Drawing 2000. (to appear)
6. Hutchinson, J., Shermer, T., Vince, A.: On Representations of some Thickness-two graphs. Computational Geometry, Theory and Applications 13 (1999) 161–171
7. Mohar, B.: Projective planarity in linear time. J. Algorithms 15 (1993) 482–502
8. Mohar, B., Rosenstiehl, P.: Tessellation and visibility representations of maps

on the torus. Discrete Comput. Geom. 19 (1998) 249–263

9. Mohar, B., Thomassen, C.: Graphs on Surfaces. Johns Hopkins Press (to appear)

10. O'Rourke, J.: Art Gallery Theorems and Algorithms. Oxford University Press, Oxford (1987)

11. Tamassia, R., Tollis, I. G.: A unified approach to visibility representations of planar graphs. Discrete and Computational Geometry 1 (1986) 321–341

12. _____: Tesselation representations of planar graphs. In: Medanic, J. V., Kumar, P. R. (eds.): Proc. 27th Annual Allerton Conf. on Communication, Control, and Computing. (1989) 48–57

13. _____: Representations of graphs on a cylinder. SIAM J. Disc. Math. 4 (1991) 139-149

14. Thomassen, C.: Planar representations of graphs. In: Bondy, J. A., Murty, U. S. R. (eds.): Progress in Graph Theory. (1984) 43–69

15. West, D.: Introduction to Graph Theory. Prentice Hall, Upper Saddle River, NJ (1996)

16. White, A.: Graphs, Groups and Surfaces. revised ed. North-Holland, Amsterdam (1984)

17. Wismath, S.: Characterizing bar line-of-sight graphs. In: Proc. 1st Symp. Comp. Geom. ACM (1985) 147–152

A Linear Time Implementation of SPQR-Trees[*]

Carsten Gutwenger[1] and Petra Mutzel[2]

[1] Max-Planck-Institut für Informatik
Saarbrücken, Germany, gutwenge@mpi-sb.mpg.de
[2] Technische Universität Wien, Austria, mutzel@ads.tuwien.ac.at

Abstract. The data structure SPQR-tree represents the decomposition of a biconnected graph with respect to its triconnected components. SPQR-trees have been introduced by Di Battista and Tamassia [8] and, since then, became quite important in the field of graph algorithms. Theoretical papers using SPQR-trees claim that they can be implemented in linear time using a modification of the algorithm by Hopcroft and Tarjan [15] for decomposing a graph into its triconnected components. So far no correct linear time implementation of either triconnectivity decomposition or SPQR-trees is known to us. Here, we show the incorrectness of the Hopcroft and Tarjan algorithm [15], and correct the faulty parts. We describe the relationship between SPQR-trees and triconnected components and apply the resulting algorithm to the computation of SPQR-trees. Our implementation is publically available in AGD [1].

1 Introduction

The data structure SPQR-tree represents the decomposition of a biconnected graph with respect to its triconnected components. SPQR-trees have been introduced by Di Battista and Tamassia [8] in a static and in a dynamic environment. In [8,10], the authors use SPQR-trees in order to represent the set of all planar embeddings of a planar biconnected graph.

Since then, SPQR-trees evolved to an important data structure in the field of graph algorithms. Many linear time algorithms that work for triconnected graphs only can be extended to work for biconnected graphs using SPQR-trees (e.g., [4,17]). Often it is essential to represent the set of all planar embeddings of a planar graph, e.g. in order to optimize a specific criteria over all planar embeddings [14,21,3,5], or for testing cluster planarity [18,6]. In a dynamic environment, SPQR-trees are useful for a variety of on-line graph algorithms dealing with triconnectivity, transitive closure, minimum spanning tree, and planarity testing [2]. Here, we restrict our attention to the static environment.

In the theoretical papers (e.g., [8,9,10]), the authors suggest to construct the data structure SPQR-tree in linear time "using a variation of the algorithm of [15] for finding the triconnected components of a graph...[10]". So far, to our knowledge, no correct linear time implementation is publically available. The only correct implementation of SPQR-trees we are aware of is part of GDToolkit [12],

[*] Partially supported by DFG-Grant Mu 1129/3-1, Forschungsschwerpunkt "Effiziente Algorithmen für diskrete Probleme und ihre Anwendungen".

J. Marks (Ed.): GD 2000, LNCS 1984, pp. 77–90, 2001.

where SPQR-trees are used in connection with a branch-and-bound algorithm to compute an orthogonal drawing of a biconnected planar graph with the minimum number of bends. However, this implementation does not run in linear time [11].

Here, we present a linear time implementation of the data structure SPQR-tree. We show the relationship between SPQR-trees and triconnected components, and show the incorrectness of the algorithm presented in [15] for decomposing a graph into its triconnected components. We develop a correct algorithm for triconnectivity decomposition by correcting and replacing the faulty parts in [15], and apply it to the computation of SPQR-trees. Our implementation (in a re-usable form) is publically available in AGD [1] (see Section 6).

The paper is structured as follows. The basics of SPQR-trees and triconnected components are described in Section 3. The algorithm for computing SPQR-trees and triconnectivity decomposition is described in Section 4, and the faulty parts of the Hopcroft and Tarjan algorithm are shown in Section 5, where we also point out the corrections we have made. We have carefully tested our implementation. Computational results concerning running time are described in Section 6.

2   Preliminaries

Let $G = (V, E)$ be an *undirected multi-graph*, that is, V is a set of vertices and E is a multi-set of unordered pairs (u, v) with $u, v \in V$. An edge (v, v) is called a *self-loop*. If an edge $(u, v) \in E$ occurs more than once in E, it is called a *multiple edge*. G is called *simple*, if it contains neither self-loops nor multiple edges. If E' is a set of edges, $V(E')$ denotes the set of all vertices incident to at least one edge in E'. A *path* $p : v \overset{*}{\Rightarrow} w$ in G is a sequence of vertices and edges leading from v to w. A path is *simple* if all its vertices are distinct. If $p : v \overset{*}{\Rightarrow} w$ is a simple path, then p plus the edge (w, v) is a *cycle*.

An undirected multi-graph $G = (V, E)$ is *connected* if every pair $v, w \in V$ of vertices in G is connected by a path. A connected multi-graph G is *biconnected* if for each triple of distinct vertices v, w, a, there is a path $p : v \overset{*}{\Rightarrow} w$ such that a is not on p. Let $G = (V, E)$ be a biconnected multi-graph and $a, b \in V$. E can be divided into equivalence classes $E_1, \ldots, E_k$ such that two edges which lie on a common path not containing any vertex of $\{a, b\}$ except as an endpoint are in the same class. The classes E_i are called the *separation classes* of G with respect to $\{a, b\}$. If there are at least two separation classes, then $\{a, b\}$ is a *separation pair* of G unless (i) there are exactly two separation classes, and one class consists of a single edge, or (ii) there are exactly three classes, each consisting of a single edge. If G contains no separation pair, G is called *triconnected*.

A *tree* T is a directed graph whose underlying undirected graph is connected, such that there is exactly one vertex (called the *root*) having no incoming edges and every other vertex has exactly one incoming edge. An edge in T from v to w is denoted with $v \rightarrow w$. If there is a (directed) path from v to w, we write $v \overset{*}{\rightarrow} w$. If $v \rightarrow w$, v is the *parent* of w, and w a *child* of v. If $v \overset{*}{\rightarrow} w$, v is an *ancestor* of w, and w a *descendant* of v. Every vertex is an ancestor and a descendant of itself. If G is a directed multi-graph, a tree T is a *spanning tree* of G if T is a subgraph of G and T contains all vertices in G.

A *palm tree* P is a directed multi-graph such that each edge in P is a either a tree arc (denoted with $v \to w$) or a frond (denoted with $v \hookrightarrow w$) satisfying the following properties:

(i) The subgraph T consisting of all tree arcs is a spanning tree of P.
(ii) If $v \hookrightarrow w$, then $w \xrightarrow{*} v$.

3   SPQR-Trees and Triconnected Components

Let $G = (V, E)$ be a biconnected multi-graph, $\{a, b\}$ a separation pair of G, and $E_1, \ldots, E_k$ the separation classes of G with respect to $\{a, b\}$. Let $E' = \bigcup_{i=1}^{\ell} E_i$ and $E'' = \bigcup_{i=\ell+1}^{k} E_i$ be such that $|E'| \geq 2$ and $|E''| \geq 2$. The two graphs $G' = (V(E'), E' \cup \{e\})$ and $G'' = (V(E''), E'' \cup \{e\})$ are called *split graphs* of G with respect to $\{a, b\}$, where $e = (a, b)$ is a new edge. Replacing a multi-graph G by two split graphs is called *splitting* G. Each split graph is again biconnected. The edge e is called *virtual edge* and identifies the split operation.

Suppose G is split, the split graphs are split, and so on, until no more split operations are possible. The resulting graphs are called the *split components* of G. Each of them is a set of three multiple edges (*triple bond*), or a cycle of length three (*triangle*), or a triconnected simple graph. The split components are not necessarily unique.

Lemma 1. *Let $G = (V, E)$ be a multi-graph.*

(i) *Each edge in E is contained in exactly one, and each virtual edge in exactly two split components.*
(ii) [15] *The total number of edges in all split components is at most $3|E| - 6$.*

Let $G_1 = (V_1, E_1)$ and $G_2 = (V_2, E_2)$ be two split components containing the same virtual edge e. The graph $G' = (V_1 \cup V_2, (E_1 \cup E_2) \setminus \{e\})$ is called a *merge graph* of G_1 and G_2. Replacing two components G_1 and G_2 by a merge graph of G_1 and G_2 is called *merging* G_1 and G_2. The *triconnected components* of G are obtained from its split components by merging the triple bonds into maximal sets of multiple edges (*bonds*) and the triangles into maximal simple cycles (*polygons*).

Lemma 2. [19,15] *The triconnected components of G are unique.*

Triconnected components of graphs are closely related to SPQR-trees. SPQR-trees were originally defined in [8] for planar graphs only. Here, we cite the more general definition given in [9], that also applies to not necessarily planar graphs.

Let G be a biconnected graph. A *split pair* of G is either a separation pair or a pair of adjacent vertices. A *split component* of a split pair $\{u, v\}$ is either an edge (u, v) or a maximal subgraph C of G such that $\{u, v\}$ is not a split pair of C. Let $\{s, t\}$ be a split pair of G. A *maximal split pair* $\{u, v\}$ of G with respect to $\{s, t\}$ is such that, for any other split pair $\{u', v'\}$, vertices u, v, s, and t are in the same split component.

Let $e = (s, t)$ be an edge of G, called the *reference edge*. The SPQR-tree $\mathcal{T}$ of G with respect to e is a rooted ordered tree whose nodes are of four types: S, P, Q, and R. Each node μ of $\mathcal{T}$ has an associated biconnected multi-graph, called the *skeleton* of μ. Tree $\mathcal{T}$ is recursively defined as follows:

Trivial Case: If G consists of exactly two parallel edges between s and t, then $\mathcal{T}$ consists of a single Q-node whose skeleton is G itself.

Parallel Case: If the split pair $\{s, t\}$ has at least three split components $G_1, \ldots,$ G_k, the root of $\mathcal{T}$ is a P-node μ, whose skeleton consists of k parallel edges $e = e_1, \ldots, e_k$ between s and t.

Series Case: Otherwise, the split pair $\{s, t\}$ has exactly two split components, one of them is e, and the other one is denoted with G'. If G' has cutvertices $c_1, \ldots, c_{k-1}$ $(k \geq 2)$ that partition G into its blocks $G_1, \ldots, G_k$, in this order from s to t, the root of $\mathcal{T}$ is an S-node μ, whose skeleton is the cycle $e_0, e_1, \ldots, e_k$, where $e_0 = e$, $c_0 = s$, $c_k = t$, and $e_i = (c_{i-1}, c_i)$ $(i = 1, \ldots, k)$.

Rigid Case: If none of the above cases applies, let $\{s_1, t_1\}, \ldots, \{s_k, t_k\}$ be the maximal split pairs of G with respect to $\{s, t\}$ $(k \geq 1)$, and, for $i = 1, \ldots, k$, let G_i be the union of all the split components of $\{s_i, t_i\}$ but the one containing e. The root of $\mathcal{T}$ is an R-node, whose skeleton is obtained from G by replacing each subgraph G_i with the edge $e_i = (s_i, t_i)$.

Except for the trivial case, μ has children $\mu_1, \ldots, \mu_k$, such that μ_i is the root of the SPQR-tree of $G_i \cup e_i$ with respect to e_i $(i = 1, \ldots, k)$. The endpoints of edge e_i are called the *poles* of node μ_i. The virtual edge of node μ_i is edge e_i of skeleton of μ. Tree $\mathcal{T}$ is completed by adding a Q-node, representing the reference edge e, and making it the parent of μ so that it becomes the root.

Each edge in G is associated with a Q-node in $\mathcal{T}$. Each edge e_i in skeleton of μ is associated with the child μ_i of μ. It is possible to root $\mathcal{T}$ at an arbitrary Q-node μ', resulting in an SPQR-tree with respect to the edge associated with μ' [9]. In our implementation, we use a slightly different, but equivalent, definition of SPQR-tree. We omit Q-nodes and distinguish between *real edges* and *virtual edges* in the skeleton graphs instead. An edge in the skeleton of μ which is associated with a Q-node in the original definition is a real edge that is not associated with a child of μ, all other skeleton edges are virtual edges associated with a P-, S-, or R-node. Using this modified definition, we can show that the skeleton graphs are the unique triconnected components of G:

Theorem 1. *Let G be a biconnected multi-graph and $\mathcal{T}$ its SPQR-tree.*

(i) *The skeleton graphs of $\mathcal{T}$ are the triconnected components of G. P-nodes correspond to bonds, S-nodes to polygons, and R-nodes to triconnected simple graphs.*

(ii) *There is an edge between two nodes $\mu, \nu \in \mathcal{T}$ if and only if the two corresponding triconnected components share a common virtual edge.*

(iii) *The size of $\mathcal{T}$, including all skeleton graphs, is linear in the size of G.*

Proof. (sketch) We remark that if $\{u, v\}$ is a separation pair, the split components of $\{u, v\}$ are the separation classes with respect to $\{u, v\}$. In the parallel, series, and rigid case of the definition of SPQR-tree, subgraphs $G_1, \ldots, G_k$ are considered. Assume that $G_1, \ldots, G_\ell$ contains more than one edge, and $G_{\ell+1}, \ldots, G_k$
contains exactly one edge. In each of the three cases, the recursive decomposition
step can be realized by performing ℓ split operations, each splitting off one G_i,
$1 \leq i \leq \ell$ and introducing a new virtual edge e' in the skeleton of node μ and the
skeleton of a child μ_i of μ. since e' remains in the skeleton of μ_i in subsequent
steps, part (ii) of the theorem follows.

The final skeleton graphs are each either a polygon, a bond, or a simple
triconnected graph, and no two polygons, and no two bonds share a common
virtual edge. Thus, the skeleton graphs are the unique triconnected components
of G proving part (i). The last part of the theorem follows directly from (i) and
Lemma 1.   □

4   The Algorithm

Let G be a biconnected multi-graph without self-loops. According to Theorem 1,
it suffices to compute the triconnected components of G, which give us enough
information to build the SPQR-tree of G. We correct the faulty parts in the
algorithm by Hopcroft and Tarjan [15] and apply this modified algorithm for
computing the triconnected components. We focus on the computation of split
pairs, because the description of this part in [15] is not only confusing but contains also severe errors. For an overview of the Hopcroft and Tarjan algorithm,
please refer to [15] or [13].

4.1   Computing SPQR-Trees

Input to the algorithm is a biconnected multi-graph $G = (V, E)$ and a reference
edge e_r. In the first step, bundles of multiple edges are replaced by a new virtual
edge as shown in Alg. 1. This creates a set of bonds $C_1, \ldots, C_k$ and results in
a simple graph G'. The required sorting of the edges in line **1.1** can be done in
$\mathcal{O}(|V| + |E|)$ time using bucket sort two times. Firstly according to the endpoint
with lower index, and secondly to the one with higher index, where we assume
that vertices have unique indices in the range $1, \ldots, |V|$. The for-loop in line **1.2**
iterates over all edges, so Alg. 1 has running time $\mathcal{O}(|V| + |E|)$.

Algorithm 1: Split off multiple edges

  1.1 *Sort edges such that all multiple edges come after each other*
  1.2 **for** *each maximal bundle of multiple edges* $e_1, \ldots, e_\ell$ *with* $\ell \geq 2$ **do**
        let $e_1, \ldots, e_\ell$ be edges between v and w
        replace $e_1, \ldots, e_\ell$ by a new edge $e' = (v, w)$
        create a new component $C = \{e_1, \ldots, e_\ell, e'\}$
  end

The second step finds the split components $C_{k+1}, \ldots, C_m$ of G'. The procedure is presented in detail in the next subsection. The triconnected components of the input graph G are created by partially reassembling the components $C_1, \ldots, C_m$. As long as two bonds or two polygons C_i and C_j containing the same virtual edge exist, C_i and C_j are merged. This is shown in Alg. 2. Removed components are marked as empty. The forall-loop in line **2.1** steps over *all* edges in C_i, i.e. those added to C_i during the loop. The test in line **2.1** can be done in constant time by precomputing for each virtual edge e the two components to which e belongs. We represent the edges in a component C_i by a list of edges, which allows to implement the set operations in lines **2.3** and **2.4** in constant time. According to Lemma 1, the total number of edges in all components is $\mathcal{O}(|E|)$, so Alg. 2 can also be implemented in time $\mathcal{O}(|V| + |E|)$.

Algorithm 2: Build triconnected components

---

    for $i := 1$ **to** m **do**

        if $C_i \neq \emptyset$ *and* C_i *is a bond or a polygon* **then**

2.1   **forall** $e \in C_i$ **do**

2.2   **if** *there exists* $j \neq i$ *with* $e \in C_j$ *and* $type(C_i) = type(C_j)$ **then**

2.3 $C_i := (C_i \cup C_j) \setminus \{e\}$

2.4 $C_j := \emptyset$

               end

           od

        end

    end

The preceding steps give enough information to build the SPQR-tree $\mathcal{T}$ of G. Applying Theorem 1, it is easy to construct the unrooted version of $\mathcal{T}$. Since we omit Q-nodes in our representation, we root $\mathcal{T}$ at the node whose skeleton contains the reference edge e_r. During the construction, we also create cross links between each tree edge $\mu \to \nu$ in $\mathcal{T}$ and the two corresponding virtual edges in skeleton of μ and skeleton of ν.

4.2   Finding Separation Pairs

Suppose we have a palm tree P for the simple, biconnected graph $G' = (V, E')$, and the vertices of G' are numbered $1, \ldots, |V|$. In the following, we identify vertices with their numbers. We introduce the following notation:

$$lowpt1(v) = \min\left(\{v\} \cup \{w \mid v \xrightarrow{*} \hookrightarrow w\}\right)$$

$$lowpt2(v) = \min\left(\{v\} \cup \left(\{w \mid v \xrightarrow{*} \hookrightarrow w\} \setminus \{lowpt1(v)\}\right)\right)$$

That is, $lowpt1(v)$ is the *lowest* vertex reachable by traversing zero or more tree arcs followed by one frond of P (or v if no such vertex exists), and $lowpt2(v)$ is the *second lowest* vertex reachable this way (or v if no such vertex exists).

We denote with $Adj(v)$ the ordered (non-cyclic) adjacency list of a vertex v, and with $D(v)$ the set of descendants of v. We seek for a numbering of the vertices and ordering of the edges in the adjacency lists satisfying the following properties:

(P1) the root of P is 1.

(P2) if $v \in V$ and $w_1, \ldots, w_n$ are the children of v in P according to the ordering in $Adj(v)$, then $w_i = w + |D(w_{i+1}) \cup \ldots \cup D(w_n)| + 1$.

(P3) the edges e in $Adj(v)$ are in ascending order according to $lowpt1(w)$ if $e = v \rightarrow w$, or w if $e = v \hookrightarrow w$, respectively.

Let $w_1, \ldots, w_n$ be the children of v with $lowpt1(w_i) = u$ in the order given by $Adj(v)$. Then there exists an i_0 such that $lowpt2(w_i) < v$ for $1 \leq i \leq i_0$, and $lowpt2(w_j) \geq v$ for $i_0 < j \leq n$. If $v \hookrightarrow u \in E'$, then $v \hookrightarrow u$ comes in $Adj(v)$ between $v \rightarrow w_{i_0}$ and $v \rightarrow w_{i_0+1}$.

It is shown in [15], how to compute such a numbering of the vertices and ordering of the adjacency lists in linear time. Unlike [15], we demand that a frond $v \hookrightarrow w$, if contained in E', must come between $v \rightarrow w_{i_0}$ and $v \rightarrow w_{i_0+1}$ in $Adj(v)$. This can easily be done by adapting the sorting function ϕ used in [15]:

$$\phi(e) = \begin{cases} 3lowpt1(w) & \text{if } e = v \rightarrow w \text{ and } lowpt2(w) < v \\ 3w + 1 & \text{if } e = v \hookrightarrow w \\ 3lowpt1(w) + 2 & \text{if } e = v \rightarrow w \text{ and } lowpt2(w) \geq v \end{cases}$$

The required ordering can be obtained by sorting the edges according to their ϕ-values using bucket sort. Using ordering ϕ and procedure PATHSEARCH as suggested in [15] will not recognize all multiple edges and thus not correctly compute the split components of G'.

Suppose we perform a depth-first-search on G' using the ordering of the edges in the adjacency lists. This divides G' into a set of paths consisting of zero or more tree arcs followed by one frond. The first path starts at vertex 1 and a path ends, when the first frond on the path is reached (see Fig. 1). Each path ends at the lowest possible vertex, and has only its initial and terminal vertex in common with previously traversed paths. From each such path $p : v \overset{*}{\Rightarrow} w$, we can form a cycle by adding the tree path from $w \overset{*}{\rightarrow} v$ to p (compare [15,16]).

Example 1. Fig. 1 shows a palm tree with a numbering that satisfies (P1)-(P3). The edges are numbered according to the generated paths. The generated paths are

1: $1 \rightarrow 2 \rightarrow 3 \rightarrow 13 \hookrightarrow 1$	7: $12 \hookrightarrow 9$
2: $13 \hookrightarrow 2$	8: $10 \rightarrow 11 \hookrightarrow 8$
3: $3 \rightarrow 4 \hookrightarrow 1$	9: $11 \hookrightarrow 9$
4: $4 \rightarrow 5 \rightarrow 8 \hookrightarrow 1$	10: $5 \rightarrow 6 \rightarrow 7 \hookrightarrow 4$
5: $8 \rightarrow 9 \rightarrow 10 \rightarrow 12 \hookrightarrow 1$	11: $7 \hookrightarrow 5$
6: $12 \hookrightarrow 8$	12: $6 \hookrightarrow 4$

We need one more definition: u_n is a *first descendant* of u_0 if $u_0 \rightarrow \cdots \rightarrow u_n$ and each $u_i \rightarrow u_{i+1}$ is the first edge in $Adj(u_i)$. In the sequel, we consider a palm

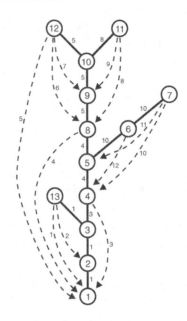

Fig. 1. Palm tree with numbered vertices and generated paths.

tree P satisfying (P1)-(P3). The following lemma gives us three easy-to-check conditions for separation pairs.

Lemma 3. (Lemma 13 in [15]) *Let $G = (V, E)$ be a biconnected graph and a, b be two vertices in G with $a < b$. Then $\{a, b\}$ is a separation pair if and only if one of the following conditions holds.*

Type-1 Case: There are distinct vertices $r \neq a, b$ and $s \neq a, b$ such that $b \to r$, $lowpt1(r) = a$, $lowpt2(r) \geq b$, and s is not a descendant of r.

Type-2 Case: There is a vertex $r \neq b$ such that $a \to r \xrightarrow{*} b$, b is a first descendant of r, $a \neq 1$, every frond $x \hookrightarrow y$ with $r \leq x < b$ has $a \leq y$, and every frond $x \hookrightarrow y$ with $a < y < b$ and $b \to w \xrightarrow{*} x$ has $lowpt1(w) \geq a$.

Multiple Edge Case: (a, b) is a multiple edge of G and G contains at least four edges.

Example 2. Consider the palm tree from Fig. 1. We have the following separation pairs:

type-1 pairs: $(1, 4), (1, 5), (4, 5), (1, 8), (1, 3)$
type-2 pairs: $(4, 8), (8, 12)$

4.3   Finding Split Components

During the algorithm, we maintain a graph G_c and a palm tree P_c of G_c. We denote with $deg(v)$ the degree of v in G_c, with $v \rightarrow w$ a tree arc in P_c, with $v \hookrightarrow w$ a frond in P_c, with $parent(v)$ the parent of v in P_c, and with $ND(v)$ the number of descendants of v in P_c. Each time we identify a split component C, we split it off, and G_c and P_c are updated. We use the following update functions:

$C := \texttt{new\_component}(e_1, \ldots, e_\ell)$: a new component $C = \{e_1, \ldots, e_\ell\}$ is created, and $e_1, \ldots, e_\ell$ are removed from G_c.

$C := C \cup \{e_1, \ldots, e_\ell\}$: the edges $e_1, \ldots, e_\ell$ are added to C and removed from G_c.

$e' := \texttt{new\_virtual\_edge}(v, w, C)$: a new virtual edge $e' = (v, w)$ is created and added to component C and G_c.

$\texttt{make\_tree\_edge}(e, v \rightarrow w)$: makes edge $e = (v, w)$ a new tree edge in P_c.

Moreover, we define the access functions

$firstChild(v) = $ first child of v in P_c according to $Adj(v)$.

$$high(w) = \begin{cases} 0 & \text{if } F(w) = \emptyset \\ \text{source vertex of first visited edge in } F(w) & \text{otherwise} \end{cases}$$

where $F(w) = \{v \mid v \hookrightarrow w \in E_c\}$, and we use two stacks for which the usual functions push, pop, and top are defined:

ESTACK contains already visited edges that are not yet assigned to a split component.

TSTACK contains triples (h, a, b) (or a special end-of-stack marker EOS), such that $\{a, b\}$ is a potential type-2 separation pair, and h is the highest numbered vertex in the component that would be split off.

The algorithm starts by calling the recursive procedure PathSearch for vertex 1, the root vertex of P (see Alg. 3). When returning from the call, the edges belonging to the last split component are on ESTACK.

Algorithm 3: Find split components

TSTACK.push(EOS)
PathSearch(1)
let $e_1, \ldots, e_\ell$ be the edges on ESTACK
3.1 $C := \texttt{new\_component}(e_1, \ldots, e_\ell)$

Procedure PathSearch is shown in Alg. 4. The testing for separation pairs applying Lemma 3 is depicted separately in Alg. 5 for type-2 and in Alg. 6 for type-1 separation pairs[1]. For a detailed description of the algorithm, please refer to [15,13]. In order to achieve linear running time, we set up the following data structures:

---

[1] The algorithm will not find *all* separation pairs, but only the separation pairs needed for dividing the graph into its split components

---

Algorithm 4: PathSearch(v)

---

forall $e \in Adj(v)$ **do**
    if $e = v \to w$ **then**
        if *e starts a path* **then**
            pop all (h, a, b) *with* $a > lowpt1(w)$ *from* TSTACK
            if *no triples deleted* **then**
                TSTACK.push($w + ND(w) - 1, lowpt1(w), v$)
            else
                $y := \max\{h \mid (h, a, b)$ *deleted from* TSTACK $\}$
                let (h, a, b) *be last triple deleted*
                TSTACK.push($\max(y, w + ND(w) - 1), lowpt1(w), b$)
            end
            TSTACK.push(*EOS*)
        end

        PathSearch(w)
        ESTACK.push($v \to w$)

        check for type-2 pairs
        check for a type-1 pair

        if *e starts a path* **then**
            remove all triples on TSTACK *down to and including EOS*
        end
        while (h, a, b) *on* TSTACK *has* $a \neq v$ *and* $b \neq v$ *and* $high(v) > h$ **do**
            TSTACK.pop()
        od
    else
        let $e = v \hookrightarrow w$
        if *e starts a path* **then**
            pop all (h, a, b) *with* $a > w$ *from* TSTACK
            if *no triples deleted* **then**
               TSTACK.push(v, w, v)
            else
               $y := \max\{h \mid (h, a, b)$ *deleted from* TSTACK $\}$
               let (h, a, b) *be last triple deleted*
               TSTACK.push(y, w, b)
            end
        end
        if $w = parent(v)$ **then**
            $C :=$ new_component($e, w \to v$)
            $e' :=$ new_virtual_edge(w, v, C)
            make_tree_edge($e', w \to v$)
        else
            ESTACK.push(e)
        end
    end
od

(4.1)

Algorithm 5: check for type-2 pairs

---

while $v \neq 1$ *and* $(((h, a, b)$ *on* TSTACK *has* $a = v)$ *or* $(deg(w) = 2$ *and*
firstChild$(w) > w))$ **do**

    if $a = v$ *and* *parent*$(b) = a$ **then**

        TSTACK.pop()

    else

        $e_{ab} := nil$

        if $deg(w) = 2$ *and* *firstChild*$(w) > w$ **then**

            $C := $ new_component()

            remove top edges (v, w) *and* (w, b) *from* ESTACK *and add to* C

            $e' := $ new_virtual_edge(v, x, C)

            if ESTACK.top() $= (v, b)$ **then** $e_{ab} := $ ESTACK.pop()

        else

            $(h, a, b) := $ TSTACK.pop()

            $C := $ new_component()

            while (x, y) *on* ESTACK *has* $a \leq x \leq h$ *and* $a \leq y \leq h$ **do**

                if $(x, y) = (a, b)$ **then** $e_{ab} := $ ESTACK.pop()

                else $C := C \cup \{$ ESTACK.pop() $\}$

            od

            $e' := $ new_virtual_edge(a, b, C)

        end

        if $e_{ab} \neq nil$ **then**

            $C := $ new_component(e_{ab}, e')

            $e' := $ new_virtual_edge(v, b, C)

        end

        ESTACK.push(e'); make_tree_edge$(e', v \rightarrow b)$; $w := b$

    end

od

Algorithm 6: check for a type-1 pair

6.1 **if** $lowpt2(w) \geq v$ *and* $lowpt1(w) < v$ *and* $(parent(v) \neq 1$ *or* v *is adjacent to*
     a not yet visited tree arc) **then**

        $C := $ new_component ()

        while (x, y) *on* ESTACK *has* $w \leq x < w + ND(w)$ *or* $w \leq y < w + ND(w)$
        do

            $C := C \cup \{$ ESTACK.pop() $\}$

        od

        $e' := $ new_virtual_edge$(v, lowpt1(w), C)$

        if ESTACK.top() $= (v, lowpt1(w))$ **then**

            $C := $ new_component(ESTACK.pop(),e')

            $e' := $ new_virtual_edge$(v, lowpt1(w), C)$

        end

        if $lowpt1(w) \neq parent(v)$ **then**

            ESTACK.push(e')

            make_tree_edge$(e', lowpt1(w) \rightarrow v)$

        else

            $C := $ new_component$(e', lowpt1(w) \rightarrow v)$

            $e' := $ new_virtual_edge$(lowpt1(w), v, C)$

            make_tree_edge$(e', lowpt1(w) \rightarrow v)$

        end

    end

- The palm tree P is represented by arrays PARENT[v], TREE_ARC[v] (the tree arc entering v), and TYPE[e] (tree arc or frond).
- The values $lowpt1(v)$, $lowpt2(v)$, and $ND(v)$ are precomputed. It is not necessary to update them.
- An array DEGREE[v] contains the degree of $v \in G_c$. It is updated each time an edge is added to or removed from G_c.
- In order to compute $firstChild(v)$, we update the adjacency lists each time an edge is added to or removed from G_c.
- In order to compute $high(v)$, we precompute the list of fronds $v_i \hookrightarrow w$ ending at w in the order they are visited. When a frond is removed from or added to G_c, the respective list is updated.
- We precompute an array START[e] which is true iff e starts a path.
- The test if "v is adjacent to a not yet visited tree arc" in line **6.1** can be done by simply counting the visited adjacent tree arcs.

5   Corrections on the Hopcroft and Tarjan Algorithm

Procedure SPLIT in [15] does not correctly split a graph into its split components. We summarize the important changes we have made in our algorithm:

- The sorting function ϕ had to be modified as described in subsection 4.2 in order to identify all multiple edges.
- The creation of the last split component (line **3.1**) was missing.
- The condition in line **4.1** was changed. The original condition could remove triples from TSTACK corresponding to real type-2 separation pairs. Such a separation pair could not be recognized by the original SPLIT procedure.
- The condition in line **6.1** was changed. The original condition could incorrectly identify separation pairs after the graph had been modified.
- The updates for $firstChild(v)$ (which is $A1(v)$ in [15]) and $DEGREE(v)$ were not sufficient.
- $high(w)$ (which is HIGHPT(w) in [15]) was not updated, which is not correct. It is necessary to update HIGHPT dynamically, when G_c is modified. We replaced HIGHPT(w) by a list of fronds ending at w, which is updated as G_c changes.

6   Computational Experiments

Our implementation is based on LEDA [20] and made publically available in AGD [1]. We tested our implementation with generated planar and non-planar biconnected graphs, and the benchmark graphs collected by Di Battista et al. [7] ranging from 10 to 100 vertices. A planar biconnected graph with n vertices and m edges is generated by n randomly chosen split-edge and $m - n$ split face operations. A general biconnected graph is generated by creating a random graph G and making G biconnected by augmenting edges. The computed SPQR-trees are automatically checked with several consistency and plausibility tests. The average running times are depicted in Fig. 2. The x-axis shows the number of vertices, and the y-axis the running time in seconds. The left side shows the results for generated graphs with n vertices and m edges applying the algorithm to 100 test instances each. Even very large instances with 40000 edges could be solved within less than 4 seconds. The right side shows the results for the benchmark graphs applying the algorithm to each biconnected component containing at least three edges. Graphs with 100 vertices took about 0.02 seconds.

Recently, Gutwenger, Mutzel, and Weiskircher [14] have presented a linear time algorithm for solving the one edge insertion problem optimally (minimum number of crossings) over all combinatorial embeddings. Their algorithm requires to compute SPQR-trees for some biconnected components of a planar input graph. In their tests, they also used the benchmark graphs from [7] and applied our implementation of SPQR-trees. The longest running time for a graph with 100 vertices was 1.15 seconds, where 38 edges had to be inserted.

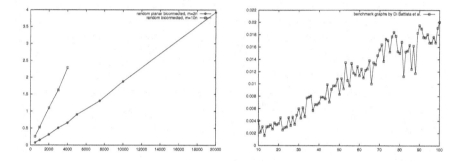

Fig. 2. Average running times (system configuration: Pentium II, 400 MHz, 128MB RAM)

References

[1] *AGD User Manual (Version 1.1.2)*, Feb. 2000. Max-Planck-Institut Saarbrücken, Technische Universität Wien, Universität zu Köln, Universität Trier. See also http://www.mpi-sb.mpg.de/AGD/.
[2] G. Di Battista and R. Tamassia. On-line graph algorithms with SPQR-trees. In M. S. Paterson, editor, *Proc. of the 17th International Colloqium on Automata, Languages and Prog ramming (ICALP)*, volume 443 of *Lecture Notes in Computer Science*, pages 598–611. Springer-Verlag, 1990.
[3] P. Bertolazzi, G. Di Battista, and W. Didimo. Computing orthogonal drawings with the minimum number of bends. In *Proc. 5th Workshop Algorithms, Data Struct.*, volume 1272 of *Lecture Notes in Computer Science*, pages 331–344, 1998.
[4] P. Bertolazzi, G. Di Battista, G. Liotta, and C. Mannino. Optimal upward planarity testing of single-source digraphs. *SIAM J. Comput.*, 27(1):132–169, 1998.

[5] D. Bienstock and C. L. Monma. On the complexity of embedding planar graphs to minimize certain distance measures. *Algorithmica*, 5(1):93–109, 1990.

[6] Elias Dahlhaus. A linear time algorithm to recognize clustered planar graphs and its parallelization. In C. L. Lucchesi and A. V. Moura, editors, *LATIN '98: Theoretical Informatics, Third Latin American Symposium*, volume 1380 of *Lecture Notes in Computer Science*, pages 239–248. Springer-Verlag, 1998.

[7] G. Di Battista, A. Garg, G. Liotta, R. Tamassia, and F. Vargiu. An experimental comparision of four graph drawing algorithms. *Comput. Geom. Theory Appl.*, 7:303–326, 1997.

[8] G. Di Battista and R. Tamassia. Incremental planarity testing. In *Proc. 30th IEEE Symp. on Foundations of Computer Science*, pages 436–441, 1989.

[9] G. Di Battista and R. Tamassia. On-line maintanance of triconnected components with SPQR-trees. *Algorithmica*, 15:302–318, 1996.

[10] G. Di Battista and R. Tamassia. On-line planarity testing. *SIAM J. Comput.*, 25(5):956–997, 1996.

[11] W. Didimo. Dipartimento di Informatica e Automazione, Università di Roma Tre, Rome, Italy, Personal Communication.

[12] *GDToolkit Online Documentation*. See http://www.dia.uniroma3.it/~gdt.

[13] C. Gutwenger and P. Mutzel. A linear time implementation of SPQR-trees. Technical report, Technische Universität Wien, 2000. To appear.

[14] C. Gutwenger, P. Mutzel, and R. Weiskircher. Inserting an edge into a planar graph. In *Proceedings of the Twelfth Annual ACM-SIAM Symposium on Discrete Algorithms (SODA '01)*. ACM Press, 2001. To appear.

[15] J. E. Hopcroft and R. E. Tarjan. Dividing a graph into triconnected components. *SIAM J. Comput.*, 2(3):135–158, 1973.

[16] J. E. Hopcroft and R. E. Tarjan. Efficient planarity testing. *Journal of the ACM*, 21:549–568, 1974.

[17] G. Kant. Drawing planar graphs using the canonical ordering. *Algorithmica, Special Issue on Graph Drawing*, 16(1):4–32, 1996.

[18] T. Lengauer. Hierarchical planarity testing. *Journal of the ACM*, 36:474–509, 1989.

[19] S. MacLaine. A structural characterization of planar combinatorial graphs. *Duke Math. J.*, 3:460–472, 1937.

[20] K. Mehlhorn and S. Näher. *The LEDA Platform of Combinatorial and Geometric Computing*. Cambridge University Press, 1999. to appear.

[21] P. Mutzel and R. Weiskircher. Optimizing over all combinatorial embeddings of a planar graph. In G. Cornuéjols, R. Burkard, and G. Woeginger, editors, *Proceedings of the Seventh Conference on Integer Programming and Combinatorial Optimization (IPCO)*, volume 1610 of *LNCS*, pages 361–376. Springer Verlag, 1999.

Labeling Points with Rectangles of Various Shapes

Shin-ichi Nakano[1], Takao Nishizeki[2], Takeshi Tokuyama[2], and
Shuhei Watanabe[2]

[1] Department of Information Engineering, Gunma University, Kiryu 376-8515,
Japan, Email: nakano@cs.gunma-u.ac.jp
[2] Graduate School of Information Sciences, Tohoku University, Aobayama, Sendai
980-8579, Japan. Emails: (nabe,nishi)@nishizeki.ecei.tohoku.ac.jp,
tokuyama@dais.is.tohoku.ac.jp

Abstract. We deal with a map-labeling problem, named LOFL (Left-part Ordered Flexible Labeling), to label a set of points in a plane with polygonal obstacles. The label for each point is selected from a set of rectangles with various shapes satisfying the *left-part ordered* property, and is placed near to the point after scaled by a scaling factor σ which is common to all points. In this paper, we give an optimal $O((n+m)\log(n+m))$ algorithm to decide the feasibility of LOFL for a fixed scaling factor σ, and an $O((n+m)\log^2(n+m))$ time algorithm to find the largest feasible scaling factor σ, where n is the number of points and m is the number of edges of polygonal obstacles.

1 Introduction

Annotating a set of points is a common task to be performed in Geographic Information Systems. It is crucial that important objects in a map have labels indicating their names or other attributes. The objects to be labeled in a map highly depend on user's interest; for example, a drainage maintainer may want to have locations and identification labels of manholes, although they are almost useless information for ordinary users. Therefore, a digital map should have a database of sets of points representing locations of objects together with labels of the objects, and should have a function to insert labels to a non-labeled map efficiently.

The problem of locating labels in a map is called the map labeling (or map lettering) [9,14,15,16]. Approximating a label (a string of characters) by its bounding rectangle, one can formulate the map-labeling problem as the problem of locating a set of n rectangles in a plane (with obstacles containing m edges) in a way that (1) each rectangle representing a label of an object should be near to the object, (2) rectangles do not overlap each other, and (3) each rectangle does not overlap any obstacle in the map. The condition (1) will be mathematically formulated in a suitable fashion.

We restrict ourselves to the *point feature label placement problem*, where each object is a point (object point) in the map. The rightmost point of an object

J. Marks (Ed.): GD 2000, LNCS 1984, pp. 91–102, 2001.
© Springer-Verlag Berlin Heidelberg 2001

is often chosen as an object point. Moreover, we only consider axis parallel rectangles as labels. See [3,7] for more complicated labeling problems.

If the size of each character is given (therefore, the size of each label is given), we want to decide whether there exists a feasible solution satisfying the conditions (1), (2), and (3) above. Such a problem is called the *decision problem*. We also want to consider the *optimization problem* in which we compute the maximum character size σ, called a *scaling factor*, for which there is a feasible solution. For convenience's sake, we assume that rectangles are closed, but we allow a rectangle to touch other rectangles or obstacles on its boundary.

The decision problem is hard in general: Formann and Wagner [5] showed that if there are four candidates of the placement for each label, it is NP-hard in general to decide the feasibility. Indeed, it is NP-hard even if each label is a unit square and must be placed in a way that the corresponding object point is at one of its four corners (four-position model); we say that such a label is *pinned* at a corner. Kato [8] showed that the problem remains to be NP-hard if there are three candidates for each label.

On the other side, if each label is a unit square pinned at one of its two left corners (two-position model), the problem is polynomial time solvable. In general, if there are at most two candidates of the placement for each label, the problem is polynomial-time solvable since it can be formulated as a 2-SAT problem [5]. Moreover, approximation algorithms with provable approximation ratios are given for several useful versions of the map labeling [9,14]. If we fix the scaling factor and measure the quality of the solution by the number of labels that can be placed without overlapping, there are PTAS algorithms for several cases [1,9].

We deal with another type of a map labeling, called a *shape-flexible labeling*, where we can flexibly choose the shape of each label from a candidate set of rectangles. The chosen labels are placed in the map after scaled by a scaling factor σ which is common for all labels. The problem of deciding the feasibility of a shape-flexible labeling problem is NP-hard in general, and the complexity of solving the problem heavily depends on features of candidate sets. If each candidate set is the set of all rectangles with a given area, the labeling is called an *elastic labeling*, and some special cases were investigated by Iturriaga and Lubiw [6].

Our motivation is as follows: Consider a rectangular label representing a character string of length l. It needs width l (character units) if it is written in a single line. However, we can fold the label to decrease the width. Moreover, in the Chinese (also Japanese or Korean) language system, we can write a character string vertically, and hence we can transpose a label (i.e. exchange its width and height). It is often seen that folding and transposition can improve the labeling layout: Suppose that we represent a character string "GSIS"[1] by using three ways: horizontal, in two lines, and vertical. Each of the first three pictures (a), (b), and (c) of Figure 1 illustrates a label placement using single-shape labels

---

[1] acronym of "Graduate School of Information Sciences"

pinned at the left-upper corner. The picture (d) shows the improvement of the scaling factor if we can use three different kinds of shapes all together.

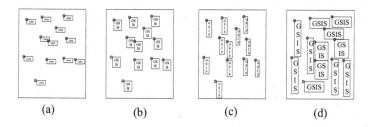

(a)   (b)   (c)   (d)

Fig. 1. Placements using three kinds of labels. Each of (a),(b), and (c) uses a single kind of labels, whereas (d) uses three kinds of labels.

Therefore, the following *fixed-corner-position shape-flexible labeling* problem naturally arises:

> Suppose that we have n points and each point has a set of candidate labels of various rectangular shapes pinned at the upper-left corner. The label for each point must be selected from the candidate set and placed after scaled by σ. How difficult is it to find the largest scaling factor σ to label all points?

In this paper, we propose a new class of shape-flexible labeling problems, named *Left-part Ordered Flexible Labeling* (LOFL), where the candidate set of rectangles given to each object point must be a *left-part ordered set*. The definition of a left-part ordered set will be given in the next section; a typical example is a set of rectangles pinned at their left-upper corners. Thus, the fixed-corner-position shape-flexible labeling problem is a special case of LOFL.

We show that the decision problem for LOFL can be solved in $O((n + m) \log(n + m))$ time by a simple plane-sweep algorithm. As a consequence, the optimization problem for LOFL can be solved in $O((n+m) \log(n+m) \log \Gamma)$ time if the coordinate values of points are represented by $\log \Gamma$-bit integers. We also give an $O((n+m) \log^2(n+m))$ time algorithm for the optimization problem. To design this algorithm, we use the parametric search paradigm in a novel way; we use parallel sort and point location query to design a "guide algorithm" required for the parametric search.

Our method can be used as a subroutine in heuristic algorithms for a practical labeling system. We have done a preliminary experiment on the ability of LOFL to enlarge the labeling size compared to single-shaped models.

2   Preliminaries

2.1   Left-Part Ordered Sets

A set $\mathcal{R}$ of rectangles in the plane is *totally ordered* with respect to inclusion if any pair R and R' of rectangles in $\mathcal{R}$ satisfies either $R \subseteq R'$ or $R' \subseteq R$.

For a rectangle L representing a label, we fix a point $q(L)$, which we call the *pinning point* of L, in the closure of L. Consider a set $\mathcal{L}$ of rectangles with pinning points as a set of candidate labels. Suppose that we place the rectangles in $\mathcal{L}$ so that their pinning points are translated to the origin. Let H^- be the closed halfplane defined by $x \leq 0$. $L \cap H^-$ (or its translated/scaled copy if L is translated/scaled) is called the *left-part* of L.

Then, $\mathcal{L}^- = \{L \cap H^- | L \in \mathcal{L}\}$ is a set of rectangles. If $\mathcal{L}^-$ is totally ordered, we say the set $\mathcal{L}$ satisfies the *left-part ordered property*, and call the set $\mathcal{L}$ a *left-part ordered set*.

We say that a left-part ordered set $\mathcal{L}$ is *degenerate* if the pinning point of each label $L \in \mathcal{L}$ is on the left edge; in other word, $L \cap H^-$ is a vertical segment. Otherwise, we say $\mathcal{L}$ is non-degenerate. Figure 2 (a) is an example of a degenerate left-part ordered set, and Figure 2 (b) is an example of non-degenerate left-part ordered set. For simplicity, we mainly consider a degenerate left-part ordered set to describe algorithms and proofs, since it is routine to generalize them for a non-degenerate case.

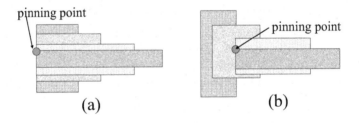

(a)   (b)

Fig. 2. (a) A degenerate left-part ordered set consisting of four rectangles, and (b) a nondegenerate left-part ordered set.

We are given a set $P = \{p_1, p_2, \ldots, p_n\}$ of n object points on the plane. Let $p_i = (x_i, y_j)$ for $i = 1, 2, \ldots, n$. We also consider a set Q of polygonal (not necessarily rectilinear) obstacles in the plane, which no label is permitted to overlap. We permit labels to touch obstacles. We assume that the obstacles do not intersect each other. Let m be the number of edges of polygons in Q. A left-part ordered set $\mathcal{L}_i$ of rectangular labels is given to each object point $p_i \in P$. $\mathcal{L}_i$ and $\mathcal{L}_j$ may be different from each other if $i \neq j$.

Let σ be a positive real value, called a *scaling factor*. We choose a suitable label L_i from $\mathcal{L}_i$ for each $p_i \in P$, scale L_i by the factor σ, and place it in the plane so that its pinning point is located at p_i. The placement $\mathbf{L} = \{L_1, L_2, \ldots, L_n\}$ is called *feasible* if (1) no two labels overlap each other and (2) no label overlaps any obstacle. Our aim is to find the largest scaling factor σ for which a feasible placement exists, and to compute the placement.

We assume that there is no pair of labels L and L' in $\mathcal{L}_i$ such that $L \subset L'$, since we need not use the larger label L' (called a *redundant label*) in our solution. If we have a label set with such pairs, we preprocess the label sets by removing all redundant labels.

We define an order $>$ on the set $\mathcal{L}_i$ as follows: $L > L'$ for two labels L and L' in $\mathcal{L}_i$ if and only if $L \cap H^- \supset L' \cap H^-$. We say that L' is *left-smaller* than L if $L > L'$.

Using the order, we can naturally give a lexicographical order among the set of feasible solutions (for a fixed σ) as follows: We sort the object points $p_1, p_2, \ldots, p_n$ in nonincreasing order with respect to the x-coordinates, and rearrange the numbering; hence, p_1 is the rightmost point and p_n is the leftmost point. Let $\mathbf{L} = (L_1, L_2, ..., L_n)$ and $\mathbf{L}' = (L'_1, L'_2, ..., L'_n)$ be two different feasible placements where $L_i \in \mathcal{L}_i$ and $L'_i \in \mathcal{L}_i$ are labels for p_i, $1 \leq i \leq n$. Then we define $\mathbf{L} > \mathbf{L}'$ if there is an index j such that $L_j > L'_j$ and $L_i = L'_i$ for every $i < j$. The minimum feasible placement with respect to this order is called the *left-minimum* solution.

Let $N = \sum_{i=1}^{n} |\mathcal{L}_i|$: N is the number of all candidate rectangles. In Geographic Information System, the cardinality of a left-part ordered set for a point is usually bounded by a constant. Therefore, we assume $N = O(n)$ in this paper, although it is not difficult to generalize our argument to the cases where N is much larger than n. Indeed, the decision problem can be solved without increasing the time complexity even if the label set of each point is an infinite set, provided that we have an efficient (in precise, $O(\log(n + m))$ amortized time) method to query the left-smallest label that does not intersect the "frontier" defined in Section 3.2. A typical example is the fixed-corner-position elastic labeling problem, where each label set consists of all rectangles with same area and each label is pinned at its upper-left corner. Another example is the *1-slider model* labeling problem proposed by van Kreveld et al.[9].

2.2 Approximation Hardness for Label Sets without Left-Part Ordered Property

Before presenting algorithms for LOFL, we remark that the left-part ordered property is crucial for designing a polynomial time algorithm. Indeed, it is NP-hard to compute a feasible placement whose scaling factor is larger than $\frac{1}{2} + \epsilon$ times the optimal scaling factor for any positive constant ϵ. The hardness result can be easily obtained by modifying the reduction of the planar 3SAT problem to a labeling problem with three candidate labels [8], and hence the proof is omitted in this version. On the other hand, one cannot construct an "alternating cycle" gadget representing a graph-edge in the reduction if only a left-part ordered set is allowed to each point.

3   Decision Problem

3.1   Algorithm for the Decision Problem

In this section, we present an $O((n + m) \log(n + m))$ time algorithm to solve the decision problem for a fixed scaling σ. Without loss of generality, we may assume $\sigma = 1$ in this section. We first sort the object points in nonincreasing order with respect to the x-coordinate values, and re-arrange the numbering as we noted before. We start with the following observation:

Lemma 1. *Let* $\mathbf{L} = (L_1, L_2, \ldots, L_n)$ *be a feasible placement, and let* $1 \le i \le n$. *If there is a label* $L \in \mathcal{L}_i$ *which is left-smaller than* L_i *and intersects none of the labels* $L_1, L_2, \ldots, L_{i-1}$ *and the obstacles, then we can replace* L_i *by* L *to obtain another feasible placement* $\mathbf{L}'$, *where* $\mathbf{L}' = (L_1, L_2, \ldots, L_{i-1}, L, L_{i+1}, \ldots, L_n)$.

Proof. Suppose for a contradiction that the placement $\mathbf{L}'$ is infeasible. Then there must be an index $j > i$ such that the label L_j assigned to p_j in the original placement $\mathbf{L}$ intersects L. Since p_j is located to the left of p_i, L_j must intersect the left-part of L. However, L_j must also intersect the left-part of L_i because L is left-smaller than L_i. This contradicts the feasibility of the original placement $\mathbf{L}$.

From this observation, we can design a simple incremental algorithm, named DECIDE, to decide the feasibility. It is clear that DECIDE is correct, and it outputs the left-minimum solution if the input instance is feasible:

Algorithm *DECIDE*
(*decide the feasibility of a LOFL instance for a given σ)
1.   **for** $i \leftarrow 1$ **to** n
2.   **do**
3.   **if** every label in $\mathcal{L}_i$ overlaps an obstacle or a label placed so far
4.   **then return** "The problem is infeasible"
5.   **else** assign p_i the left-smallest label L_i that overlaps
                         neither any obstacle nor any label placed so far
6.   **fi**
7.   **end**
8.   **return** $\mathbf{L} = (L_1, L_2, \ldots, L_n)$ as a feasible solution

3.2   Implementation and Analysis of the Algorithm

We give an implementation of DECIDE by using a standard plane sweep method, and show that it takes $O((n + m) \log(n + m))$ time. We remark that the plane sweep method is widely used in the rectangle placement and labeling problems; See for example, van Kreveld *et al.* [9].

First, we assume for simplicity that there is no obstacle, and give an algorithm for the case; we will briefly explain later how to modify it for the case with obstacles. For a label L, its *right-part width* is the horizontal distance between its pinning point and its right edge. As a preprocessing, we sort the rectangles in $\mathcal{L}_i$ in descending order of right-part width for each $i = 1, 2, \ldots, n$. It takes $O(N \log N) = O(n \log n)$ time. For each set $\mathcal{L}_i$ and a positive real number w, let $L_i(w)$ be the left-smallest label in $\mathcal{L}_i$ whose right-part width is at most w.

Let $\mathbf{U}$ be a set of geometric objects (in our case, placed labels) in the plane, and let l be a vertical line, then we say that a point q in an object in $\mathbf{U}$ is *left-visible* from l if q is on the right of l and the horizontal half-line emanating from q to the left does not intersect any objects of $\mathbf{U}$ until it meets l. The union of all left-visible points of objects in $\mathbf{U}$ is called the *frontier* of $\mathbf{U}$ at l.

We denote by $\mathbf{L}[i]$ the labeling of points $p_1, p_2, \ldots, p_i$ obtained by the algorithm DECIDE. Recall that $p_i = (x_i, y_i)$. For $t, x_i > t \geq x_{i+1}$, the frontier of $\mathbf{L}[i]$ at the vertical line $x = t$ is a union of all left-visible segments on left edges of labels (rectangles) in $\mathbf{L}[i]$. The frontier has $O(i)$ segments, and its orthogonal projection onto the y-axis induces a partition of the y axis into $O(i)$ intervals (in precise, at most $2i + 1$ intervals). The sorted list $Proj(\mathbf{L}[i])$ of these intervals with respect to the y-coordinate values of the endpoints is called the *projected frontier*. To each interval I in the projected frontier, we assign the x-coordinate value x_I of the segment in the frontier whose projected image is I. The value x_I is set to be ∞ if there is no label in $\mathbf{L}[i]$ whose projection contains I. We implement the list $Proj(\mathbf{L}[i])$ of intervals by using a suitable dynamic binary-search data structure [11]. Thus, we can find the interval I containing y_{i+1} in time $O(\log n)$, and scan the list of intervals in $O(\log n)$ time per interval.

Our plane sweep algorithm moves the sweep line $x = t$ to the left from $t = \infty$ to $t = -\infty$. While $x_i > t \geq x_{i+1}$, we maintain the projected frontier $Proj(\mathbf{L}[i])$ together with the values x_I for all intervals I. When the sweep line comes to $t = x_{i+1}$, we insert a label of p_{i+1} to $\mathbf{L}[i]$, and update $Proj(\mathbf{L}[i])$ to $Proj(\mathbf{L}[i + 1])$. We omit details in this version because of space limitation.

Proposition 1. *The time complexity of the algorithm is $O(n \log n)$.*

The decision problem is at least as difficult as the element uniqueness problem [13], and hence the $O(n \log n)$ time complexity is optimal on the algebraic decision tree model.

If there are obstacles, the framework of the algorithm is the same as one described above. However, the frontier contains "parts" of obstacles as well as left edges of placed labels. Each of such parts may be a segment of an edge of an obstacle or a connected component of the intersection of obstacles and the sweep line. If a part is a segment of an edge, the corresponding interval in the projected frontier should contain the equation of the edge.

A major difficulty is in handling of obstacles intersecting the sweep line. For each connected component of the intersection of obstacles with the sweep line, we have an interval, which linearly changes in the parameter t, in the projected frontier. Such an interval or its adjacent interval may shrink and be eliminated. Therefore, we need to consider a new type of events where an interval in the projected frontier is eliminated. However, the number of such events is $O(n+m)$. We maintain a priority queue to query the earliest elimination time of intervals in the projected frontier. This priority queue can be updated in $O(\log(m+n))$ time per each of insertion, deletion and cut-down operations of intervals. Analogously to the analysis in the previous subsection, we can prove that $O(n+m)$ intervals in the projected frontier are inserted, deleted, or cut down in the whole plane sweep procedure. Hence, we can maintain the priority queue in $O((n+m)\log(n+m))$ time in total.

4   Optimization Problem

4.1   A Precision-Dependent Algorithm

We consider the problem of finding the maximum feasible value of the scaling factor σ. A simple binary search algorithm on σ works; the algorithm for the decision problem can decide whether we should try a larger σ or a smaller one than the current scaling factor for the next search. If coordinate values of all points are integral and Γ is the maximum of their absolute value, then it suffices to run the algorithm given in Section 3.1 an $O(\log \Gamma)$ number of times (note that $\Gamma^2 \geq n$). Thus, we have:

Theorem 1. *The optimization problem of LOFL can be solved in* $O((n+m)\log(n+m)\log \Gamma)$ *time.*

4.2   Precision-Independent Algorithms

The binary search algorithm above is efficient for practical inputs for which $\log \Gamma = O(\log(n+m))$ holds. However, an efficient algorithm with time complexity independent of Γ is desirable from the theoretical point of view. We design such an $O((n+m)\log^2(n+m))$ time algorithm for the optimization problem.

One possible method is to consider the *conflict graph* of candidate labels, and first list up all possible critical values of σ, and then apply the binary search to the list. For several labeling problems [5], a list of size $O(n)$ of the critical values can be found, and the above method is efficient. This is because the size of the conflict graph at $\sigma = \sigma_{opt}$ is reduced to $O(n)$ for those problems if "clearly useless" labels are omitted. Unfortunately, the property does not hold for the LOFL, and we do not know how to obtain such a list of size $o(n^2)$. Thus, we apply another approach.

Meggido's parametric search [10] is a famous method to transform a precision-dependent binary search algorithm into a precision-independent algorithm. Especially, the method is quite useful in computational geometry [12].

We give a brief introduction to the parametric search paradigm (see [12] for details). Suppose that F is a 0-1 valued *monotone* function on a parameter θ: there is a value θ_{opt} such that $F(\theta) = 1$ if $\theta \leq \theta_{opt}$ and $F(\theta) = 0$ if $\theta > \theta_{opt}$. Our aim is to compute the value θ_{opt}. Parametric search assumes that the following two algorithms, A and D, for computing $F(\theta)$ for a given value θ are available: The algorithm D is called a *decision algorithm*, which is the fastest available algorithm to compute $F(\theta)$. Assume that D takes $O(T_D)$ time. The other algorithm A is called a *guide algorithm*. We simulate the behavior of A for $\theta = \theta_{opt}$ without knowing the value θ_{opt} in cooperation with the decision algorithm D, and find the value θ_{opt} in the course of the simulation. It is advantageous to use a guide algorithm that has a parallel structure, although we do not use a parallel machine in our computation. If A takes $O(t_A)$ parallel time with M processors, then we can simulate A for $\theta = \theta_{opt}$ without inputting the value θ_{opt} in $O(t_A M \log M + t_A T_D \log M)$ sequential time. Cole's acceleration method [4] can often improve the time complexity to $O(t_A M \log M + t_A T_D)$.

Let us consider our LOFL problem. We define a monotone function F as follows: $F(\sigma) = 1$ if and only if there is a feasible placement for the scaling factor σ. We can use the parametric search paradigm regarding σ as the parameter. We use DECIDE for the decision algorithm. Unfortunately, for our problem, a guide algorithm with $t_A = O(\log(n + m))$ and $M = O(n + m)$ seems to be difficult to design. To overcome the difficulty, we adopt a "heterogeneous" version of parametric search. The heterogeneous parametric search paradigm uses a "weaker" guide algorithm A that cannot compute $F(\sigma)$ itself even if σ is given as an input. Instead, A computes another function $G(\sigma)$, where the range of $G(\sigma)$ is not $\{0,1\}$ but is a much larger category. The required condition is that $G(\sigma) = G(\sigma')$ always implies $F(\sigma) = F(\sigma')$ for any σ and σ'. Intuitively, G gives a refinement of F. In particular, we will use a guide algorithm consisting of parallel sort and point location query algorithms.

The idea of the heterogeneous parametric search was implicitly given in Megiddo's paper [10], in which he solved a problem on the parametric minimum spanning tree of a graph by using a parallel sorting algorithm as its guide algorithm. Cole [4] dealt with the heterogeneous parametric search in which the guide algorithm is a parallel sort using a sorting network. However, to the author's knowledge, this is the first time that a heterogeneous parametric search algorithm using a guide algorithm involving a computational geometric procedure is proposed.

4.3   Parametric Search Algorithm for LOFL

As preprocessing of our parametric search algorithm, we prepare a point location data structure from the set Q of polygonal obstacles as follows: We first construct a triangulation $\mathcal{D}(Q)$ of the plane into $O(m)$ triangles so that each triangle is either contained in an obstacle or completely outside obstacles. All vertices, edges, and triangles in $\mathcal{D}(Q)$ are called *faces* of $\mathcal{D}(Q)$. Then, we prepare a point location data structure so that we can find the face of $\mathcal{D}(Q)$ containing a query point in $O(\log m)$ time. The triangulation and the point location data structure can be constructed in $O(m \log m)$ time (e.g. [13]), and we do not need to construct it in parallel since it is independent of the value of the scaling factor.

Let $\mathcal{L} = \cup_{i=1}^{n} \mathcal{L}_i$ be the set of all candidate labels, and let $S(\sigma)$ be the set of corner points of rectangles in $\mathcal{L}$ after scaled by σ and placed so that the pinning points come to their corresponding object points in P. Let $V(Q)$ be the set of all vertices of polygonal obstacles in Q.

Our guide algorithm first computes the sorting lists $X(S(\sigma) \cup V(Q))$ and $Y(S(\sigma) \cup V(Q))$ of the point set $S(\sigma) \cup V(Q)$ with respect to x- and y-coordinate values, and then locates all points of $S(\sigma)$ in $\mathcal{D}(Q)$ in parallel.

A pair τ and σ of parameter values are called *equivalent* to each other if (1) $X(S(\sigma) \cup V(Q)) = X(S(\tau) \cup V(Q))$, (2) $Y(S(\sigma) \cup V(Q)) = Y(S(\tau) \cup V(Q))$ and (3) each point in $S(\tau)$ is contained in the same face of $\mathcal{D}(Q)$ as the corresponding point in $S(\sigma)$ is.

Lemma 2. *Let σ and τ be equivalent, then there is a feasible solution of LOFL for the scaling factor τ if and only if there is a feasible solution for σ. Moreover, there is no value $\tau \neq \sigma_{opt}$ such that τ is equivalent to σ_{opt}.*

Hence, by simulating our guide algorithm, we can compute σ_{opt}. Sorting of $O(n + m)$ elements can be done by applying AKS sorting network [2] in $O(\log(n+m))$ parallel time using $O(n+m)$ processors. The point location query can be done in parallel for each of $O(n)$ points in $O(\log m)$ time. Thus, the guide algorithm runs in $O(\log(n + m))$ time using $O(n + m)$ processors, Moreover, we can apply Cole's acceleration method [4]. Hence, our parametric search algorithm runs in $O((n + m) \log^2(n + m))$ sequential time. Thus, we have obtained the following theorem:

Theorem 2. *In $O((n+m) \log^2(n+m))$ time, we can find the maximum scaling factor permitting a feasible labeling of LOFL of n points in a plane with polygonal obstacles of m edges.*

5   Heuristics by Using LOFL

In a practical GIS system, a map labeling problem is often given in a form that is theoretically NP-hard. Therefore, heuristics methods or hybrid methods are often effective in practice [14,15,16]. LOFL can be used as a powerful weapon to design heuristics combined with other methods. Suppose we have a feasible labeling with a scaling factor σ given by some method, and want to improve the factor by changing the shape of labels. Let L_i be the label for $p_i \in P$ in the labeling. In place of the single label L_i, to each $p_i \in P$, we assign an appropriate left-part ordered set $\mathcal{L}_i$ such that $\mathcal{L}_i \ni L_i$. Thus, we have an instance of LOFL. The scaling factor in the solution of this LOFL instance is larger than or equal to σ, and is often much larger than σ. This can be considered as a "local improvement routine," which is an important tool in meta-heuristics.

5.1   A Heuristic Algorithm for the Two-Position LOFL

Suppose that we are given an instance for which the set of candidate labels for each $p_i \in P$ is a union of two left-part ordered sets $\mathcal{L}_i$ and $\mathcal{M}_i$. We call this model *two-position LOFL*, since it can be regarded as a combination of LOFL and the two-position model [5,9]. As we have noted before, the problem is NP-hard to approximate the optimal scaling factor within a ratio $\frac{1}{2} + \epsilon$; therefore, we need a heuristic. We solve the two-position LOFL by using a combination of solutions of LOFL and 2LABEL, where 2LABEL is an algorithm for solving the two-position labeling problem where we have (at most) two candidate rectangles for each object point. Forman and Wagner [5] gave an efficient implementation of 2LABEL.

If we have an oracle to determine from which of $\mathcal{L}_i$ or $\mathcal{M}_i$ the label for p_i, $i = 1, 2, \ldots, n$, should be selected, we can reduce the problem to LOFL. In our heuristic, we use 2LABEL to substitute for such an oracle. We can also use LOFL to construct an instance of 2LABEL. We omit details in this version. Thus, we alternately apply 2LABEL and LOFL until the increase of the scaling factor stops. We can similarly combine the one-slider (vertical slider) model [9] and LOFL.

6   Preliminary Experimental Results

We have done a preliminary experiment to see the ability of LOFL to enlarge the scaling factor. We compared four different labeling models: (1) fixed-position model, (2) LOFL, (3) two-position model , and (4) two-position LOFL.

In precise, for each object point, we assign the following candidate labels for the respective labeling models: (1) A left-upper pinned rectangle with height 3σ and breadth 4σ. (2) A set of six kinds of left-upper pinned rectangles of area $12\sigma^2$ whose height-breadth ratios are 12, 3, 4/3, 3/4, 1/3, and 1/12 (they correspond to factorizations of 12 to 12×1, 6×2 and 4×3). (3) A pair of rectangles with height 3σ and breadth 4σ, one of which is left-upper pinned and the other is left-lower pinned[2]. (4) A set of rectangles consisting of those in (2) and their reflected copies pinned at the left-lower corner.

We randomly generate n integral object points in a square region of size 50000×50000 for each of $n = 20, 40, 60$, and 80. We did not place obstacles. Table 1 shows the average scaling factor over 1000 instances for each of (1), (2), (3), and (4) for each of n. Note that the table does not indicate the quality of labeling outputs: Although (2) can use labels with a larger area than (3), it does not say that it is better than (3), since a labeling with various shapes is often less beautiful than a labeling with a single shape. Indeed, in a practical map labeling instance, only a portion of the point set should be given labels with various shapes.

7   Concluding Remarks

Our current implementation of LOFL in the experiment is rather naive; We are preparing for an experiment by using larger and practical instances, and performance of algorithms will be reported there.

In practical applications, we often want to have an algorithm to place as many labels as possible for a given instance of LOFL which is infeasible for a

---

[2] It is also a LOFL if we rotate the instance by $90°$.

Table 1. Scaling factors of the labeling models

	fixed-position	LOFL	two-position	two-position LOFL
n=20	447	1146	1044	1477
n=40	235	643	550	835
n=60	155	445	359	598
n=80	115	333	250	467

fixed scaling factor σ. Design of efficient algorithms or heuristics for this problem is an important future problem.

References

1. P. Agarwal, M. van Kreveld, and S. Suri, Label placement by maximum independent set in rectangles, *Computational Geometry, Theory and Applications,* **11** (1998) 209–218.
2. M. Ajtai, J. Komlos, and E. Szemeredi, Sorting in $c\log(n)$ parallel steps, *Combinatorica,* **3** (1983), pp.1–19.
3. H. Aonuma, H. Imai, K. Imai, and T. Tokuyama, Maximin locations of convex objects in a polygon and related dynamic Voronoi diagrams, *Proc. 6th ACM Symp. on Computational Geometry* (1990) 225–234.
4. R. Cole, Slowing down sorting network to obtain faster sorting algorithms, *J. ACM,* **34** (1987) 200–208.
5. M. Formann and F. Wagner, A packing problem with applications to lettering of maps, *Proc. 7th ACM Symp. on Computational Geometry* (1991) 281–290.
6. C. Iturriaga and A. Lubiw, Elastic labels around the perimeter of a map, *Proc. WADS'99* (1999) 306–317
7. K. Kakoulis and I. Tollis, A unified approach to labeling graphical features, *Proc. 14th ACM Symp. on Computational Geometry* (1998) 347–356.
8. K. Kato, Studies on the Geometric Location Problems, L1 Approximation and Character Placing, Master Thesis, Kyushu University (February 1989).
9. M. van Kreveld, T. Strijk, and A. Wolff, Point set labeling with sliding labels, *Computational Geometry, Theory and Applications,* **13** (1999) 21–47.
10. N. Megiddo, Applying parallel computation algorithms in the design of serial algorithms, *J. ACM,* **30** (1983) 852–865.
11. K. Mehlhorn, *Data Structures and Algorithms 1: Sorting and Searching,* ETACS Monograph 1, Springer Verlag, 1984.
12. J. Salowe, Parametric search, Section 37 of *Handbook of Discrete and Computational Geometry,* 683–695, (ed. J. Goodman and R. Polack), (1997) CRC Press.
13. I. Shamos and F. Preparata, *Computational Geometry – An Introduction,* Springer Verlag, 1985.
14. F. Wagner and A. Wolff, A practical map labeling heuristics algorithm *Computational Geometry, Theory and Applications,* **7** (1997) 387–404.
15. F. Wagner and A. Wolff, A combinatorial framework for map labeling, *Proc. Graph Drawing '98,* LNCS 1547 (1998) 316–331.
16. M. Yamamoto, G. Camara, L. Lorena, Tabu search heuristics for point-feature cartographical label placement, *GeoInformatica* (2000) (also see http://www.lac.inpe.br/~lorena/missae/index.html)

How to Draw the Minimum Cuts
of a Planar Graph
(Extended Abstract)

Ulrik Brandes, Sabine Cornelsen, and Dorothea Wagner

University of Konstanz, Department of Computer and Information Science
{Ulrik.Brandes, Sabine.Cornelsen, Dorothea.Wagner}@uni-konstanz.de

Abstract. We show how to utilize the cactus representation of all minimum cuts of a graph to visualize the minimum cuts of a planar graph in a planar drawing. In a first approach the cactus is transformed into a hierarchical clustering of the graph that contains complete information on all the minimum cuts. We present an algorithm for c-planar orthogonal drawings of hierarchically clustered planar graphs with rectangularly shaped cluster boundaries and the minimum number of bends. This approach is then extended to drawings in which the two vertex subsets of every minimum cut are separated by a simple closed curve.

1 Introduction

The edge connectivity is a fundamental structural property of a graph. Dinitz et al. [1] discovered that the set of all minimum cuts of a connected graph G with positive edge weights has a tree-like structure. It can be represented by a cactus, i.e. by a connected graph in which every edge is contained in at most one cycle. Although the number of minimum cuts in a graph is in $\mathcal{O}(n^2)$, the size of the cactus is linear in the number n of vertices of G. From the cactus representation, the bipartitions of the vertex sets can easily be extracted, but it contains almost no information about the edges in G. We want to visualize a graph G together with the cactus representation of its minimum cuts in one drawing.

A simple closed curve divides the plane into two connected components. A minimum cut divides the set of vertices of a graph into two connected subsets. Thus, it is natural to visualize a minimum cut in a drawing of a graph by a simple closed curve separating the two subsets.

Feng et al. [4] introduced the model of hierarchically clustered graphs. In a drawing of a hierarchically clustered graph, a set of vertices of a graph is represented by a region that is bounded by a simple closed curve. The set of subsets of the vertex set that is represented simultaneously in this way has to have tree structure. In terms of cuts, this means that we can represent a set of pairwise non-crossing cuts as a hierarchically clustered graph. Graphs having no crossing minimum cuts are, for example, maximal planar graphs and chordal graphs.

J. Marks (Ed.): GD 2000, LNCS 1984, pp. 103–114, 2001.

If there are crossing cuts, the structure of the set of minimum cuts implies that they are represented at least implicitly in a drawing of the pairwise non-crossing cuts. We show, however, that the model of hierarchically clustered graphs can be extended to cactus-clustered graphs such that the goal of visualizing every minimum cut by a simple closed curve is achieved. This extension is mainly based on the fact that for two crossing minimum cuts, the four corner cuts are also minimum.

The contribution of this paper is as follows. In Sect. 2 and 3, we provide some background on the cactus representation and on hierarchically clustered graphs, respectively. In Sect. 4, we show how to construct a hierarchically clustered graph from a cactus representation and prove an interesting property of crossing cuts in planar graphs that enables us to represent all minimum cuts by closed curves. Finally, our two methods for drawing planar graphs that are clustered according to their minimum cuts are presented in Sect. 5.

2   The Cactus of the Set of Minimum Cuts

Let G be an undirected connected graph. With $E(G)$ we denote the set of edges of G and with $V(G)$ the set of vertices of G. A graph G together with a positive weight function $\omega : E(G) \to \mathbb{R}^+$ is a *weighted graph*. For two subsets S and T of $V(G)$, let $E(S,T) := \{\{v,w\}; v \in S \text{ and } w \in T\}$ be the set of edges between S and T, and let $\omega(S,T) := \sum_{e \in E(S,T)} \omega(e)$ be the sum of the weights of the edges between the two subsets.

A *cut* is an unordered pair $\{S, \overline{S}\}$ where $\emptyset \subsetneq S \subsetneq V(G)$ and $\overline{S} := V(G) \setminus S$. A set S *induces* the cut $\{S, \overline{S}\}$. The *weight* of this cut is $\omega(S, \overline{S})$. With $\lambda := \min_{\emptyset \subsetneq S \subsetneq V(G)} \omega(S, \overline{S})$ we denote the minimum of all these weights and a cut $\{S, \overline{S}\}$ of G satisfying $\omega(S, \overline{S}) = \lambda$ is called a *minimum cut*. With $\mathcal{C}(G)$ we denote the set of minimum cuts of G. By $G(S)$ we denote the subgraph of G induced by a set S.

A *cycle* $c : v_1, \ldots, v_k$ is a sequence of $k \geq 3$ distinct vertices, such that $E(c) := \{\{v_1, v_2\}, \ldots, \{v_{k-1}, v_k\}, \{v_k, v_1\}\} \subset E(G)$. For a subset $E' \subset E(G)$, we denote by $G - E'$ the graph $(V(G), E(G) \setminus E')$.

Definition 1. *A* representation *for a set $\mathcal{C}$ of cuts of a graph G is a pair $(\mathcal{G}, \varphi)$ such that $\mathcal{G}$ is a weighted graph and $\varphi : V(G) \longrightarrow V(\mathcal{G})$ is a mapping such that $\mathcal{C} = \varphi^{-1}(\mathcal{C}(\mathcal{G})) := \{\{\varphi^{-1}(S), \varphi^{-1}(\overline{S})\}; \{S, \overline{S}\} \in \mathcal{C}(\mathcal{G})\}$. A node $\nu \in V(\mathcal{G})$ is called* empty *if $\varphi^{-1}(\nu) = \emptyset$.*

Definition 2. *Two cuts $\{S, \overline{S}\}$ and $\{T, \overline{T}\}$ are* crossing, *if none of the* corner sets *$S \cap T$, $S \cap \overline{T}$, $\overline{S} \cap T$, and $\overline{S} \cap \overline{T}$ is empty. A cut induced by a corner set is a* corner cut *and the cut induced by $S \Delta T := S \setminus T \cup T \setminus S$ is the* diagonal cut.

A cut is a *crossing cut* of a family $\mathcal{C}$ of cuts, if it crosses any cut in $\mathcal{C}$. If $\mathcal{C}(G)$ contains no crossing cuts, the set of minimum cuts of G can be represented by a tree. Dinitz et al. [1] showed that the set of minimum cuts of an arbitrary weighted connected graph can be represented by a cactus where cycles correspond to sets of crossing cuts. More precisely:

Definition 3 (Cactus). *A* cactus *is a connected graph in which every edge belongs to at most one cycle. An edge that belongs to no cycle is called a* tree edge. *An edge that belongs to one cycle is called a* cycle edge.

In what follows, we assume that a weighted cactus is *uniform*, i.e. that all cycle edges have the same weight and that every tree edge has twice the weight of a cycle edge.

Theorem 1 ([1]). *The set $C(G)$ of all minimum cuts of a weighted connected graph G has a representation $(\mathcal{G}, \varphi)$ such that $\mathcal{G}$ is a uniform cactus with $\mathcal{O}(n)$ nodes.*

Figure 1 shows an example of a weighted graph and its cactus. Dinitz and Nutov

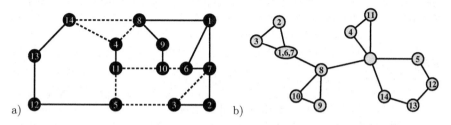

a)   b)

Fig. 1. a) A weighted connected graph and b) the cactus representation of its minimum cuts. In a), solid edges have weight 2 and dashed edges have weight 1. In b), φ is represented by the labels of the nodes.

[2] characterized all sets of cuts that can be represented by a cactus.

Theorem 2. *A set C of cuts can be represented by a cactus if and only if for any two crossing cuts in C*

- *the four corner cuts are in C and*
- *the diagonal cut is not in C.*

In this case, there is always a cactus representation with $O(n)$ nodes.

In what follows, we denote by $\mathcal{G}$ the cactus representation of all minimum cuts of G. Note that there is a bijection between the set of minimum cuts of $\mathcal{G}$ and the set of tree edges and pairs of cycle edges belonging to the same cycle. Thus, we can also say that a cut in G is represented by a tree edge or a pair of cycle edges of $\mathcal{G}$. An important property of a cactus representation that we will use later is, that "edges in G do not cross a cycle of $\mathcal{G}$". More precisely:

Lemma 1. *For a cycle $c : \nu_1, \ldots, \nu_k$ in $\mathcal{G}$ let $\mathcal{V}_i, i = 1, \ldots, k$ be the set of vertices in the connected component of $\mathcal{G} - E(c)$ that contains ν_i and $V_i := \varphi^{-1}(\mathcal{V}_i)$. Then*

$$\omega(V_i, V_j) = 0 \quad \text{if} \quad i - j \not\equiv \pm 1 \mod k.$$

Fleischer [5] showed that the cactus of all minimum cuts of a weighted graph can be constructed in $\mathcal{O}(mn \log \frac{n^2}{m})$ time. For an unweighted graph, it can be computed in $\mathcal{O}(\lambda n^2)$ time [12]. Using the linear-time shortest-path algorithm of Henzinger et al. [7] for max-flow computations, the cactus of a weighted planar graph can be obtained in $\mathcal{O}(n^2)$ time with the construction described in [5].

3   Hierarchically Clustered Graphs

Feng et al. [4] introduced the hierarchically clustered graph model and characterized graphs that have a planar drawing with respect to the clustering. In this section, we summarize definitions and results of [4] and [3] that we will use later.

A *hierarchically clustered graph* (G, T) consists of a graph $G = (V, E)$ and a rooted tree T such that the set of leaves of T is exactly V. Vertices of T are called *nodes*. Each node ν of T represents the *cluster* $V(\nu)$ of leaves in the subtree of T rooted at ν. T is called the *inclusion tree* of (G, T). An edge e of G is said to be *incident* to a cluster $V(\nu)$, if $|e \cap V(\nu)| = 1$.

A hierarchically clustered graph (G, T) is *connected*, if each cluster induces a connected subgraph of G.

A *drawing* $\mathcal{D}$ of a hierarchically clustered graph (G, T) includes the drawing of the underlying graph G and of the inclusion tree T in the plane. Each vertex v of G is represented as a point $\mathcal{D}(v)$ and each edge $e = \{v, w\}$ as a simple curve $\mathcal{D}(e)$ between $\mathcal{D}(v)$ and $\mathcal{D}(w)$. Each non-leaf node ν of T is drawn as a simple closed region $\mathcal{D}(\nu)$ bounded by a simple closed curve $\partial\mathcal{D}(\nu)$ such that

- $\mathcal{D}(\mu) \subset \mathcal{D}(\nu)$ for all descendents μ of ν.
- $\mathcal{D}(\mu) \cap \mathcal{D}(\nu) = \phi$ if μ is neither a descendent nor an ancestor of ν.
- $\mathcal{D}(e) \subset \mathcal{D}(\nu)$ for all edges e of G with $e \subset V(\nu)$.
- $\partial\mathcal{D}(e) \cap \partial\mathcal{D}(\nu)$ is a single point if $|e \cap V(\nu)| = 1$.

Roughly speaking, T is drawn in the inclusion representation and edges of G may only cross cluster boundaries if necessary.

The drawings of an edge e and a cluster ν have an *edge-cluster-crossing*, if $e \cap V(\nu) = \phi$ but $\mathcal{D}(e) \cap \mathcal{D}(\nu) \neq \phi$. A drawing of a hierarchically clustered graph is *c-planar*, if there are no crossing edges and no edge-cluster-crossings. A graph is *c-planar* if it has a c-planar drawing.

Theorem 3 (Characterization of c-Planar Graphs [4]). *A connected hierarchically clustered graph* $C = (G, T)$ *is c-planar if and only if there exists a planar drawing of* G, *such that for each node* ν *of* T *all vertices of* $V - V(\nu)$ *are in the outer face of the drawing of* $G(\nu)$.

In an *OGRC (orthogonal grid rectangular cluster)* drawing of a hierarchically clustered graph (G, T), curve $\mathcal{D}(e)$ is a sequence of horizontal and vertical segments for every edge e of G and $\mathcal{D}(\nu)$ is an axis-parallel rectangle for every non-leaf node ν of T.

Theorem 4 ([3]). *For a c-planar connected clustered graph with* n *vertices of degree at most 4, a c-planar OGRC drawing with* $\mathcal{O}(n^2)$ *area and with at most 3 bends per edge can be constructed in* $\mathcal{O}(n)$ *time.*

4   From the Cactus Representation to a Hierarchically Clustered Graph

Both, the cactus representation of the minimum cuts of a graph and the inclusion tree of a hierarchically clustered graph, represent structural information of a graph. We show how to transform the cactus representation into an inclusion tree such that all minimum cuts can be recognized in a drawing of the corresponding hierarchically clustered graph.

Let $(\mathcal{G}, \varphi)$ be the cactus structure of the minimum cuts of a weighted connected planar graph G with n vertices.

1. For every cycle $c : \nu_1, \ldots, \nu_k$ in $\mathcal{G}$, delete all edges in c and add a new (empty) node ν_c and edges $\{\nu_i, \nu_c\}$, $i = 1, \ldots, k$.
2. Replace every empty node of degree 2 and its incident edges by a single edge.
3. For every vertex v of G, add a new node ν_v and an edge $\{\varphi(v), \nu_v\}$.
4. Find a suitable root.

We call the thus constructed rooted tree $\mathcal{T}(G)$. Note that $(G, \mathcal{T}(G))$ is now a hierarchically clustered graph. The number of nodes in $\mathcal{T}(G)$ is in $\mathcal{O}(n)$: By Theorem 1, we have $|V(\mathcal{G})| \in \mathcal{O}(n)$. In Step 1, we add a new node for every cycle in $\mathcal{G}$ and in Step 3, we add n new nodes. Thus $|V(\mathcal{T}(G))|$ remains in $\mathcal{O}(n)$.

Figure 2 shows the inclusion tree $\mathcal{T}(G)$ of the graph G from Fig. 1. There are

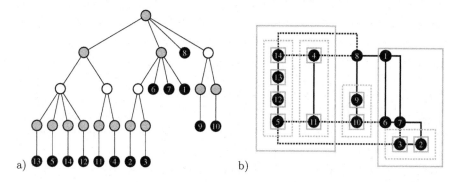

Fig. 2. a) White nodes in the inclusion tree $\mathcal{T}(G)$ of the graph G in Fig. 1 represent the nodes that were added for a cycle in $\mathcal{G}$. b) The corresponding cluster boundaries are drawn as dashed grey rectangles in the c-planar OGRC drawing of $(G, \mathcal{T}(G))$.

several options for choosing a root. We have chosen the root such that $|V(\nu)| \leq |\overline{V(\nu)}|$ for every inner node ν of $\mathcal{T}(G)$. This has the advantage that the most balanced minimum cut $\{S, \overline{S}\}$, i.e. the cut such that $||S| - |\overline{S}||$ is minimal, is seen on the top level. Another possibility is to take the center of the tree, i.e. to minimize the height. In both cases, the root can be computed in linear time.

To visualize the minimum cuts in G we want to construct a c-planar drawing of the hierarchically clustered graph $(G, \mathcal{T}(G))$. The next lemma guarantees that we can fix an arbitrary embedding of G and add cluster boundaries.

Lemma 2. *Every planar embedding of a weighted connected planar graph G can be extended to a c-planar drawing of the hierarchically clustered graph $(G, \mathcal{T}(G))$.*

Proof. To see this, recall the following facts about minimum cuts.

Remark 1. Let $\{S, \overline{S}\}$ be a minimum cut in a weighted connected graph. Then both $G(S)$ and $G(\overline{S})$ are connected.

Remark 2. Let $\{S, \overline{S}\}$ be a minimum cut in a weighted connected planar graph G. In any embedding of G, the dual of the graph induced by $E(S, \overline{S})$ is a cycle.

These two facts guarantee that for every embedding of the weighted connected planar graph G the hierarchically clustered graph $(G, \mathcal{T}(G))$ fulfills the preconditions of Theorem 3 and thus has a c-planar drawing.   □

In Step 1, we replace each cycle of the cactus by a star. Thus, the information about the cyclic order of the edges in a cycle of $\mathcal{G}$ is not preserved in $\mathcal{T}(G)$. However, this order can be reconstructed from a c-planar drawing of $(G, \mathcal{T}(G))$ as follows.

Let $c : \nu_1, \ldots, \nu_k$ be a cycle in $\mathcal{G}$ and let $\mathcal{V}_i$ and V_i be defined as in Lemma 1. Consider the dual of the graph induced by $E(V_i, \overline{V_i})$. Since $\{V_i, \overline{V_i}\}$ is a minimum cut, its dual is a cycle, denoted by d_i^*. Note that the cycles d_i^* correspond to cluster boundaries in a c-planar drawing of $(G, \mathcal{T}(G))$. For each edge e in G let $\omega(e^*) = \omega(e)$. An immediate consequence of Lemma 1 is

Lemma 3. $\displaystyle\sum_{e^* \in E(d_i^*) \cap E(d_j^*)} \omega(e^*) = 0$ if $i - j \not\equiv \pm 1 \bmod k$.

Lemma 4. *The set of edges $E(d_i^*) \cap E(d_{i+1}^*)$ is consecutive in d_i^* and d_{i+1}^*.*

Proof. Suppose not. Let $e_1^*, \ldots, e_l^*$ be the sequence of edges in a path in d_i^* such that $e_1^*, e_l^* \in E(d_{i+1}^*)$, $e_2^*, \ldots, e_{l-1}^* \notin E(d_{i+1}^*)$. Let $e^* \in E(d_i^*) \setminus (E(d_{i+1}^*) \cup \{e_2^*, \ldots, e_{l-1}^*\})$ be another edge in d_i^* and let $e = \{v, w\}$ with $w \notin V_i$. For $j = i, i+1$ let p_j be a path from $e_1 \cap V_j$ to $e_l \cap V_j$ in the graph induced by V_j. Let c' be the cycle in G that is induced by edge e_1, path p_i, edge e_l and path p_{i+1}. Without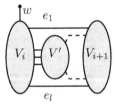
loss of generality we can assume that edges $e_2^*, \ldots, e_{l-1}^*$ are inside cycle c'. Let $V' \subset V(G)$ be the set of vertices that are incident to $e_2, \ldots, e_{l-1}$ and that are not in V_i. Then, by Lemma 1, $w \in V_{i-1}$ and $V' \subset V_{i-1}$. Thus, V_{i-1} is not connected, contradicting Remark 1.   □

Lemma 5. $\displaystyle\left| \bigcap_{i=1}^{k} V(d_i^*) \right| = 2$

Proof. Let $d_1^* : f_1, \ldots, f_l, f_{l+1}, \ldots, f_r$ and $d_2^* : f_1, \ldots, f_l, f_{l+1}', \ldots, f_s'$ be such that $f_{l+1} \neq f_{l+1}'$ and $f_r \neq f_s'$. From Lemma 3 we conclude that $E(d_3^*) \cap E(d_2^*) = E(d_2^*) \setminus E(d_1^*)$ and with Lemma 4 we get $\cap_{i=1}^3 V(d_i^*) = \{f_1, f_l\}$. It follows inductively that $\cap_{i=1}^k V(d_i^*) = \{f_1, f_l\}$.   □

Thus, we can choose one face f where all the cycles in G^* that correspond to the cuts of a cycle c in $\mathcal{G}$ intersect. The cyclic order in f displays the cyclic order of c. Moreover, every cut that is represented by two edges of c can be reconstructed from a c-planar drawing of $(G, \mathcal{T}(G))$ in the following way. Divide every cluster boundary corresponding to a cycle d_i^* into two paths between f_1 and f_l. This results in k disjoint paths. The cycle consisting of any two of these paths is the dual of a cut represented by two edges of c and every cut that is represented by a two-cut of c is of this form.

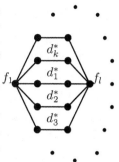

5   Two Models for Drawing Cuts

5.1   C-Planar Drawings of Hierarchically Clustered Graphs

As mentioned in Sect. 3, Eades et al. [3] introduced a method for drawing hierarchically clustered planar graphs orthogonally with rectangularly shaped cluster boundaries. To do this, they made the undirected graph directed and allowed edges to cross the cluster boundary only at the top or bottom of the rectangle. This might introduce unnecessary bends into the drawing. We introduce a different way of drawing a connected c-planar graph (G, T). We add edges and vertices to G such that the newly constructed graph G' remains planar and each cluster boundary corresponds to a cycle in G'. Now any embedding preserving algorithm can be applied to draw graph G' and thus to obtain a c-planar drawing of (G, T). In case G has maximum degree 4, using the model of Tamassia [13] with some additional constraints on the flow, this leads to a c-planar OGRC drawing with the minimum number of bends.[1]

Let (G, T) be a hierarchically clustered c-planar graph with an embedding in the plane that fulfills the conditions of Theorem 3. Recall, that for $T = \mathcal{T}(G)$, by Lemma 2, every embedding of a planar graph is suitable.

For every cluster, we add a cycle of new edges and new vertices to G in the following way:

Proceeding from the leaves to the root of T, for every non-leaf node ν of T let $e_1, \ldots, e_k$ be the edges incident to cluster $V(\nu)$ in their cyclic order around $V(\nu)$. Let $e_{k+1} = e_1$ and $e_i = \{v_i, w_i\}, i = 1, \ldots, k+1$. For $i = 1, \ldots, k$, we *split* edge e_i, i.e. we add a vertex v_{e_i} to $V(G)$ and replace edge e_i by edges $\{v_i, v_{e_i}\}$ and $\{w_i, v_{e_i}\}$. Finally, we add edges $\{v_{e_i}, v_{e_{i+1}}\}$. These k edges are called *boundary edges* of $V(\nu)$. They form a cycle, called the *boundary cycle* of $V(\nu)$, that model the cluster boundary of $V(\nu)$.

---

[1] While implementing the extension of Tamassia's model to hierarchically clustered graphs we learned that it was independently described in [9] and is now part of the AGD library [11].

A special case occurs, if there are only one or two edges incident to a cluster. In that case, two or one additional vertices are inserted in this cycle to avoid loops and multiple edges. The resulting edges are also called boundary edges. Let the resulting graph be G'. Let $n = |V(G)|$ and $h(T)$ be the height of the inclusion tree T.

Lemma 6. $|V(G')| \in \mathcal{O}(n \cdot h(T))$

Proof. Let $e = \{u, v\} \in E(G)$ and let k be the number of vertices on the path in T between u and v. Then $k - 3 \le 2 \cdot h(T)$ vertices are inserted into e. Thus, $|V(G')| \le n + 2|E(G)| \cdot h(T) \in \mathcal{O}(n \cdot h(T))$. □

Note that in case $T = \mathcal{T}(G)$ and $\omega(e) \ge 1$ for every edge $e \in E(G)$ it is also true, that $|V(G')| \in \mathcal{O}(\lambda \cdot n)$: Every cluster is incident to at most λ edges and the number of clusters is in $\mathcal{O}(n)$. For unweighted planar graphs we have $\lambda \le 5$ and this implies $|V(G')| \in \mathcal{O}(n)$.

Lemma 7. G' *can be constructed in* $\mathcal{O}(|V(G')|)$ *time.*

Proof. Proceeding for each edge $\{u, v\} \in E(G)$ along the path in T between u and v, splitting the edges can be done in $\mathcal{O}(|E(G)| + |\text{added vertices}|) = \mathcal{O}(|V(G')|)$.

From the leave to the root of T, add the boundary edges along the outer face of each cluster. Doing this, every edge can be touched at most twice. Thus, inserting the boundary edges is in $\mathcal{O}(|E(G')|) = \mathcal{O}(|V(G')|)$. □

In the flow network for an orthogonal [13] or quasi-orthogonal drawing [14,8] of G', we restrict the flow over a boundary edge to be zero, if it goes from outside the corresponding boundary cycle into it. This guarantees that the boundary cycles are rectangularly shaped in any resulting orthogonal drawing. This restriction is necessary. Even in the case of unweighted graphs with the root chosen in such a way, that $|V(\nu)| \le |\overline{V(\nu)}|$, there are examples of planar graphs G such that the bend minimum solution without restriction of the clustered graph $(G, \mathcal{T}(G))$ have non-rectangularly shaped cluster boundaries. See for example Fig. 3. Theorem 4 guarantees that there is a feasible flow for the restricted flow network. The resulting drawing is a bend minimum c-planar OGRC drawing. Moreover, all inserted vertices have degree 4 and split edges alternate with boundary edges. Thus, the corresponding original edges in G have no bends at cluster boundaries.

Lemma 8. *The area requirement of the constructed bend–minimum c–planar OGRC drawing of* (G, T) *is in* $\mathcal{O}(n^2)$.

Proof. There are $\mathcal{O}(n)$ clusters and each cluster boundary requires two horizontal and two vertical lines.

Those edges in G' that are not boundary edges correspond to $\mathcal{O}(n)$ original edges in G. As the constructed drawing is bend–minimum, by Theorem 4, there are at most $3 \cdot |E(G)|$ bends on those edges. Thus, the non–boundary edges require at most $4 \cdot |E(G)| \in \mathcal{O}(n)$ horizontal and vertical lines. □

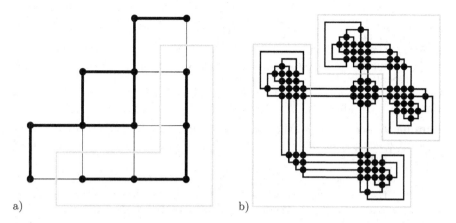

a)   b)

Fig. 3. Drawings of the non-trivial minimum cuts of a) a weighted and b) an unweighted graph without rectangularity restriction on the cluster shape. Grey edges are boundary edges, thin edges have weight 1 and thick edges have weight 6.

The running time of our algorithm is as follows:

- Constructing the cactus $\mathcal{G}$ of a planar connected weighted graph G is in $\mathcal{O}(n^2)$.
- Constructing the inclusion tree $\mathcal{T}(G)$ from the cactus is in $\mathcal{O}(n)$.
- Constructing G' from (G, T) is in $\mathcal{O}(n \cdot \mathrm{h}(T))$.
- Constructing the orthogonal drawing of G' with $N := |V(G')|$ vertices is in $\mathcal{O}(N^{7/4}\sqrt{\log N})$ [6].

We can finally summarize that the running time is dominated by the construction of the cactus and the orthogonal drawing and is in $\mathcal{O}(n^2 + N^{7/4}\sqrt{\log N})$ time. Figure 2 shows an OGRC drawing of the hierarchically clustered graph $(G, \mathcal{T}(G))$ where G is the graph in Fig. 1.

5.2 Planar Cactus-Clustered Drawings

The information about the cyclic order of the edges in a cycle of $\mathcal{G}$ is preserved in a c-planar drawing of $(G, \mathcal{T}(G))$. However, crossing cuts are given only implicitly and have to be reconstructed from the others as mentioned at the end of Sect. 4. In this subsection we show how we can modify the c-planar drawing in such a way that on one hand, for every minimum cut $\{S, \overline{S}\}$ there is a cycle c added to G separating S from $\overline{S}$ and, on the other hand, for every cycle c added to G the cut defined by the inside and the outside of c is a minimum cut. We achieve this, roughly speaking, by merging the cluster boundaries corresponding to pairs of incident nodes on a cycle in the cactus.

Let $\mathcal{T}(G)$ be defined as in Sect. 4, except that we do not replace nodes of degree 2. Let again $c : \nu_1, \dots, \nu_k$ be a cycle in $\mathcal{G}$ and let $\mathcal{V}_i$ and V_i be defined as in Lemma 1. Let ν_c be the node added for c in $\mathcal{T}(G)$. Note that $\overline{V_j} = V(\nu_c)$ and

$V(\nu_j) \notin \{V_j, \overline{V_j}\}$ if ν_j is the predecessor of ν_c. In this case, node ν_c represents cut $\{V_j, \overline{V_j}\}$ and in the following, we substitute ν_j by ν_c in c. Now, every node in $\mathcal{T}(G)$ is associated with at most one cycle in $\mathcal{G}$.

By Lemma 4, we already know that duals of the edges between V_i and V_{i+1} are consecutive on the cycles corresponding to their cluster boundaries. We now show that those vertices that were inserted into an edge of G for a drawing of these duals are adjacent.

Lemma 9. *Let* $e \in E(V_i, V_{i+1})$ *and let* v_i *and* v_{i+1} *be the vertices that were inserted into* e *for the cluster corresponding to* ν_i *and* ν_{i+1}, *respectively. Then* $\{v_i, v_{i+1}\} \in E(G')$.

Proof. Let μ be the root of the smallest subtree of $\mathcal{T}(G)$ containing ν_i and ν_{i+1}. If $\nu_c = \mu$, the vertices corresponding to the two descendents ν_i and ν_{i+1} of μ are inserted consecutively in e. If not, one of the two nodes, say ν_{i+1}, was the node that was substituted by ν_c. The vertices corresponding to ν_i and ν_c are inserted consecutively in e.   □

Thus, we can merge vertices that were inserted into the same edge for clusters corresponding to nodes of the same cycle. Finally, for each cycle c, we add a vertex in both of the faces f_1 and f_l identified in Lemma 5 and we split each edge that is incident to one of the faces and is a boundary edge of a cluster corresponding to c at the added vertex. This results in k paths between f_1 and f_l.

Let the constructed graph be G''. As the number of cycles in $\mathcal{G}$ is in $\mathcal{O}(n)$, we add $\mathcal{O}(n)$ vertices to G'. Thus, $|V(G'')| \in \mathcal{O}(|V(G')|)$ and G'' can be constructed in $\mathcal{O}(|V(G')|)$ time.

As in the previous subsection, we can now apply any embedding preserving algorithm to draw graph G'' and thus to visualize all minimum cuts in G. We call the resulting drawing a planar *cactus-clustered* drawing.[2] An example using the quasi-orthogonal drawing method is shown in Fig. 4. The advantage of this drawing is obviously that every minimum cut is represented by a simple closed curve. Moreover, parallel boundary edges are avoided.

6   Conclusion

We outlined two methods for representing the minimum cuts of a weighted planar graph in a planar drawing of the graph. Utilizing the cactus representation, all

---

[2] In analogy to c-planar drawings of hierarchically clustered graphs, we define a *planar cactus-clustered drawing* $\mathcal{D}$ of a planar graph G with a cactus representation of a set $\mathcal{C}$ of cuts as follows. Again, each vertex of G is represented as a distinct point and each edge $\{v, w\}$ as a simple curve between $\mathcal{D}(v)$ and $\mathcal{D}(w)$ such that no two edges cross. Each cut $C = \{S, \overline{S}\} \in \mathcal{C}$ is represented by a simple closed curve such that (1) $\mathcal{D}(S)$ and $\mathcal{D}(\overline{S})$ are in different connected components of $\mathbb{R}^2 \setminus \mathcal{D}(C)$ and such that (2) for every simple closed curve c in $\bigcup_{C \in \mathcal{C}} \mathcal{D}(C)$ there is a cut $\{T, \overline{T}\} \in \mathcal{C}$ such that (a) $\mathcal{D}(T)$ and $\mathcal{D}(\overline{T})$ are contained in different connected components of $\mathbb{R}^2 \setminus c$ and (b) for an edge $e \in E$ we have $|\mathcal{D}(e) \cap c| = 1$, if e is incident to both, a vertex in T and a vertex in $\overline{T}$ and $\mathcal{D}(e) \cap \mathcal{D}(C) = \emptyset$, if not.

Moreover, for two cuts $C, C' \in \mathcal{C}$ it shall hold that $|\mathcal{D}(C) \cap \mathcal{D}(C')| \neq 1$.

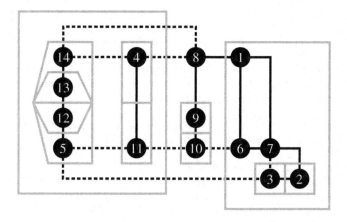

Fig. 4. A cactus-clustered drawing of the graph in Fig. 1.

minimum cuts can be shown in a c-planar drawing of a hierarchical clustering of the graph. This approach was then extended to cactus-clustered drawings that visualize all minimum cuts by simple closed curves. Both approaches have been demonstrated to work for bend-minimum orthogonal drawings, but can be used with any drawing algorithm that preserves the embedding of cluster boundaries.

Moreover, our methods can be extended to any set $\mathcal{C}$ of, not necessarily minimum, cuts of a planar graph G that has a cactus representation $(\mathcal{G}, \varphi)$ and the following additional properties:

1. For each cut $\{S, \overline{S}\} \in \mathcal{C}$, the graphs $G(S)$ and $G(\overline{S})$ are connected.
2. No edge of G crosses a cycle of $\mathcal{G}$.

If $\mathcal{T}(\mathcal{G})$ is the inclusion tree constructed from $\mathcal{G}$ as described in Sect. 4, it holds that $(G, \mathcal{G}, \varphi)$ has a planar cactus-clustered drawing if and only if $(G, \mathcal{T}(\mathcal{G}))$ is c-planar.

Acknowledgments. We thank Christian Fieß for providing part of the implementation used to compute the cactus, determine a clustering, and draw the clustered graphs. Our implementation uses LEDA [10] and an older version of the AGD library [11].

References

1. Y. Dinitz, A. V. Karzanov, and M. Lomonosov. On the structure of a family of minimal weighted cuts in a graph. In A. Fridman, editor, *Studies in Discrete Optimization*, pages 290–306. Nauka, 1976. (in Russian).
2. Y. Dinitz and Z. Nutov. A 2-level cactus model for the system of minimum and minimum+1 edge–cuts in a graph and its incremental maintenance. In *Proceedings of the 27th Annual ACM Symposium on the Theory of Computing (STOC '95)*, pages 509–518. ACM, The Association for Computing Machinery, 1995.

3. P. Eades, Q. Feng, and H. Nagamochi. Drawing clustered graphs on an orthogonal grid. *Journal on Graph Algorithms and Applications*, 3(4):3–29, 1999.
4. Q. Feng, R. F. Cohen, and P. Eades. Planarity for clustered graphs. In P. Spirakis, editor, *Proceedings of the 3rd European Symposium on Algorithms (ESA '95)*, volume 979 of *Lecture Notes in Computer Science*, pages 213–226. Springer, 1995.
5. L. Fleischer. Building chain and cactus representations of all minimum cuts from Hao–Orlin in the same asymptotic run time. *Journal of Algorithms*, 33(1):51–72, 1999.
6. A. Garg and R. Tamassia. A new minimum cost flow algorithm with applications to graph drawing. In S. C. North, editor, *Proceedings of the 4th International Symposium on Graph Drawing (GD '96)*, volume 1190 of *Lecture Notes in Computer Science*, pages 201–213. Springer, 1996.
7. M. R. Henzinger, P. Klein, S. Rao, and S. Subramanian. Faster shortest-path algorithms for planar graphs. *Journal of Computer and System Sciences*, 55:3–23, 1997. Special Issue on Selected Papers from STOC 1994.
8. G. W. Klau and P. Mutzel. Quasi orthogonal drawing of planar graphs. Technical Report MPI-I-98-1-031, Max-Planck-Institut für Informatik, Saarbrücken, Germany, 1998.
9. D. Lütke-Hüttmann. Knickminimales Zeichnen 4-planarer Clustergraphen. Master's thesis, Universität des Saarlandes, 1999. (Diplomarbeit).
10. K. Mehlhorn and S. Näher. *The LEDA Platform of Combinatorial and Geometric Computing*. Cambridge University Press, 1999. Project home page at http://www.mpi-sb.mpg.de/LEDA/.
11. P. Mutzel, C. Gutwenger, R. Brockenauer, S. Fialko, G. W. Klau, M. Krüger, T. Ziegler, S. Näher, D. Alberts, D. Ambras, G. Koch, M. Jünger, C. Buchheim, and S. Leipert. A library of algorithms for graph drawing. In S. H. Whitesides, editor, *Proceedings of the 6th International Symposium on Graph Drawing (GD '98)*, volume 1547 of *Lecture Notes in Computer Science*, pages 456–457. Springer, 1998. Project home page at http://www.mpi-sb.mpg.de/AGD/.
12. H. Nagamochi and T. Kameda. Constructing cactus representation for all minimum cuts in an undirected network. *Journal of the Operations Research*, 39(2):135–158, 1996.
13. R. Tamassia. On embedding a graph in the grid with the minimum number of bends. *SIAM Journal on Computing*, 16:421–444, 1987.
14. R. Tamassia, G. Di Battista, and C. Batini. Automatic graph drawing and readability of diagrams. *IEEE Transactions on Systems, Man and Cybernetics*, 18(1):61–79, 1988.

2D-Structure Drawings of Similar Molecules [*]

J.D. Boissonnat, F. Cazals, and J. Flötotto

Projet Prisme, INRIA Sophia-Antipolis, 2004, route des Lucioles - BP 93, F-06902
Sophia Antipolis Cedex, France, {Firstname.Lastname}@sophia.inria.fr

Abstract. A common strategy in drug design and pharmacophore identification consists of evaluating large sets of molecular structures by comparing their 2D structure drawings. To simplify the chemists' task, the
drawings should reveal similarities and differences between drugs. Given
a family of molecules all containing a common template, we present an
algorithm to compute standardised 2D structure drawings. The molecules being represented as a graph, we compute a structure called supertree
in which all molecules can be embedded. Using the correspondences between atoms provided by the supertree, we are able to coordinate the
drawings performed by a breadth-first traversal of the molecular graphs. Both parts of the problem are NP-hard. We propose algorithms of
heuristic nature.

1 Introduction

1.1 Problem Statement

A *Molecule* is defined as an undirected graph of bounded degree with labelled
nodes and edges. We restrict this definition to the case where molecules contain
only induced cycles of length at most α —typically $\alpha = 6$, and two such cycles
share at most one edge. This way, a molecule reduces to a tree with local cycles
and blocks of cycles. [1] We define a *molecular family* as a set of molecules $\mathcal{M} =$
$\{M_1, M_2, .., M_m\}$ sharing a connected subgraph T called the *template*. For a
molecule M_k, the connected subgraphs of $M_k \backslash T$ are called its *appendices* and
are denoted T_{t_i}, i.e. $M_k = T \cup \{T_{t_1}, T_{t_2}, .., T_{t_k}\}$. We further assume that each
appendix is rooted at the unique node connecting it to the template. At last, all
the molecules processed are assumed to have a planar drawing. Notice, that in
the context of drug design, the assumptions are mostly satisfied.

The question we are interested in is related to the pharmacophore identification problem and was posed by a French pharmaceutical company as a follow-up
to [4]. Using the previous definitions, we can state it as follows:

---

[*] A poster version of this work has been presented at JOBIM 2000 –Jounées Ouvertes
  Biologie Informatique Mathématique, France, Mai 2000.
[1] Two cycles are adjacent if they share an edge. A block is a connected subgraph
  consisting of adjacent cycles. For further definitions concerning basic graph theory,
  we refer to [5, Chapter 1].

J. Marks (Ed.): GD 2000, LNCS 1984, pp. 115–126, 2001.

> *Given a molecular family and its template, draw the molecules*
> *in the plane in a standardised way to make the comparisons and*
> *interpretations of the chemist easier.*

This is exemplified on Figure 1 which depicts two "similar" molecules with the drawings as they are stored in the database as well as the drawings automatically generated by our algorithm. The dotted edge served as template. [2]

Naturally, $2D$ drawings cannot capture all the subtleties of a molecule. (In particular the stereo-chemistry is not taken into account.) These $2D$ representations should therefore be considered as a necessary but not sufficient information: whenever they fall short from providing the precision required, they can be enriched by more precise facts such as the $3D$ Connolly surface, chemical measurements, etc.

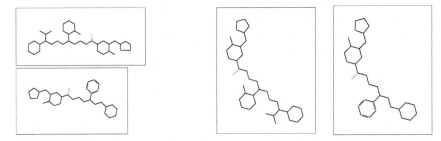

Fig. 1. (left) Database drawings, (right) Drawing algorithm "longest chain"

1.2   Paper Overview

In Section 2, we review relevant previous work. Section 3 outlines our drawing method. Its main ingredients are developed in Sections 4, 5, and 6. Implementation issues and results are presented in Section 7. Finally, Section 8 lists the future plans.

2   Previous Work

The bibliographic work consists of three parts: literature concerning the drawing of chemical structures, graph drawing methods in general, and the graph theoretic question of subgraph isomorphisms.

To the best of our knowledge, there exist very few publications concerning automatic drawing of $2D$ chemical structures. Several chemists confirmed that no automatic drawing software exists on the market. Nevertheless, in the Seventies and Eighties several approaches were proposed. First attempts by Zimmerman

---

[2] Examples are taken from our partner's database. For confidentiality reasons, they are slightly modified.

and Feldman [8] use a library of basic substructures and combine them according to a set of rules. The main criticism to this approach is lack of generality. In [3], a three-dimensional model is projected in the plane, but cyclic structures are not drawn with correct angles. In [14], Shelley presents a promising heuristic method. Similar to our approach, it is based on a breadth-first traversal of the molecule with the successive positioning of atoms. However, the heuristic is quite involved, and the final choice for atom positions is not completely described. A more recent genetic algorithm by D.B.Hibbert [10] generates and displays chemical structures. Angles are fixed following the valence of the atoms, and a drawing is evaluated with respect to the distance between non-adjacent atoms.

Concerning graph drawing literature and to the best of our knowledge, efficient algorithms for planar drawing of trees with unitary edge length and several fixed angles do not exist. However, hardness results for drawing trees are particularly interesting. We use the logic machine approach of Whitesides and Eades [5, Chapter 11.2] to show the hardness of our drawing problem.

The comparison of graphs and trees are well known questions in graph theory. Problems involving isomorphism testing of graphs are in general NP-hard problems [15, Chapter 2.6]. For molecular graphs, i.e. node labelled graphs, a simplification is shown in [13]. The computational complexity is lowered down by an important factor, but the problem still remains NP-complete. Polynomial-time algorithms exist for trees: The embedding problems for subgraph isomorphism and topological embedding can be solved in $O(n^{2.5} \log n)$ time for two trees of size $O(n)$ [9]. The problem is NP-hard for more than two trees, as shown in [1].

3   Algorithm Outline

3.1   Drawing Conventions and Constraints

A chemical drawing must satisfy several properties. The drawing is planar, and the degree of the node determines the angle between its adjacent edges. However, assuming one edge is fixed, the positioning of its adjacent nodes is not fully determined. The possible choices are shown in Figure 2. The angle between two edges incident to a node of degree 3 and a node of degree 2 that does not have two double bonds is $\alpha = \pm\frac{2\pi}{3}$. For a node of degree 2 with two double bonds there is no choice, $\alpha = \pi$. There are six possibilities to draw the adjacent nodes of a node of degree 4: The angles are multiples of $\alpha = \frac{\pi}{2}$, and the order among the edges is not fixed. If the edge belongs to a cycle, the angle is given by the type of cycle (i.e. for a hexagon $\alpha = \frac{2\pi}{3}$).

3.2   Automated Molecular Drawing: A Three Stages Process

From the previous discussion, a satisfactory solution to our problem needs to be able to:

1. identify the sub-structures common to several molecules,
2. perform the drawing of the molecules.

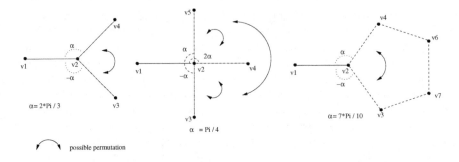

Fig. 2. Angles between bonds

To solve the first part of the problem, we construct a graph, the *supertree*, which "includes" all the molecules of the family. It is a fictive molecule into which each molecule is embedded. In general, the supertree itself does not admit a planar drawing. It is simply used to record and coordinate the drawings of the molecules.

The second stage of the drawing process consists of effectively drawing the molecules. The criteria to be respected are twofold. First, the constraints of Section 3.1 must hold. Second, in order to improve readability, some attention has to be devoted to aesthetic aspects. In particular, we shall be concerned with the diameter of the drawing — the maximum distance between two atoms, as well as the total area of the drawing.

To summarise, the outline of the algorithm is the following. First, cycles and blocks are detected in all the molecules. Next, a representation of the whole family of molecules is computed, the so-called supertree. At last, the drawing algorithm is applied to each molecule separately, but drawing decisions are coordinated using the supertree. The heuristics applied in the drawing procedure are based on the supertree.

4   Preprocessing Step: Cycle and Block Detection

The first step is the detection of cycles and blocks of cycles in all molecules. For each appendix, a Breadth-First Search (BFS) algorithm stores parent-child relations and *non-tree edges*, i.e. the last edge of each cycle to be traversed by the BFS. The corresponding cycle to a non-tree edge is found by collecting the ancestors of the nodes incident to the non-tree edge until a common ancestor is found. Adjacent cycles are grouped to a block. The first node of a block in BFS order is called its *entry node*. It is easy to see that the detection of cycles in a molecular graph of size $O(n)$ with c induced cycles of maximum size α has $O(n + \alpha c)$ running time.

5   Identification of Common Substructures

5.1   Supertree under Topological Embedding

The first part of the problem is to identify substructures that are isomorphic in several molecules of the family. This is necessary to be able to draw them the same way, up to a rigid motion. To give an illustration, Molecule 1 and Molecule 2 shown in Figure 3 have the template (dotted) but also the dashed parts in common. An embedding of a graph G into a graph S is a one-to-one

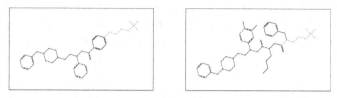

Fig. 3. Identified substructures, (a) Molecule 1 (b) Molecule 2

mapping of all nodes of G to the nodes (not necessarily all) of S. The type of embedding defines the condition under which two nodes can be mapped.[3] The smallest common embeddable supertree (SCES), thus, contains all atoms of all molecules, but several identical atoms may be associated to a single node.

We choose topological embedding which allows matching an edge to a path. This way, similar substructures are matched even if they are connected by paths of different length. This is, for example, employed in Figure 3. Figure 4 presents a simple supertree example.

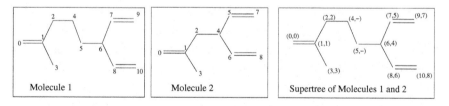

Fig. 4. Simple example to demonstrate the supertree

5.2   Supertree Computation

In this section, we show how to compute the supertree $S(r(M), r(M'))$ of two rooted trees M and M' with roots $r(M)$ and $r(M')$. In the sequel, $S(r(M), r(M'))$

---

[3] For definitions in this section, we refer to [9].

stands for the supertree as well as its size. Recall, that the roots are provided by the template. M and M' are appendices of the molecules rooted at the same template node.

Algorithm of Gupta and Nishimura. We adapt the algorithm proposed by Gupta and Nishimura in [9] which is based on a dynamic programming scheme. Without giving the details of the proofs, the results of [9] are briefly illustrated and later revisited for the treatment of cycles and blocks.

The algorithm is based on the following observation. An embedding can either map the two roots $r(M)$ and $r(M')$ to the root of the supertree and some children of $r(M)$ to distinct children of $r(M')$, or one of the roots can be mapped to a descendant of the other root. In [9, Lemma 7.4], the three possibilities are formalised as follows:

For any node a of M with children $b_1, ..., b_k$ and for any node u of M' with children $v_1, ..., v_l$, three quantities are defined:

$$M_1 = \min_{1 \leq i \leq l} \left\{ S(a, v_i) + 1 + \sum_{j \neq i} S(\emptyset, v_j) \right\};$$

$$M_2 = \min_{1 \leq i \leq k} \left\{ S(b_i, u) + 1 + \sum_{j \neq i} S(b_j, \emptyset) \right\}; \text{ and}$$

$$M_3 = MinWM(\{b_1, ..., b_k\}, \{v_1, ..., v_l\}) + 1.$$

Then $S(a, u) = \min \{M_1, M_2, M_3\}$.

As in [9], $MinWM(\{b_1, ..., b_k, n_1, ..., n_l\}, \{v_1, ..., v_l, m_1, ..., m_k\})$ is the minimum weight perfect matching of the bipartite graph built from the two sets of nodes adjacent to a and u and a copy of each node in the party node set. The weight of the edge (b_i, v_j) is the size of the minimum supertree $S(b_i, v_j)$. Edges (b_i, m_i) and (n_j, v_j) correspond to a matching of a node with its copy. They have weight $S(b_i, \emptyset)$ and $S(\emptyset, v_j)$, respectively.

The running time of the algorithm is $O(n^{2.5} \log n)$ for two trees of size $O(n)$. This cost is dominated by the bipartite matching calculations. For details, we refer to [9].

Dealing with Cycles and Blocks. Unfortunately, a dynamic programming scheme does not apply to general graphs. A reduction from trees with blocks of cycles to trees is necessary in order to avoid NP-hardness. Two solutions come to mind: The reduction of a block to one node, or the destruction of the cycles by "hiding" an edge.

The first solution has the disadvantage that only complete blocks of cycles can be matched. The matching of a path that is not part of a cycle to some edges of a cycle is, yet, desirable for some chemical structures. The second one has a different drawback: since the choice of the hidden edge, e.g. the non-tree

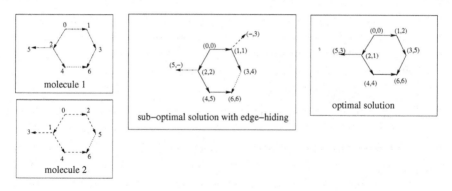

Fig. 5. Hiding non-tree edges (the node labels follow the BFS-order)

edge, is not uniquely determined by the graph-structure, the result might not be optimal as illustrated in Figure 5.

We choose a hybrid solution to avoid these inconveniences: For the matching of two blocks with entry nodes a or u, we do not regard their internal structure but consider a block as an atomic node —solution 1. In the case where either a or u is not part of a block, the non-tree edges of the blocks are hidden as proposed in the second solution, and the normal procedure applies.

Adaptation of the Algorithm to Labelled Trees. Only a slight modification is necessary to take atom types into account. The matching of nodes that are not of the same type is punished by counting the corresponding supertree node twice.

Data-Structures and Implementation. For every node a of M proceeding from leaves to root and for every node u of M' proceeding from leaves to root, $S(a, u)$ can be recursively computed following the scheme:

$$
S(a, u) = \begin{cases}
1 & \text{if they are leaves and type}(a) = \\
& \text{type}(u), \\
2 & \text{if they are leaves and type}(a) \neq \\
& \text{type}(u), \\
\max(\text{size}(block_1), \text{size}(block_2)) & \text{if } a \text{ root of } block_1 \text{ and } u \text{ root of} \\
& block_2, \text{ and the blocks are leaves}, \\
\min\{M_1, M_2, M_3\} & \text{otherwise}.
\end{cases}
$$

Of course, to be able to actually construct the supertree, one has to keep track of the subtrees as well as their sizes. To do so, for each pair of nodes (a, u), a potential supertree node is created. Depending on whether $S(a, u)$ was the result of M_1, M_2, or M_3, the corresponding edges are inserted. For example, in the case of $S(a, u) = M_3$, the edges inserted are $((a, u), (b_j, v_i))$ for each pair (b_j, v_i) matched in the computation of M_3, $((a, u), (b_l, \emptyset))$ and $((a, u), (\emptyset, v_k))$ for the non-matched nodes (matched with its own copy).

The tree rooted at the node $(r(M), r(M'))$ is the smallest supertree of M and M'. This is due to the fact that at each node insertion, we insert the edges that connect the node to its children in the supertree.

5.3 Supertree for the Family of Molecules

Theorem 1. *The computation of the smallest common embeddable supertree of m rooted trees of bounded degree is NP-hard and hard to approximate within a factor of $n^{1-\epsilon}$, for any $\epsilon > 0$.*

Proof. The problem is reduced from the INDEPENDENT SET problem following the result of [1]. For details see [2]. ∎

In this context, we content ourselves to an approximate solution and compute the supertrees two by two in a binary fashion. At first, the family is divided into pairs of molecules for which the supertree is computed. This is repeated recursively until we are left with a single supertree. Notice that the optimality of the result over several recursive calls is not guaranteed.

The size of the supertrees is in the worst case the sum of the size of the molecules —a rough upper bound. Thus, for m molecules of size $O(n)$, the size of the supertree of the family is in the worst case of order $O(m \times n)$. The following property is easily checked.

Proposition 1. *The computation of the supertree of a family of m molecules of sizes $O(n)$ has a worst case running time of $O((mn)^{2.5} \log(mn))$*

6 Drawing Molecules

The algorithm is based on the observation that all possible drawings of the molecule can be enumerated. Once an edge has been drawn, there is only a limited number of possibilities to draw the adjacent edges – see Section 3.1. The drawing algorithm consists of determining the choice of angles that leads to a permissible drawing which, in addition, fulfils as much as possible the aesthetic criteria.

6.1 NP-Hardness Proof

Before describing the algorithm, we state that it is an NP-hard problem to decide whether there exists a permissible drawing of a molecule following the constraints described in Section 3.1. The proof mimics the "logic engine" of [7]. See [2] for details.

Theorem 2. *It is an NP-hard problem to decide whether there exists a permissible drawing of a molecule.*

6.2   Basic Drawing Procedure

The basic procedure to draw a molecule consists of traversing the nodes in breadth-first-order starting at a template node and assigning a coordinate to each adjacent node. In case of a cycle or a block of cycles, the breadth-first order is not followed, and the whole block is processed at once. Before assignment, each position is checked for conflicts, i.e. intersections of the new edge with already drawn nodes and edges. If the assignment of coordinates to an adjacent node is impossible, we backtrack to the preceding node (in the breadth-first traversal), and try another permutation of node/position pairs. In case of failure, we backtrack even further.

To complete the description, one needs to define an order of preference:

P1.   on the adjacent nodes of the last node drawn.
Two options referred to as "largest" and "longest" subtree are implemented. Nodes are ordered with respect to the depth or size of the subtree they are root of.

P2.   on the possible positions of the nodes to be drawn.
The positions maximising the sum over the distances to already drawn nodes or the continuation of the longest chain are preferred.

The combination of the possibilities for P1 and P2 yield 4 heuristics. (But we disregard the combination "largest subtree" – "development of longest chain' which does not make sense.)

6.3   Rigid Blocks

The analysis of the decision tree which describes all possible drawings immediately leads to the notion of rigid blocks. We define a *rigid block* to be a part of the molecule whose embedding depends on only one decision. In case of a block, for example, the choice of the position of one node determines the placement of the whole block and of its adjacent edges. In the drawing procedure, this is taken into account by processing an entire block immediately after the placement of the first node, as already mentioned above. For the adjacent edges, the angle is determined by the angle inside the cycle. An example of a molecule with a block of 5 cycles is given in Figure 6(c).

6.4   Coordination of the Drawings

This section describes how drawing decisions are coordinated using the supertree. The first molecule is drawn as described above, with the only difference that the heuristics (depth or size of subtrees) are based on the supertree structure. For the following molecules, the heuristic consists of trying the choice that has been successful in the majority of the molecules already processed. If it fails, another angle is chosen with the usual heuristics. Already drawn molecules are not corrected in order to avoid further backtracking.

The data-structure used for the coordination of the angles consists of a priority queue for each edge that contains all angles used for the edge. The priority is the number of molecules that employed this solution. It is updated each time a molecule has been processed. In fact, it is a persistent data structure as proposed in [6] using the "fat node model".

7   Implementation and Results

The implementation consists of about 10000 lines of C++ code. It uses the Graph Template Library GTL (http://infosun.fmi.uni-passau.de/GTL) and the STL library (Standard Template Library) [12]. The GTL library implements a graph data-structure together with basic algorithms such as graph traversal algorithms. It is based on the generic data-structures of STL. Molecules are given

as molfiles or structure-data files, two connection table (CTAB) formats that are standard in the chemical software industry (see http://www.mdli.com). The list of atoms and bonds in CTAB format is transformed into the graph description language GML [11] by a perl script. It is possible to display the former drawing of the molecule (that was done by the chemist) to be able to compare the results. The graphical output is written in OpenGL (http://reality.sgi.com/opengl). A CTAB file is produced for further use in standard chemical set-ups. For printing, a Postscript file is dumped.

The drawing algorithm can be applied to single molecules separately from the supertree computation. Some of the results are shown as examples in the paper. Usually, there is not much difference between the different heuristics, neither in running time nor in the drawings. For the supertree program, some molecules are entered – either with specified templates or without (default: the template consists of the first edge). Again, the options for the drawing algorithm are valid.

The running time on a SUN computer with a SPARC processor at 300 MHz and 250 MB of RAM is about 25 seconds for 12 molecules of average size of 35 atoms. So far, the code has not been optimised. More experimental facts will be provided in [2].

We tested examples from two molecular databases. The examples of database 1 were selected by the pharmacologist as good and typical examples for their application. In the second molecular database, the drawings are given as 3D projections of the molecular structures –see Figure 6(b).

8   Conclusion and Future Work

The calculation of the supertree is a promising solution to coordinate drawing decisions in the layout of $2D$-chemical structures of several similar molecules. In addition to the fact that the drawings reveal similarities and differences between several molecules, the supertree speeds up the drawing procedure. Whenever a

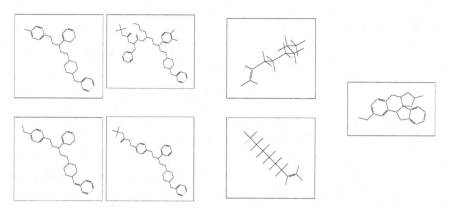

Fig. 6. (a) Family of 4 molecules, (b) Example from database 2, (c) Molecule with block of cycles

backtracking is necessary in the first molecule, copying its result for the other molecules prevents from testing the misleading decisions again.

Unfortunately, the method is restricted to the case where a template or at least one common atom is known in all molecules of the family. The choice of the template is crucial to the quality of the result. For now, a generalisation of the supertree algorithm to the case of non-rooted trees seems difficult to realize without a significant slow down in computation time (a factor of n^2 per call). Notice, however, that template queries to data-bases yield in general the necessary information. Second, the optimality of the size of the supertree after successive recursive calls is not guaranteed. It might be interesting to derive bounds on the error and to propose more sophisticated strategies to avoid it. Last but not least, one might look at a further use of the supertree representation in this context. The size of the supertree as a measurement of similarity, for example, could be considered.

The results of the drawing algorithm are in general correct and readable to the chemist. The heuristics avoid backtracking in most cases which makes the computation very fast. Problems occur when no permissible drawing can be found. Should this happen, we return the original database drawings. A mechanism to allow compromises in the drawing such as the changing of angles or variances of the edge length would allow more flexibility. Another problem is the orientation of ring systems such as steroids. Such chemical structures must be drawn with a certain orientation. In order to draw them in the right manner, we propose to supply a dictionary of critical ring systems with their predefined drawings.

Acknowledgements. The authors wish to thank Sue Whitesides for interesting discussions on rigid blocks and on the complexity of the drawing algorithm during her visit at INRIA in May 1999.

References

1. Tatsuya Akutsu and Magnús M. Halldórsson. On the approximation of largest common subtrees and largest common point sets. *Theoretical Computer Science*, to appear 2000.

2. Jean-Daniel Boissonnat, Fréderic Cazals, and Julia Flötotto. 2d-structure drawings of similar molecules. Technical report, INRIA Sophia Antipolis, to appear 2000.

3. Raymond E. Carhart. A model-based approach to the teletype printing of chemical structures. *Journal of Chemical Information and Computer Sciences*, 16:82–88, 1976.

4. F. Cazals. Effective nearest neighbours searching on the hyper-cube, with applications to molecular clustering. In *14th ACM Symposium on Computational Geometry*, pages 222–230, 1998.

5. G. di Battista, P. Eades, R. Tamassia, and I. Tollis. *Graph Drawing*. Prentice Hall, 1999.

6. J. R. Driscoll, N. Sarnak, D. D. Sleator, and R. E. Tarjan. Making data structures persistent. *Journal of Computer and System Sciences*, 38:86–124, 1989.

7. Peter Eades and Sue Whitesides. The logic engine and the realization problem for nearest neighbor graphs. *Theor. Comput. Sci.*, 169(1):23–37, 1996.

8. R. J. Feldmann, S. R. Heller, E. Hyde, and W. T. Wipke (Ed.). *Computer Representation and Manipulation of Chemical Information*, pages 55–60. Wiley, New York, 1974.

9. A. Gupta and N. Nishimura. Finding largest subtrees and smallest supertrees. *Algorithmica*, 21:183–210, 1998.

10. D. Brynn Hibbert. Generation and display of chemical structures by genetic algorithms. *Chemometrics and Intelligent Laboratory Systems*, 20:35–43, 1993.

11. M. Himsolt. Gml: A portable graph file format. Technical report, Universität Passau, 1997. cf. http://www.fmi.uni-passau.de/himsolt/Graphlet/GML.

12. D.R. Musser and Atul Saini. *STL Tutorial and Reference Guide*. Addison-Wesley Publishing Company, 1995.

13. V. Nicholson, C.-C. Tsai, M. Johnson, and M. Naim. A subgraph isomorphism theorem for molecular graphs. In *Graph theory and topology in chemistry, Collect. Pap. Int. Conf.*, volume 51 of *Stud. Phys. Theor. Chem.*, pages 226–230, Athens/GA, 1987.

14. Craig A. Shelley. Heuristic approach for displaying chemical structures. *J. Chem. Inf. Comput. Sci.*, 23:61–65, 1983.

15. J. van Leeuwen (Ed.). *Handbook of Theoretical Computer Science*, volume A. Elsevier, 1990.

Fast Layout Methods for Timetable Graphs[*]

Ulrik Brandes[1], Galina Shubina[2], Roberto Tamassia[2], and Dorothea Wagner[1]

[1] University of Konstanz, Dept. of Computer & Information Science,
Box D 188, 78457 Konstanz, Germany.
{Ulrik.Brandes,Dorothea.Wagner}@uni-konstanz.de
[2] Brown University, Dept. of Computer Science, Center for Geometric Computing,
Providence, Rhode Island 02912-1910, USA. {gs,rt}@cs.brown.edu

Abstract. Timetable graphs are used to analyze transportation networks. In their visualization, vertex coordinates are fixed to preserve the underlying geography, but due to small angles and overlaps, not all edges should be represented by geodesics (straight lines or great circles).

A previously introduced algorithm represents a subset of the edges by Bézier curves, and places control points of these curves using a force-directed approach [5]. While the results are of very good quality, the running times make the approach impractical for interactive systems.

In this paper, we present a fast layout algorithm using an entirely different approach to edge routing, based on directions of control segments rather than positions of control points. We reveal an interesting theoretical connection with Tutte's barycentric layout method [18], and our computational studies show that this new approach yields satisfactory layouts even for huge timetable graphs within seconds.

1 Introduction

We consider timetables comprised of transportation schedules, which may be originating from, e.g., trains, flights, or product shipments. A large amount of such timetable data is provided to us by our industrial partner.[1] These are mostly train schedules from companies all over Europe, but may contain timetables from other public transport authorities (running busses, ferries, etc.) as well. Due to the size of the data (e.g., more than 140,000 trains serving a combined number of 28,000 stations for schedules of trains in Europe), visualization has proven to be a valuable tool for data inspection and maintenance.

The main purpose of the analysis of such data is quality management, since schedules may vary between time periods, and are exchanged in a variety of formats, including hand-written. The basis for several aspects of the analysis [19,

---

[*] Research partially supported by the U.S. National Science Foundation under grants CCR-9732327 and CDA-9703080, by the U.S. Army Research Office under grant DAAH04-96-1-0013, and by the German Academic Exchange Service (DAAD/Hochschulsonderprogramm III).
[1] *TLC/EVA* the IT subsidiary of *Deutsche Bahn AG* that is, among other things, responsible for collecting, analyzing, and publishing timetable information.

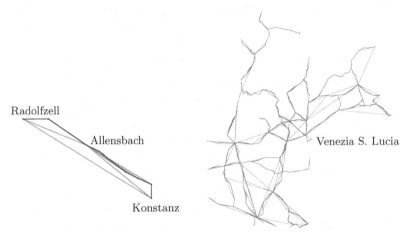

Fig. 1. All edges represented by straight-line segments

17,14] are graphs generated from the timetables. For the purpose of this paper, we assume that each station that any train stops at corresponds to a vertex, and an undirected edge is introduced for each pair of stations for which there is a non-stop service in either direction. Consequently, the timetable graphs considered here are undirected and simple.

The links between points of departure and arrival are tied to geographic locations, thus providing an intuitive vertex placement for the corresponding connection graph, leaving us with the problem of routing its edges. In order for the visualization to be effective, the edge routing algorithm has to produce clear and helpful drawings, and do so quickly to make it usable in interactive tools.

In the most common form of geographic network visualization, edges are shown as geodesics (straight lines or great circles, depending on whether the graph is shown in the plane or on the sphere) [2,16]. While such drawings can be produced very quickly, small angles and overlap of edges often hinder their unambiguous identification, as can be seen in Fig. 1.

A method to produce effective timetable graph visualizations is presented in [5]. It uses an automatically classification of the edges into *minimal* and *transitive*, where minimal edges are assumed to correspond to railroads directly connecting pairs of stations, while transitive edges typically correspond to regional or long-distance services that do not stop at each station they pass. Since railroads can be expected to cover a geographic region efficiently, this method represents minimal edges by straight lines. Transitive edges, by their very nature, are bound to cause small angles and overlap. Therefore they are represented by cubic Bézier curves. An elaborate force-directed model places the control points of these curves. While according to the data analysts, the output is very satisfactory, running times are not acceptable (in the range of several minutes for graphs of realistic size). See Fig. 2 for an example.

Since we plan to integrate visualization into an existing interactive query engine that allows to generate timetable graphs from the complete data set based

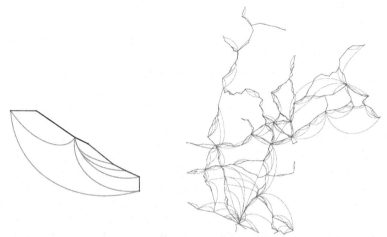

Fig. 2. Force-directed placement of Bézier control points for transitive edges [5]

on a variety of attributes (coordinates, train classes, traveling times, service frequencies, etc.), faster, yet still effective, layout methods are sought.

A recently introduced alternative approach to automatic routing of cubic Bézier curves is based on the angles between control segments rather than the positions of control points (*rotation approach* [4]). Though it is extremely fast, its application to timetable graphs produces drawings with several deficiencies, such as excessive crossings and S-shaped curves.

Building on the underlying principle of the general rotation approach, we present a new edge routing model that yields better layouts without sacrificing too much of the running-time advantage. It uses properties of timetables to preprocess the graph so as to make it more susceptible to a rotation-like method. A new objective function is introduced, that combines three criteria: *angular resolution*, *straightness*, and *roundness*. A theoretically interesting aspect of this model is its close relation to the barycentric model of Tutte [18], even though it is entirely based on angles rather than coordinates.

Though several other graph drawing techniques explicitly consider curved or polyline edges, none of them seems applicable in our case, because they either position vertices [12,13,7] or are mainly designed to route edges around obstacles [11,9,1,10]. Our goal here is to disentangle a straight-line drawing in the simplest way possible without moving vertices.

The main features of the angle-based approach for cubic-curve routing are outlined in Sect. 2. Our new model for timetable graphs is introduced in Sect. 3. We conclude with some real-world examples and running time experiments.

2   An Alternative Approach to Curved Edge Layout

In this section, we outline some aspects of a recently introduced approach [4] for edge layout, when a graph has fixed vertex positions.

A cubic Bézier curve [3] is determined by its two endpoints, b_0, b_3, and two inner *control points* b_1, b_2 (see Fig. 3). Note that the same curves are obtained when the order of control points is reversed, while other permutations in general define different curves. We thus call segments $\overline{b_0 b_1}$ and $\overline{b_2 b_3}$ the *initial control segments*, while $\overline{b_1 b_2}$ is called the *inner control segment*. Two other important properties of Bézier curves

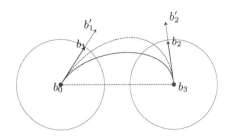

Fig. 3. Cubic Bézier curves [3] defined by rotation and length of initial segments

are *i)* that the entire curve is contained in the convex hull of its defining points, and *ii)* that the tangents at its endpoints are collinear with the first and last control segment. The second property provides an immediate generalization of angles between edges represented by straight lines to angles between edges represented by Bézier curves. From now on, we will neglect the distinction between an edge or a vertex and its graphical representation.

Since vertex positions are fixed, so are the endpoints of each edge. Instead of placing control points directly, the layout can be divided into the following two steps, also illustrated by Fig. 3:

1. determine a direction for each initial control segment
2. determine a length for each initial control segment

The *(local) angular resolution* at some vertex is defined as the smallest angle formed by the edges incident to that vertex. The *global angular resolution* of a drawing is the minimum local angular resolution. In the first of the above steps, the angles between incident edges are determined, and therefore the angular resolution at all vertices. The second step can be used to ensure additional properties of a curve, e.g. to reduce its curvature. For our purposes, however, the simple heuristic of choosing a fixed proportion of the distance between the endpoints proved sufficient.

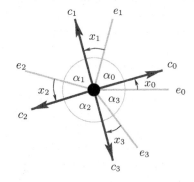

Fig. 4. Angles α_i between straight-line edges, and angular differences x_i

Cubic curves are the simplest representation that allows to maximize the angular resolution for all vertices. Note that we can treat control segments incident to one vertex independently from control segments incident to other vertices. Therefore, let $e_0, \ldots, e_{d_G(v)-1}$, be a counterclockwise ordering of the edges around some $v \in V$ in the given drawing (with ties broken arbitrarily), and denote by α_i, $i = 0, \ldots, d_G(v)-1$, the angle between e_i and its counterclockwise neighbor.

Accordingly, we define $c_0, \ldots, c_{d_G(v)-1}$ to be the corresponding ordering of control segments incident to v. The angles between neighboring control segments equal $\frac{2\pi}{d_G(v)}$ because of the optimal angular resolution constraint. Denote by x_i the angle between e_i and c_i, $i = 0, \ldots, d_G(v) - 1$, where $x_i > 0$, if e_i comes before, and $x_i < 0$, if e_i comes after c_i in the counterclockwise order around v. We call these deviations from straight-line directions the *angular errors*. See Fig. 4 for an illustration and note that $x_3 < 0$.

Any set of control segments satisfying the angle constraints is called a *rotation* at v, but arbitrary rotations usually lead to unpleasant "spaghetti" drawings. The *straightness* of a rotation is the degree to which a rotation succeeds in keeping the angular errors small. Several penalty functions can be used to quantify straightness. In particular, the rotation minimizing the squared angular errors $\sum_{i=0}^{d_G(v)-1}(x_i)^2$ is unique, and can be computed in time $\mathcal{O}(d_G(v))$ simply by averaging over the angular errors. It is called *balanced rotation*, because it satisfies $\sum_{i=0}^{d_G(v)-1} x_i = 0$. Figure 5 shows that optimal local angular resolution at all vertices is too strong a criterion. Note, however, that angles between control segments can be specified arbitrarily. Much better drawings are obtained, when angles between control segments at vertices v with a *dominant angle* angle $\alpha_i \geq \pi$ are constrained to π (for the dominant angle) and to $\frac{\pi}{d_G(v)-1}$ for the other angles. This set of angle constraints is called the *half-sided template*. For further details we refer to [4].

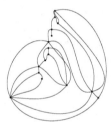

Fig. 5. Result of a standard planar graph drawing algorithm [8], and balanced rotations with and without half-sided templates for dominant angles

The rotation approach applied to timetable graphs is very fast (see Sect. 4). Though an obvious improvement over straight-line drawings, the examples in Fig. 6 also show that the results are not entirely satisfactory for our present application. Despite optimal angular resolution and straightness, the drawings in general display sharp turns, S-shaped curves and appear cluttered.

3   A New Layout Model for Timetable Graphs

Though the rotation approach described in the previous section yields drawings that are much more readable than straight-line representations, they are still far from the quality obtained by the force-directed layout approach. In this section, we present a new layout model based on the same strategy as the rotation approach but tailored to timetable graphs.

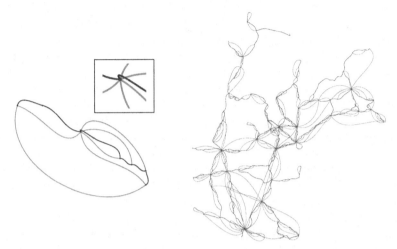

Fig. 6. Balanced rotations with the half-sided template for vertices with a dominant angle. Note the unfortunate ordering on the left.

Edges of a timetable graph are routinely classified into minimal and transitive. Since minimal edges typically correspond to actual railways, they seldom cause readability problems and can hence be drawn straight, both to reduce the size of the input and to visually emphasize their role as a support of the network.

We are going to customize the rotation approach to our specific application in several ways. Most importantly, we introduce a preprocessing step that determines a good ordering of edges around a vertex, and our new objective function aims at rounding out S-shaped curves. In consequence, angular resolution must be relaxed from a constraint to an optimization criterion. We show that there still is a unique optimum solution, which, however, is no longer computed as easily as in the rotation approach.

3.1   Preprocessing

Since minimal edges are represented by straight lines, the ordering of minimal edges around a vertex is fixed. Transitive edges have an inherent tendency to short-cut paths of minimal edges. Small deviations in the course of the latter cause the straight-line ordering of transitive edges to be somewhat arbitrary (as was demonstrated in Fig. 6). To reduce the crossing problem illustrated in Fig. 6, an ordering of transitive edges is determined in two stages: in the first stage, control segments with similar target directions are grouped together, and in the second stage, the control segments of each group are put in order.

Grouping. The initial control segments incident to a vertex are grouped according to the minimal edge their straight-line representation is closest to, and according to the side toward which they depart (i.e. whether they are counterclockwise before or after their closest minimal edge). As illustrated in Fig. 7(a), each initial control segment of a transitive edge is assigned to the left or right hand side of its closest minimal edge.

Crossing reduction. The assignment is then refined to reduce the number of crossings among transitive and minimal edges. If the two initial control segments of a single edge are both classified to lie on, say, the right hand side of their respective minimal edge, the assumption that these minimal edges are linked by a path of minimal edges (the railroad that the long-distance trains inducing this edge travel along) suggests that the inner control segment of the transitive edge will cross this path. We therefore reassign the control segment incident to the vertex of smaller degree (presumably a less important station), or, in case of equal degrees, to the one that deviates further from the straight-line connecting the endpoints, to another group as depicted in Fig. 7(b).

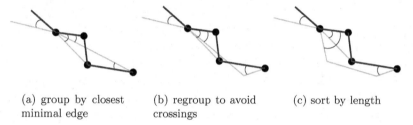

(a) group by closest   (b) regroup to avoid   (c) sort by length
   minimal edge   crossings

Fig. 7. Initial ordering of edges. Arcs indicate the side of minimal edge a control segment is assigned to

Sorting. Within a group, control segments are sorted according to the length of the straight-line representation of their corresponding edge, such that the shortest ones are closest to their assigned minimal edge. See Fig. 7(c). This order is likely to avoid crossing of adjacent transitive edges.

Thus the initial control segments of transitive edges incident to a common vertex are grouped into a number of groups that is twice the number of minimal edges incident to that vertex. Each group is assigned a wedge that is an angle between a minimal edge and a bisector of the angle between this edge and its clockwise or counterclockwise minimal neighbor and space out evenly. The overall running time of the preprocessing step is $\mathcal{O}\left(\sum_{v \in V} d_G(V) \log d_G(v)\right) = \mathcal{O}\left(|E| \log |E|\right)$.

3.2 Layout Objectives

With this heuristic ordering, we are now able to state our objective function formally. It combines the criteria of local angular resolution, straightness, and roundness, subject to straight-line representation of minimal edges. Considering the straight minimal edges to be control segments with fixed direction, let $c_0, \ldots, c_{d_G(v)-1}$ be the directions of control segments incident to a vertex v, in the order resulting from the preprocessing.

Angular Resolution. The optimal angular differences a_i, $i = 0, \ldots, a_{d_G(v)-1}$, between consecutive control segments (which have been a constraint in the rotation approach) are determined by equally dividing up the wedge assigned to the group. Satisfaction of the angular resolution criterion can then be expressed in terms of the squared angular error with respect to the target values,

$$A_v(c) = \sum_{i=0}^{d_G(v)-1} (c_{i+1} - c_i - a_i)^2, \qquad (angular\ resolution\ criterion)$$

where indices are modulo $d_G(v)$, and pairs c_i, c_{i+1} of control segments in different groups are omitted. Recall that control segments of minimal edges lie in two groups.

Straightness. For reasons mentioned in the discussion of the rotation approach, the deviation of control segments from straight edges should be penalized. We use the squared angular errors with respect to straight-line directions, i.e. the objective function of the balanced rotation approach,

$$S_v(c) = \sum_{i=0}^{d_G(v)-1} x_i^2 = \sum_{i=0}^{d_G(v)-1} (c_i - e_i)^2. \qquad (straightness\ criterion)$$

Roundness. Much of the clarity in timetable graph layouts produced with force-directed placement stems from the prevailing symmetry, or *roundness*, of those edges represented by Bézier curves. We measure the roundness of a Bézier curve by the squared difference in deviation of the two initial control segments from the straight-line edges connecting the endpoints. Note that, for a highly desirable curve, the magnitudes of deviation are the same at both ends, but with opposite sign.

As a result of the preprocessing, one initial control segment of a transitive edge is assigned to a group associated with the right hand side of a minimal edge, while the other is assigned to a group associated with the left hand side of a minimal edge. Reversing the sense of direction within each group associated with, say, a left hand side changes the sign of all their angular differences, but does not affect the other two criteria (if the sign of optimal angular differences a_i is reversed as well). Non-roundness is thus defined as

$$R_v(c) = \sum_{i=0}^{d_G(v)-1} ((c_i - e_i) - (c_i' - e_i'))^2, \qquad (roundness\ criterion)$$

where c_i' is the direction of the initial control segment at the opposite end of e_i, and e_i' is the reverse straight-line direction of the edge.

The objective function for edge layout of preprocessed timetable graphs is now defined as a weighted sum of the above criteria, $U(c) = \sum_{v \in V} w_a \cdot A_v(c) + w_s \cdot S_v(c) + w_r \cdot R_v(c)$.

3.3   Optimizing the Objective Function

Next we show that the layout objective function $U(c)$ defined in the previous section is a generalized version of the objective function of the barycentric layout model, and has a unique minimum under equivalent assumptions.

Consider the following transformation of a preprocessed timetable graph. We construct a new undirected graph $G = (V, E)$ that has a vertex for each of the two initial control segments, and for each of the two directions of a transitive edge. Two vertices in G are adjacent, if one is a straight-line direction and the other is the corresponding control segment (e_i and c_i of some vertex), if they are consecutive control segments in some group (recall that groups and order are defined in the preprocessing), or if they are initial control segments of the same transitive edge.

Assume that for each edge $e = \{u, v\} \in E$ there are weights $\omega_e > 0$ and target differences $\theta_{uv} = -\theta_{vu}$. Then it is easy to see that our objective function can be restated as

$$U(c) = \sum_{e=\{u,v\}\in E} \omega_e \cdot (c_v - c_u - \theta_{uv})^2.$$

Since the essential properties of this function are the same as those of the barycentric layout model $\sum_{e=\{u,v\}\in E}(c_v - c_u)^2$, the following parallels Tutte's analysis [18]. For a vector $c = (c_v)_{v\in V}$ minimizing this function, the partial derivatives

$$\frac{\partial}{\partial c_v} U(c) = \sum_{u\,:\,e=\{u,v\}\in E} 2\omega_e \cdot (c_v - c_u - \theta_{uv})$$

must equal zero for all $v \in V$. Cancelling the constant factor of 2, this system of linear equations can be reordered into the form

$$(D(G) - A(G)) \cdot c = L(G) \cdot c = b,$$

where $D(G)$ is a diagonal matrix with weighted degrees $d_{vv} = \sum_{u\,:\,e=\{u,v\}\in E} \omega_e$ on the diagonal, $A(G)$ is the weighted adjacency matrix with entries $a_{uv} = \omega_{\{u,v\}}$ if $\{u, v\} \in E$ and $a_{uv} = 0$ otherwise, and b is a vector with constant entries $b_v = \sum_{u\,:\,e=\{u,v\}\in E} \omega_e \theta_{uv}$. The resulting matrix $L(G)$ is called the Laplacian of the graph.

Lemma 1 ([6]). *The determinant of any submatrix of $L(G)$ obtained by omitting any pair of a row and a column corresponding to a vertex in G equals*

$$\sum_T \prod_{e\in E(T)} \omega_e$$

where the sum is over all spanning trees of G, and $E(T)$ denotes the edge set of a tree T.

Note that fixing any entry in c corresponds to omitting its row and column from $L(G)$ and adjusting b. Fixing the entries of more than one vertex of G corresponds to contracting these vertices, and then omitting the row and column of this vertex and adjusting b. Consequently, the determinant of the resulting submatrix is positive, if the value of at least one vertex in each connected component of G is fixed. Since, by definition, no station is incident only to transitive edges, every component of G has at least one vertex that corresponds to a fixed control segment of a minimal edge.

Theorem 1. *The timetable graph layout objective function $U(c)$ has a unique minimum that can be determined by solving a system of linear equations with twice as many unknowns as there are transitive edges.*

Due to the size of typical systems (cf. Tab. 1), we cannot afford to solve it exactly in time still acceptable for an interactive system. Since the matrix $L(G)$ is weakly diagonally dominant, we instead use Gauss-Seidel iteration to quickly approximate the optimal directions. Note that this nicely corresponds to a one-dimensional spring embedder, that does an optimal move at each step.

Initial directions are determined by equally dividing the angle formed by the bounding pair of a minimal edge and an angle bisector for each group. Clearly, these layouts optimize the angular resolution criterion subject to the heuristic ordering and grouping constraint.

4   Results and Discussion

Figure 8 gives the result of our new approach as applied to the running example. The larger examples given in the appendix show that our new method clearly outperforms the general rotation approach in terms of visual quality, though it still does not quite match the quality of force-directed placements.

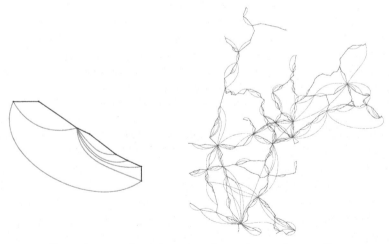

Fig. 8. Layout after 7 iterations ($\omega_a = 2$, $\omega_s = 0$, $\omega_r = 1$)

By necessity, the angular error of control segments with respect to straight edges is usually much larger than the angle between neighboring control segments, so that w_s should be chosen significantly smaller than the other two weights. Since it turns out that straightness is sufficiently taken care of by the preprocessing step, we generally omit this criterion altogether. The relative choice of angular resolution vs. roundness depends on personal preferences. The examples in the appendix use $w_a = 2 \cdot w_r$. We also found that the initial directions do fairly well for any reasonable choice of weights, and since the system is rather sparse, the maximum rotation of a control segment is below 0.01 radians after 3–10 iterations.

All three approaches (force-directed layout, rotation approach, and the approach of this paper) have been implemented in C++ using LEDA [15]. Since we compare proof-of-concept implementations, our running-time experiments should be understood as qualitative. The indication is nevertheless quite clear. Though most of the time is spent on a preprocessing step that determines the "neighborhood" of each control point [5], the force-directed approach is very slow, and will probably remain so even with a sophisticated implementation. While the rotation approach is the fastest, even our current implementation of the approach presented in this paper performs at interactive speed,[2] but produces drawings of much better quality.

Table 1. Running times on a Sun Ultra 5 workstation (360 Mhz, 192 MBytes). Times given in parentheses are without preprocessing

instance	nodes	edges (transitive)	force-directed	rotation	new
switzerland	2218	3203 (536)	53 (10) sec	0.36 sec	**1.31 (0.17) sec**
italy	2386	4370 (1849)	309 (42) sec	0.51 sec	**2.21 (0.57) sec**
france	4551	7793 (2408)	621 (54) sec	0.80 sec	**3.44 (0.73) sec**
germany	7083	9713 (1956)	582 (38) sec	1.18 sec	**4.21 (0.60) sec**

There are several avenues for future work. With respect to the present application, we have yet no way of modeling the second most effective feature after roundness, i.e. *binding*. By introducing dummy edges between a pair of control points when their initial segments are incident to the same vertex, the force-directed approach succeeds in dragging consecutive or nested transitive edges to the same side of a path of minimal edges. Is there a way to integrate this feature in the present approach?

Similar problems are encountered for geographic networks whose vertices are placed on a globe. We are working on a three-dimensional interpretation in the mold of [4] that would take into account lengths of geodesics to better untangle the edges.

We are also investigating useful strategies for control segment length assignments that would satisfy certain properties of the resulting curves, like low curvature, but potentially also to preserve certain features, like planarity.

We devised a fast and effective layout method for timetable graphs by taking a different view on edge routing and utilizing the underlying network structure.

---

[2] Note that the runnings times compare favorably with the time that LEDA's graph editor needs to render the results.

It will be interesting to devise similar extensions of the rotation approach for other applications like internet traffic, flight routes, or non-geographic networks.

Acknowledgment. We thank Michael Baur and Marc Benkert for implementing the preprocessing step, and Stina Bridgeman for help in additional experiments.

References

1. J. Abello and E.R. Gansner. Short and smooth polygonal paths. *Proc. LATIN '98*, Springer LNCS 1380, pp. 151–162, 1998.
2. R.A. Becker, S.G. Eick, and A.R. Wilks. Visualizing network data. *IEEE Transactions on Visualization and Graphics*, 1(1):16–28, 1995.
3. P. Bézier. *Numerical Control.* John Wiley & Sons, 1972.
4. U. Brandes, G. Shubina, and R. Tamassia. Improving angular resolution in visualizations of geographic networks. *Data Visualization 2000. Proc. VisSym '00*, pp. 23–32. Springer, 2000.
5. U. Brandes and D. Wagner. Using graph layout to visualize train interconnection data. *Journal of Graph Algorithms and Applications*, 2000. To appear.
6. R.L. Brooks, C.A.B. Smith, A.H. Stone, and W.T. Tutte. The dissection of rectangles into squares. *Duke Mathematical Journal*, 7:312–340, 1940.
7. C.C. Cheng, C.A. Duncan, M.T. Goodrich, and S.G. Kobourov. Drawing planar graphs with circular arcs. *Proc. GD '99*, Springer LNCS 1731, pp. 117–126, 1999.
8. H. de Fraysseix, J. Pach, and R. Pollack. How to draw a planar graph on a grid. *Combinatorica*, 10:41–51, 1990.
9. D.P. Dobkin, E.R. Gansner, E. Koutsofios, and S.C. North. Implementing a general-purpose edge router. *Proc. GD '97*, Springer LNCS 1353, pp. 262–271, 1997.
10. E.R. Gansner and S.C. North. Improved force-directed layouts. *Proc. GD '98*, Springer LNCS 1547, pp. 364–373, 1998.
11. E.R. Gansner, S.C. North, and K.-P. Vo. DAG – A program that draws directed graphs. *Software—Practice and Experience*, 17(1):1047–1062, 1988.
12. M.T. Goodrich and C.G. Wagner. A framework for drawing planar graphs with curves and polylines. *Proc. GD '98*, Springer LNCS 1547, pp. 153–166, 1998.
13. C. Gutwenger and P. Mutzel. Planar polyline drawings with good angular resolution. *Proc. GD '98*, Springer LNCS 1547, pp. 167–182, 1998.
14. A. Liebers, D. Wagner, and K. Weihe. On the hardness of recognizing bundels in time table graphs. *Proc. WG '99*, Springer LNCS 1665, pp. 325–337, 1999.
15. K. Mehlhorn and S. Näher. *The LEDA Platform of Combinatorial and Geometric Computing.* Cambridge University Press, 1999.
16. T. Munzner, E. Hoffman, K. Claffy, and B. Fenner. Visualizing the global topology of the MBone. *Proc. IEEE InfoVis '96*, pp. 85–92, 1996.
17. F. Schulz, D. Wagner, and K. Weihe. Dijkstra's algorithm on-line: An empirical case study from public railroad transport. *Proc. WAE '99*, Springer LNCS 1668, pp. 110–123, 1990.
18. W.T. Tutte. How to draw a graph. *Proceedings of the London Mathematical Society, Third Series*, 13:743–768, 1963.
19. K. Weihe. Covering trains by stations or the power of data reduction. Electronic *Proc. ALEX '98*, pp. 1–8, 1998. http://rtm.science.unitn.it/alex98/proceedings.html.

An Algorithmic Framework for Visualizing Statecharts*

R. Castelló, R. Mili, and I. G. Tollis

Department of Computer Science
The University of Texas at Dallas
Box 830688, Richardson, TX 75083-0688, USA
email: {castello, rmili, tollis}@utdallas.edu

Abstract. Statecharts [9] are widely used for the requirements specification of reactive systems. In this paper, we present a framework for the automatic generation of layouts of statechart diagrams. Our framework is based on several techniques that include hierarchical drawing, labeling, and floorplanning, designed to work in a cooperative environment. Therefore, the resulting drawings enjoy several important properties: they emphasize the natural hierarchical decomposition of states into substates; they have a low number of edge crossings; they have good aspect ratio; and require a small area. We have implemented our framework and obtained drawings for several statechart examples. The preliminary drawings are very encouraging.

1 Introduction

Statecharts [9] is a graphical notation widely used for the requirements specification of reactive systems. Because of their hierarchical property, statecharts are prime candidates for visualization. Nice and intuitive drawings of statecharts would be invaluable aids to software engineers who would like to check the correctness of their design visually. In this paper, we study the problem of visualizing statecharts and present an algorithmic framework for producing clear and intuitive drawings.

Several visualization tools for reactive system specification and design are available in the market [10,19,18,26]. Eventhough these tools are helpful in organizing a designer's thoughts, they are mostly sophisticated graphical editors, and therefore are severely inadequate for the modeling of complex reactive systems. For example, the Rational Rose tool [21] provides a feature to layout UML [3] statechart diagrams. Figure 1 shows an example of a statechart after the Rational Rose layout feature is applied. We notice that transition labels overlap; transition edges overlap with state boxes; and there is a large number of unnecessary edge bends and edge crossings. Figure 5 in Section 5 shows a drawing of the same diagram using our algorithmic framework.

---

* Research supported in part by Sandia National Labs and by the Texas Advanced Research Program under grant number 009741-040.

J. Marks (Ed.): GD 2000, LNCS 1984, pp. 139–149, 2001.

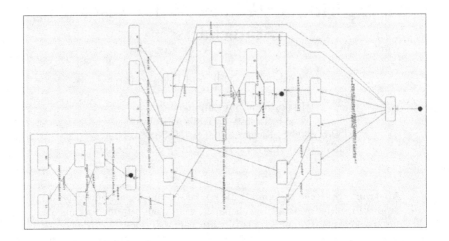

Fig. 1. An example of a drawing of a Statechart generated by Rational Rose (drawing rotated by 90 degrees due to space limitations).

A comprehensive approach to hierarchical drawings of directed graphs is described in Sugiyama et al. [25]. Several extensions and variations of this approach have been introduced in the literature. A comprehensive survey is given in [1]. A first extension that takes into consideration cycles and dummy nodes for large edges (i.e., edges that span more that one level) was introduced by Rowe et al. [22]. Gansner et al. [8,7] provide a technique to draw directed graphs using a simplex-based algorithm that assigns vertices to layers; at the same time, they provide an extension to the basic algorithm of Sugiyama et al. by drawing edge-bends as curves. A divide-and-conquer approach is described by Messinger et al. [17] to improve the layout-time performance for large graphs consisting of several hundreds of vertices. More recently, a combination of the algorithm of [25] with incremental-orthogonal drawing techniques was proposed by Seemann [23] to automatically generate a layout of UML class diagrams. In [11], Harel and Yashchin discuss an algorithm for drawing edgeless highgraph-like structures. The problem of drawing clustered graphs without crossings was studied in [5,6]. Most of the research on the *Edge Labeling Problem* (ELP) has been done on labeling graphs with fixed geometry, such as geographical and technical maps [14]. Kakoulis and Tollis [13] present an algorithm for the ELP problem that can be applied to hierarchical drawings with fixed geometry. Gansner et al. [7] use a simple approach to solve the ELP problem for hierarchical drawings: they assign labels to the middle position of edge lines. However, they assume that edge labels are small and do not consider the possibility of overlap with other drawing components.

In this paper, we present a framework for the automatic generation of layouts of statechart diagrams. Our framework is based on several techniques that include hierarchical drawing, labeling, and floorplanning. Our algorithm for hierarchical drawings is a variant of the algorithm by Sugiyama et al. [25] that is tailored to statecharts. Since edge labels are crucial in describing transitions in statecharts, we have developed edge labeling techniques. Previously, edge labeling techniques were described for graph drawings, and geographical and technical maps with fixed geometry [14,13]. In our work, we address the problem of graph drawings with flexible geometry. Finally, in order to reduce the area and improve the aspect ratio of the statechart drawings we apply floorplanning techniques inspired by the ones used for the area minimization of VLSI layouts [24, 16]. In our approach, the hierarchical, labeling, and floorplanning techniques are designed to work in a cooperative environment. Therefore, the resulting drawings enjoy several properties: they emphasize the natural hierarchical decomposition of states into substates; they have a low number of edge crossings; and require a small area. We have implemented our framework and have obtained drawings for several statechart examples. The preliminary drawings are very encouraging.

2   Statecharts

Statecharts [9] are extended finite state machines used to describe control aspects of reactive systems. They provide mechanisms to describe synchronization and concurrency, and manage exponential explosion of states by using state decomposition. In the statechart notation, a state is denoted by a box labeled in the upper left corner. Directed arcs are used to denote transitions between states. A transition label has the form $E[C]/A$, where E is a boolean combination of external stimuli; C is a boolean combination of conditions; and A is an action that is executed when the transition is active, E occurs, and C is true. A *superstate* is a state that can be used to aggregate sets of states with the same transitions. A state can be repeatedly decomposed into substates in two ways, through the *OR* or the *AND* decomposition. The *OR* decomposition reflects the hierarchical structure of a state machine and is represented by encapsulation. The *AND* decomposition reflects concurrency of independent state machines and is represented by splitting a box with lines.

In our approach, a statechart is treated as a graph. Nodes [1] in the graph correspond to states, and arcs correspond to transitions between states. A node includes the following information: its name; its width and height; the coordinates of its origin point; a pointer to its parent; the list of its children; its decomposition type (e.g., *AND*, *OR* or *leaf*); the list of incoming arcs; the list of outgoing arcs; a list of attributes; and finally its aliases.

The underlying structure of a statechart is an *AND/OR* tree where the leaves are called *basic* states. We call this structure a *decomposition tree*. The root of a decomposition tree corresponds to the system state; leaves correspond to atomic states. Each object in the tree can be decomposed through the *AND* or *OR* decomposition. In the remainder of the paper, we assume that relevant information is extracted from a textual description of requirements, and stored in a decomposition tree.

---

[1] In the remainder of this paper we will use the words *node* and *object* interchangeably.

3   Automatic Layout of Statecharts

In this section, we describe our statechart drawing algorithm. Our algorithm proceeds as follows: first, the decomposition tree is traversed in order to determine the dimensions (and origin point) of every node in a recursive manner. If a node v is a leaf then a drawing procedure is called. This procedure produces a labeled rectangle and returns the dimensions of the rectangle. If v is an *AND* node then a recursive algorithm constructs the drawings of each child of v and places the drawings next to each other. If v is an *OR* node then a recursive algorithm constructs the drawings of v's children, then assigns each child to a specific layer. For the sake of simplicity, we generate our drawings horizontally, from left to right. A similar approach can be used to generate vertical drawings.

An *AND* node reflects concurrency of independent state machines. The children of an *AND* node are drawn as adjacent rectangles. The height of an *AND* node is equal to the maximum height of its children's rectangles; its width is equal to the sum of the widths of its children's rectangles. This algorithm is very simple and thus not very efficient in terms of area. As the size of each node depends on the recursive drawings of the substate nodes that are nested in it, and these drawings depend also on the size of the edge labels, it becomes clear that drawing an *AND* node should be done more carefully. More area-efficient drawings can be obtained by applying techniques similar to floorplanning as used in VLSI layout [15,24,27]. We will revisit this topic in Section 5.

An *OR* node reflects the decomposition of states into substates. The substates of an *OR* node are drawn as rectangles. The drawing (and hence the dimensions of the enclosing rectangle) of an OR node is obtained by recursively performing a hierarchical drawing algorithm [2] on the node and each of its substates. The algorithm that constructs the drawing of an OR node has the following characteristics: (i) substates are drawn recursively; (ii) substates are assigned to layers by using a modified version of Sugiyama's algorithm [25] (procedure *realDimensionHierarchyDrawing*).

Procedure *realDimensionHierarchyDrawing* (see Figure 2) consists of two steps:

1. We construct a hierarchy of substates by treating each substate as a point by calling procedure *hierarchyDrawing*, which proceeds as follows:

   a) We assign the node that corresponds to the initial state to the first layer.

   b) We apply a *depth-first* search to identify those edges that form graph-cycles; then we temporarily remove them.

   c) Once the cycles are removed, we assign every node v to a specific layer which is determined by the length of a longest path from the start node to v. At this stage, every node is assigned an x coordinate.

   d) We add dummy vertices to deal with edges whose initial and final states are not in adjacent layers.

   e) Finally, we apply a node ordering procedure whose purpose is to minimize edge crossings within each layer. This ordering provides the y coordinate for each node.

```
realDimensionHierarchyDrawing(ObjectList o.children)
Begin
    hierarchyDrawing(o.children);
    hierarchy.height = 0;
    hierarchy.width = 0;
    for i = 1 to depth(hierarchyDrawing of o.children) do
    begin_do

    1. layer[i].largestWidth = largest width among the objects in layer[i];
    2. if (layer[i+1] ≤ depth(hierarchyDrawing of o.children)) then add
       layer[i].largestWidth as an offset to the origin_x of every object in
       layer[i+1];
    3. layer[i].height = summation of each object's height at layer[i];
    4. if (hierarchy.height < layer[i].height) then hierarchy.height =
       layer[i].height;
    5. hierarchy.width = hierarchy.width + layer[i].largestWidth;
    6. Increase the origin_y of each object in layer[i] in order to deal with
       the height of each object and avoid overlapping;

    end_do;
End
```

Fig. 2. Procedure that generates the final hierarchy of an OR node.

2. We incorporate into the hierarchy the dimensions (i.e., height and width) of each node in the drawing, as described in Figure 2. The resulting hierarchy is used to determine the height and width of the parent object/state, as well as the coordinates of the origin of the object's rectangle.

Most of the steps of the algorithm have linear time-complexity with respect to the number of edges of the graph. The last step of procedure *hierarchyDrawing* attempts to beautify the obtained drawing by reducing the number of edge crossings. Our approach is based on the general *layer by layer sweep* paradigm [2]. The time-complexity of this step of the algorithm depends on the number of vertices that exist on each layer. If layer L contains $|L|$ nodes, then the time required the algorithm is $O(|L|^2)$. Clearly, the total time for this step depends upon the distribution of nodes into layers. Any step of the above framework can be replaced by any algorithm that achieves results that are acceptable for the next step. Due to space limitations, we cannot provide more details in this paper. For more details please see [4].

4   Labeling

In the labeling literature, it is common to distinguish between *node label placement* (NLP) and *edge label placement* (ELP). In the Statecharts [9] notation, NLP depends primarily on the node type. Hence, the label placement for nodes in statecharts is rather simple: if a node is a *leaf*, then the label size will deter-

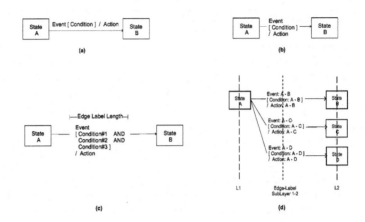

Fig. 3. Edge label placement in statecharts: (a) label on a single line, (b) one label component per line, (c) label with fixed length, (d) edge label placement.

mine the node size. If a node is an AND or an OR, then the label is placed in the top left corner of the enclosing rectangle.

Now we discuss our solution to the ELP problem for statecharts. In cartography, the placement of an edge label must satisfy the following criteria [12, 28,13]:

1. A label cannot overlap with any other graphical component except with its associated edge.
2. The placement of a label has to ensure that it is identified with just one edge in the drawing. Therefore it must be very close to its associated edge.
3. Each label must be placed in the best possible position among all acceptable positions.

In the statecharts notation, an edge label consists of three components: *event*, *condition* and *action* (see Figure 3(a)). In order to satisfy the labeling criteria discussed above we have defined the following steps:

1. We fix the maximum length of the label to a constant, and we write the transition's three components (i.e., *events*, *conditions* and *actions*) on three separate lines (see Figure 3(b)). If the size of a component is greater than the maximum length of the label, then we write it on several lines (see Figure 3(c)).
2. At the beginning of the execution of the drawing algorithm (see Section 3), we assign labels to sublayers (see Figure 3(d)).
3. We traverse the hierarchy from left to right, considering two adjacent layers L_1 and L_2 at a time (see Figure 3(d)). For each vertex a in L_1, we identify the set of edges E_a between a and the vertices in L_2. We order E_a in such a way that potential label crossings are removed.

The time complexity of this step is linear with respect to the number of edges in the graph.

5 Floorplanning Heuristics

Because of the representation of statecharts, it is possible that certain *AND* nodes of the decomposition tree are very large in one dimension or the other. Recall that our algorithm places all the subnodes (vertically) next to each other. The height of the resulting drawing of an *AND* node is equal to the maximum of the heights of the subnodes and the width is equal to the sum of the widths of the subnodes. This implies that a bad combination of two subnode rectangles (one with large height and one with large width) will result in a drawing of the *AND* node that occupies a very large area. This is clearly undesirable. Additionally, the aspect ratio of the drawing, another important aesthetic criterion, is not controllable. We tackle this problem by applying a technique similar to the one used for the minimization of VLSI chip areas [15,24,27], namely *floorplanning*. Floorplanning partitions a floor rectangle into *floorplans* using line segments called *slices*. Floorplans are combined in such a way that the enclosing rectangle covers a minimum area. A floorplan is *slicing* whenever the floorplan is an atomic rectangle or there exists slice that divides the rectangle into two. The floorplanning problem has an efficient solution when the floorplan is slicing [24, 16].

Although one could apply the slicing floorplanning technique for drawing the *AND* nodes [24], due to the special representation of statecharts, we have simplified this technique. We apply the slicing floorplanning concept to derive a set of heuristics that can be applied to statecharts. To this effect, we define the following drawing criteria for statecharts:

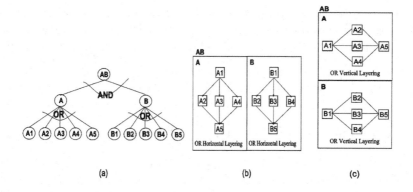

(a)   (b)   (c)

Fig. 4. AND-OR combination: (a) AND/OR decomposition tree, (b) AND vertical slicing with OR horizontal layering, (c) AND horizontal slicing with OR vertical layering.

– Leaves are used to represent atomic states whose size depends solely on their labels. Since labels are usually written horizontally (for readability purposes), we will draw leaves horizontally.

- The AND decomposition reflects concurrency, and is represented by splitting an AND-state box into a number of concurrent substates. Since one of the most important aesthetic criteria in graph drawing is *symmetry* [20], we choose to slice AND-state boxes either horizontally or vertically.
- OR states can be drawn in a hierarchical fashion using either a horizontal or a vertical layering depending on the slicing type of the parent node (i.e., horizontal / vertical slicing).

Our goal is to generate drawings that use the horizontal and vertical dimensions in a uniform way, in order to optimize the drawing area. To this effect we define several heuristics: The AND/OR heuristic applies to the case where the parent is an AND node and the children are OR nodes (see Figure 4(a)). There are two cases:

1. The parent node (AND) is sliced vertically. Then the children nodes (OR) are drawn on horizontal layers (see Figure 4(b)). In this case, the height of the parent object is the height of the highest child node; and the total width of the parent is the sum of the children's widths.
2. The parent node (AND) is sliced horizontally. Then the children nodes (OR) are drawn on vertical layers (see Figure 4(c)). In this case the height of the parent node is the sum of children's heights; and the width of the parent node is the width of the widest child.

Heuristics that handle the other cases (OR/AND, AND/AND, and OR/OR) are defined similarly, and are omitted due to space limitation. For more details please see [4].

Figure 5 shows the statecharts diagram after we applied our improvement drawing techniques (i.e., edge-crossing and edge-bend reduction, edge labeling, and floorplaning) to the diagram of Figure 1. We observe that both, the horizontal and vertical dimensions, grow in a uniform manner; edges do not overlap with any other drawing component; every edge crossing has been removed; and the number of edge bends has been reduced considerably.

6   Conclusions and Experimental Results

In this paper we presented an algorithmic framework for the automatic generation of layouts of statechart diagrams. Our framework is based on hierarchical drawing, labeling, and floorplanning techniques. Clearly any algorithm used for any step can be replaced with an improved algorithm thus resulting in an improved tool. We implemented a tool using the algorithms described in this paper, and ran the tool on four statechart examples. We generated drawings using first, the basic version (without the optimized algorithms), then the optimized version. Due to space limitations, the drawings produced by our tool are available at http://www.utdallas.edu/~rmili/GD2000/. Our results are described in Table 1. We notice that, after the application of the optimized algorithms, (1)

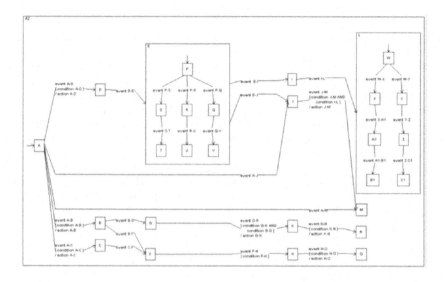

Fig. 5. Same statechart as in Figure 1, generated by our drawing algorithm with optimization techniques.

edge-crossings are completely eliminated; (2) the number of edge-bends is considerably reduced; (3) the drawings enjoy a good aspect ratio. This optimization improves considerably the readability of the diagrams. Therefore, it constitutes an invaluable tool to the specifier who will shift his/her focus from organizing the mental or physical structure of the requirements to its analysis.

Table 1. Comparison of four examples of statecharts drawn by our algorithms

Aesthetic Criteria	Drawing 1		Drawing 2		Drawing 3		Drawing 4	
	Without Improve	With Improve	Without Improve	With Improve	Without Improve	With Improve	Without Improve	With Improve
Edges Crossings	33	0	23	0	22	0	34	0
Edge Bends	53	19	39	18	24	18	70	32
Width	1,953	875	2,733	1,029	1,794	861	2,820	1,632
Height	548	1,259	735	1,722	736	1,593	651	1,424
W/H Ratio	3.5227	0.695	3.718	0.5975	2.4375	0.54	4.33	1.44
Area		1,059,284	1,102,884	2,008,755	1,771,938	1,320,384	1,371,573	1,835,820	2,323,968

References

1. G. Di Battista, P. Eades, R. Tamassia, and I. G. Tollis. Algorithms for drawing graphs: an annotated bibliography. *Comput. Geom. Theory Appl.*, (4):235–282, 1994.
2. G. Di Battista, P. Eades, R. Tamassia, and I. G. Tollis. *Graph Drawing: Algorithms for the Visualization of Graphs*. Prentice Hall, 1999.

3. G. Booch, I. Jacobson, and J. Rumbaugh. *The Unified Modeling Language User Guide*. Addison-Wesley, 1998.
4. R. Castelló, R. Mili, and I. G. Tollis. Automatic layout of statecharts. Technical Report UTD-04-00, University of Texas at Dallas, 2000.
5. P. Eades and Q. Feng. Drawing clustered graphs on an orthogonal grid. In G. Di Battista, editor, *Graph Drawing (Proceedings GD'97)*, pages 146–157. Springer-Verlag, 1997. Lecture Notes in Computer Science 1353. .
6. P. Eades, Q. Feng, and X. Lin. Straight-line drawing algorithms for hierarchical graphs and clustered graphs. In S. North, editor, *Graph Drawing (Proceedings GD'96)*, pages 113–128. Springer-Verlag, 1997. Lecture Notes in Computer Science 1190.
7. E. R. G., E. Koutsofios, S. C. North, and K. Vo. A technique for drawing directed graphs. *IEEE Transactions on Software Engineering*, 19(32):214–230, March 1993.
8. E. R. Gansner, S. C. North, and K. P. Vo. Dag—a program that draws directed graphs. *Software Practice and Experience*, 18(11):1047–1062, November 1988.
9. D. Harel. Statecharts: A visual formalism for complex systems. *Science of Computer Programming*, 8(3):231–274, June 1987.
10. D. Harel, H. Lachover, A. Naamad, A. Pnueli, M. Politi, R. Sherman, A. Shtull-Trauring, and M. Trakhtenbrot. Statemate: A working environment for the development of complex reactive systems. *IEEE Transactions on Software Engineering*, 16(4):403–414, May 1990.
11. D. Harel and G. Yashchin. An algorithm for blob hierarchy layout. In *Proceedings of International Conference on Advanced Visual Interfaces*, AVI'2000, Palermo, Italy, May 1990.
12. E. Imhof. Positioning names on maps. *The American Cartographer*, 2(2):128–144, 1975.
13. K. G. Kakoulis and I. G. Tollis. An algorithm for labeling edges of hierarchical drawings. In G. Di Battista, editor, *Graph Drawing (Proceedings GD'97)*, pages 169–180. Springer-Verlag, 1997. Lecture Notes in Computer Science 1353.
14. K. G. Kakoulis and I. G. Tollis. On the edge label placement problem. In S. North, editor, *Graph Drawing (Proceedings GD'96)*, pages 241–256. Springer-Verlag, 1997. Lecture Notes in Computer Science 1190.
15. E. S. Kuh and T. Ohtsuki. Recent advances in VLSI layout. *Proceedings of the IEEE*, 78(2):237–263, 1990.
16. T. Lengauer. *Combinatorial Algorithms for Integrated Circuit Layout*. John Wiley & Sons, 1990.
17. E. B. Messinger, L. A. Rowe, and R. R. Henry. A divide-an-conquer algorithm for the automatic layout of large graphs. *IEEE Transactions on Systems, Man, and Cybernetics*, 21(1):1–11, February 1991.
18. R. O'Donnel, B. Waldt, and J. Bergstrand. Automatic code for embedded systems based on formal methods. Available from Telelogic over the Internet. http://www.Telelogic.se/solution/techpap.asp. Accessed on April 1999.
19. J. Peterson. Overcoming the crisis in real-time software development. Available from Objectime over the Internet. http://www.Objectime.on.ca/otl/technical/crisis.pdf. Accessed on April 1999.
20. H. Purchase. Which aesthetic has the greatest effect on human understanding. In G. Di Battista, editor, *Graph Drawing (Proceedings GD'97)*, pages 248–261. Springer-Verlag, 1997. Lecture Notes in Computer Science 1353.
21. Rational. Rose java. Downloaded from Rational over the Internet. http://www.rational.com. Accessed on November 1999.

22. L. A. Rowe, M. Davis, E. Messinger, and C. Meyer. A bowser for directed graphs. *Software Practice and Experience*, 17(1):61–76, January 1987.

23. J. Seeman. Extending the sugiyama algorithm for drawing UML class diagrams: Towards automatic layout of object-oriented software diagrams. In G. Di Battista, editor, *Graph Drawing (Proceedings GD'97)*, pages 415–424. Springer-Verlag, 1997. Lecture Notes in Computer Science 1353.

24. L. Stockmeyer. Optimal orientations of cells in slicing floorplan designs. *Information and Control*, (57):91–101, 1983.

25. K. Sugiyama, S. Tagawa, and M. Toda. Methods for visual understanding of hierarchical system structures. *IEEE Transactions on Systems, Man, and Cybernetics*, 11(2):109–125, February 1981.

26. Artisan Software Tools. Real-time studio: The rational alternative. Available from Artisan Software Tools over the Internet.
http://www.artisansw.com/rtdialogue/pdfs/rational.pdf. Accessed on April 1999.

27. S. Wimer, I. Koren, and I. Cederbaum. Floorplans, planar graphs and layout. *IEEE Transactions on Circuits and Systems*, pages 267–278, 1988.

28. P. Yoeli. The logic of automated map lettering. *The Cartographic Journal*, 9(2):99–108, 1972.

Visualization of the Autonomous Systems Interconnections with HERMES*

Andrea Carmignani, Giuseppe Di Battista, Walter Didimo,
Francesco Matera, and Maurizio Pizzonia

Dipartimento di Informatica e Automazione,
Università di Roma Tre, via della Vasca Navale 79,
00146 Roma, Italy.
{carmigna,gdb,didimo,matera,pizzonia}@dia.uniroma3.it

Abstract. HERMES is a system for exploring and visualizing Autonomous Systems and their interconnections. It relies on a three-tiers architecture, on a large repository of routing information coming from heterogeneous sources, and on sophisticated graph drawing engine. Such an engine exploits static and dynamic graph drawing techniques.

1 Introduction and Overview

Computer networks are an endless source of problems and motivations for the Graph Drawing and for the Information Visualization communities. Several systems aim at giving a graphical representation of computer networks at different abstraction levels and for different types of users. To give only some examples (an interesting survey can be found in [17]):

1. Application level: Visualization of Web sites structures, Web maps [18,16], and Web caches [10].
2. Network level: Visualization of multicast backbones [22], internet traffic [23], routes, and interconnection of routers.
3. Data Link level: interconnection of switches and repeaters in a local area network [1].

We deal with the problem of exploring and visualizing the interconnections between *Autonomous Systems*. An Autonomous System (in the following AS) is a group of networks under a single administrative authority. Roughly speaking, an AS can be seen as a portion of Internet, and Internet can be seen as the totality of the ASes and their interconnections. Each AS is identified by an integer number.

A *route* is a path on the network that can be used to reach a specific set of (usually contiguous) IP addresses. A route is described by its IP addresses, its cost, and by the set of ASes that are traversed. Routes are "announced" from an

---

* Research supported in part by the Murst Project: "Algorithms for Large Data Sets".
HERMES is presented at http://www.dia.uniroma3.it/~hermes.

J. Marks (Ed.): GD 2000, LNCS 1984, pp. 150–163, 2001.

AS with messages like "through me you can reach a certain set of IP addresses, with a certain cost and traversing a certain set of other ASes".

In order to exchange route's information, the ASes adopt a network protocol called *BGP* (Border Gateway Protocol) [31]. Such a protocol is based on a distributed architecture where border routers that belong to distinct adjacent ASes exchange information about the routes they know. The exchange of routing information between ASes is subject to *routing policies*; for example an AS can say "I do not want to be traversed by packets going to a certain other AS".

Several tools have been developed for analyzing and visualizing the Internet topology at the ASes level [9,12,23,30]. However, in our opinion, such tools are still not completely satisfactory both in the interaction with the user and in the effectiveness of the drawings. Some of them have the goal of showing large portions of Internet, but the maps they produce can be difficult to read (see e.g. [9]). Other tools point the attention on a specific AS, only showing that AS and its direct connections (see e.g. [23]).

In this work we describe a new system, called HERMES, which allows to get several types of information on a specific AS and to explore and visualize the ASes interconnections. The main features of HERMES are the following.

HERMES has a three tiers architecture. The user interacts with a top-tier client which collects the user requests and forwards them to a middle-tier server. The server translates the requests into queries to a repository (the bottom tier). With the top-tier the user can explore and visualize the ASes interconnections, and several information about ASes and the BGP routing policies. The user interacts with a subgraph (called *map*) of the graph of all ASes interconnections. Each exploration step enriches the map with new ASes and connections (vertices and edges).

HERMES handles a large repository (about 50 MB). The repository is updated off-line from a plurality of sources [24]: APNIC, ARIN, BELL, CABLE&WIRE, CANET, MCI, RADB, RIPE, VERIO. The data in the repository are used from HERMES to construct the ASes interconnection graph.

The middle-tier server of HERMES encapsulates a graph drawing module that computes the drawing of the ASes interconnections already explored at a specific time. Such a module is based on the GDToolkit library [21] and has the following main features:

- Its basic drawing convention is the podevsnef [20] model for orthogonal drawings having vertices of degree greater than four. However, since the handled graphs have often many vertices (ASes) of degree one connected with the same vertex, the podevsnef model is enriched with new features for representing such vertices.
- It is equipped with two different graph drawing algorithms. In fact, at each exploration step the map is enriched and hence it has to be redrawn. Depending on the situation, the system (or the user) might want to use a static or a dynamic algorithm. Of course, the dynamic and the static algorithms have advantages and drawbacks. The dynamic algorithm allows to preserve the mental map of the user [19,27] but can lead, after a certain number of

exploration steps, to drawings that are less readable than those constructed with a static algorithm.

- The static algorithm is based on the topology-shape-metrics approach [14] and exploits recent compaction techniques that can draw vertices with any prescribed size [13]. The topology-shape-metrics approach has been shown to be very effective and reasonably efficient in practical applications [15].
- The dynamic algorithm is a new dynamic graph drawing algorithm that has been devised according to three main constraints. (1) It had to be completely integrated within the topology-shape-metrics approach in such a way to be possible to alternate its usage with the usage of the static algorithm. (2) It had to be consistent with the variation of the podevsnef model used by HERMES. (3) It had to allow vertices of arbitrary size. Several algorithms have been recently proposed in the literature on dynamic graph drawing algorithms. In [4] a linear time algorithm for orthogonal drawings is presented, where the position of the vertices cannot be changed after the initial placement. In [29] four different scenarios for interactive orthogonal graph drawings are studied. Each scenario defines the changes allowed in the common part of two consecutive drawings. In [7] it is described an interactive version of GIOTTO [33]; it allows to incrementally add vertices and edges to an orthogonal drawing so that the shape of the common part of two consecutive drawings is preserved and the number of bends is minimized under this constraint. In [5] it is presented a dynamic algorithm for orthogonal drawings that allows to specify the relative importance of the number of bends vs. the number of changes between two consecutive drawings. Other algorithms for constructing drawings of graphs incrementally, while preserving the mental map of the user, are for example [26,11,28]. Also, in [6] it is presented a study on different metrics that can be used to evaluate the changes between drawings in an interactive scenario. However, as far as we know, none of the cited dynamic algorithms enforces all the constraints (1)–(3).

The paper is organized as follows. In Section 2 we explain how the user interacts with HERMES and provide a high level description of the functionalities of the system. In Section 3 we give some details about the three tiered architecture of HERMES. Section 4 shows the results of a study on the ASes interconnection graph we have performed before choosing the graph drawing algorithms to apply in HERMES. In particular we give measures on the *local density* and on the average degree of the vertices. In Section 5 we describe the drawing convention and the algorithms.

2   Using HERMES

The user interacts with HERMES through an *ASes subgraph*. An ASes subgraph (also called *map*) is a subgraph of the graph of all ASes and their interconnections. A map is initially constructed using two possible starting primitives. **AS selection**: An AS is chosen. The obtained map consists of such an AS plus all the ASes that are connected to it. See Fig. 1(a). **Routes selection**: A set of routes is selected. The user has two possibilities: (1) selection of all the routes traversing a specific AS (see Fig. 2); (2) selection of all the routes traversing a specific pair of ASes. The obtained map consists of all the ASes and connections traversed by the selected routes. The user has also the option of cutting the map pruning all the ASes that are after (before) the selected AS in the routes propagation (see Figs. 2(a) and 2(b)).

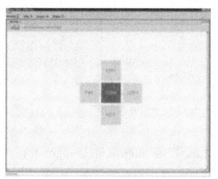

(a) Selection of AS 12300.

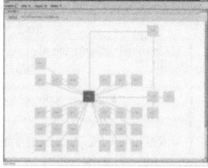

(b) Exploration of AS 5583.

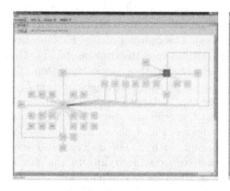

(c) Exploration of AS 5484.

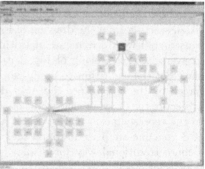

(d) Exploration of AS 6715.

Fig. 1. Exploration steps in the ASes graph. The selected AS is always drawn red.

The user can explore and enrich the map by using the following primitive. **AS exploration**: an AS u among those displayed in the current map is selected. The current map is augmented with all the ASes that are connected to u. Further, for each AS v connected to u an edge (u, v) is added. Observe that, according

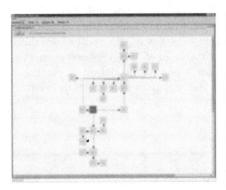 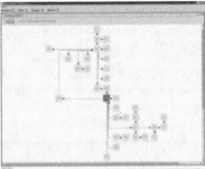

(a) Map containing only the ASes af-   (b) Map containing all the traversed
ter AS 137.   ASes.

Fig. 2. Selection of the routes traversing AS 137.

to this definition, a map is a subgraph of the ASes interconnection graph but it
is not an induced subgraph.

Fig. 1 shows a sequence of exploration primitives applied to the map of
Fig. 1(a). ASes 5583, 5484, and 6715 are explored in Fig's. 1(b), 1(c), and 1(d),
respectively. Fig. 1 highlights several features of HERMES. HERMES can construct
new drawings either using a static or a dynamic graph drawing algorithm. Observe how the drawing of Fig. 1(b) has been constructed with a dynamic algorithm starting from the drawing of Fig. 1(a). The drawing of Fig. 1(d) has been
constructed with a dynamic algorithm starting from the drawing of Fig. 1(c).
Conversely, the drawing of Fig. 1(c) is obtained with a static algorithm. The
choice of the algorithm to apply can be done by the system (see Section 5) or
forced by the user. Since the ASes degree can be large (see Section 4), in the
project of the drawing algorithms of HERMES, special attention has been devoted
to the representation of vertices of high degree. Fig. 1 shows how the vertices
of degree one are placed around their adjacent vertices. The user can also select
a more traditional way for displaying such vertices, according to the simple-
podevsnef model described in [13] (see Fig. 2). A more complex map obtained
with HERMES is depicted in Fig. 3. It contains more than 150 ASes.

Working on a map, independently on the way it has been obtained, the user
can get several information on any AS:

General Info : Name, Maintainers, Connections, and Description. See Fig. 4.
Routing Policies : For each connected AS, an expression describing the policy
and its cost. See Fig. 4. This is possible both for in and for out policies. The
default AS is also displayed.
Internal Routers : List of the known border routers with the IP-numbers of
the interfaces. Peering sessions with other routers are displayed.

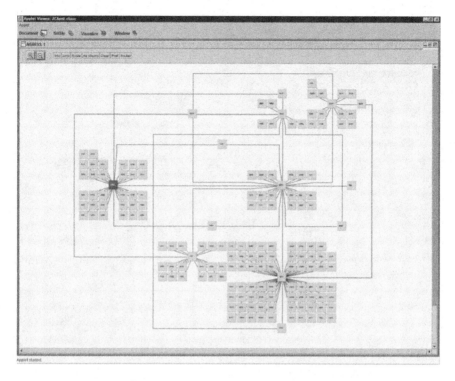

Fig. 3. A map obtained with several exploration steps.

Routes : List of the routes originated by the AS. It is also possible to visualize the propagation of a given route in the ASes composing the map.

AS Macros : List of the macros [2] including the AS.

3   A Three Tiers Architecture

The architecture of HERMES is three tiered. The user interacts with a top-tier client which is in charge of collecting user requests and showing results. The requests are forwarded by the client to a middle-tier server which is in charge to process the raw data extracted from a repository (bottom tier).

The client is a multi-document GUI-based application. It allows the user to carry-on multiple explorations of the ASes graph at the same time. The Java technology has been used to ensure good portability. Snapshots of the GUI have been shown in Section 2.

The repository is updated off-line from a plurality of sources. At the moment we access the following databases adopting, for representing data, the RIPE-181 language [2,8]: ANS, APNIC, ARIN, BELL, CABLE&WIRE, CANET, MCI,

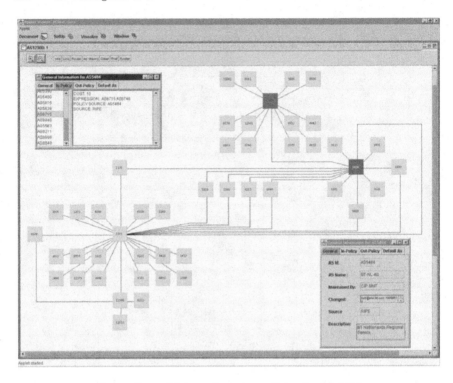

Fig. 4. General Info and routing policies for AS 5484.

RADB, RIPE, VERIO. Further, we access the routing BGP data provided by the Route views project of the Oregon University [25]. However, the repository is easily extensible to other data sources.

Data are filtered so that only the information used by HERMES are stored in the database, but no consistency check or ambiguity removal is performed in this stage. The overall size of the repository is about 50 MB. The adopted DBMS technology is currently mysql.

The crucial part of the system is mainly located in the middle-tier. The top tier requests two types of service to the middle tier. **General info services**: the top tier queries about ASes, routes, and path properties. **Topology services**: the top tier queries for a new exploration and gets back a new map.

Info services requests are independent each other and hence are independently handled by the middle-tier. On the contrary, topology services requests are always part of a *drawing session*. Each client may open one or more drawing sessions. Each drawing session is associated with a map that can be enriched by means of exploration requests.

Info services requests are directly dispatched to a *mediator*. The mediator module is in charge to retrieve the data from the repository and to remove ambiguities on-the-fly.

Topology services requests are handled by the kernel of the system. It gets information from the mediator and inserts new edges and vertices into the map. The drawing is computed by the *drawing engine* module (see Section 5). The drawing engine encapsulates GDToolkit [21].

4   AS Interconnection Data from a Graph Drawing Perspective

In order to devise effective graph drawing facilities for HERMES, we have analyzed the ASes and their interconnections considering them as a unique large graph G. The data at our disposal show the following structure for G.

The number of vertices of G is 6,849, while the number of edges is 27,686. Fig. 5(a) illustrates the distribution of the degree of the vertices. The figure shows that while there are many vertices (about 75%) with degree less or equal than 4, there are also several vertices whose degree is more than 100. For improving the readability of the chart, we have omitted two vertices with degree 862 and 1,044, respectively. Further, consider that G contains 473 isolated vertices.

The density of G is 4.04. However, the "local" density can be much greater. In order to estimate such a local density, we have computed, for each vertex v, the density of the subgraph induced by the vertices adjacent to v. We call such graphs *local graphs*. Fig. 5(b) illustrates the distribution of the densities of the local graphs. From the figure it is possible to observe that about 5% of the local graphs have density greater than 10.

We have also tried to estimate the probability, for a user that explores G, to encounter a portion of G that is locally dense. Fig. 5(c) shows, for each value d of density, what is the percentage of vertices that are adjacent to a vertex whose local graph has density at least d. Note that more than 30% of the vertices are adjacent to a vertex whose local graph has density at least 10.

Concerning connectivity, the graph has 480 connected components, including the above mentioned 473 isolated vertices. One of them has 6,360 vertices; each of the remaining 6 components has less than 6 vertices.

5   Drawing Conventions and Algorithmic Issues

At each exploration step, HERMES computes a new drawing. Namely, when the user selects in the map a new vertex v, all the vertices and edges connected to v are added to the map, and such a map is redrawn. We use, depending on the specific situation, two different drawing algorithms.

Static algorithm The current map is completely redrawn, after the new vertices and edges have been added.

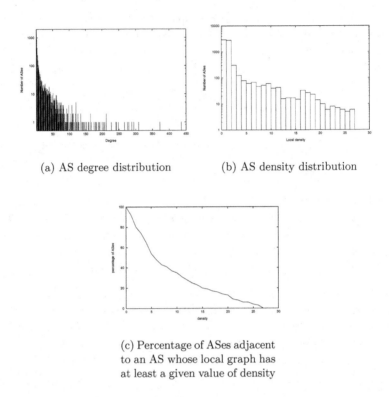

(a) AS degree distribution   (b) AS density distribution

(c) Percentage of ASes adjacent
to an AS whose local graph has
at least a given value of density

Fig. 5. Several aspects of the density of the ASes graph.

Dynamic algorithm The new vertices and edges are added to the current
drawing in such a way that the shape of the existing edges and the position
of the existing vertices and bends are preserved "as much as possible".

Of course, the dynamic and the static algorithms have advantages and drawbacks. The dynamic algorithm allows to preserve the mental map of the user but
can lead, after a certain number of exploration steps, to drawings that are less
readable than those constructed with the static algorithm. In fact, the dynamic
algorithm makes use of local optimization strategies. The optimization strategies
of the static algorithm are global and more effective. However, with the static
algorithm, the new drawing can be quite different from the previous one and the
user's mental map can be lost.

Because of the above motivations, we automatically choose between the static
algorithm and the dynamic algorithm, depending on the number and the kind of
egdes and vertices that are added when a new vertex is explored. However, the
user can always force the system to apply one of the two algorithms. Namely,
suppose v is the vertex the user wants to explore. The choice of the drawing

algorithm to apply is done by calculating an *exploration cost* for v and comparing such a cost with a treshold that can be set-up in a configuration menu. The exploration cost is computed as follows. For any new edge the cost is 1. For any vertex that had degree 1 in the old graph and whose degree is increased in the new graph the cost is 0.5. An extra cost is also added as a function of the number of times the dynamic algorithm has been invoked before. The exploration cost is a very rough estimate of the efficiency and effectiveness of the dynamic algorithm with respect to the new exploration. The reason why vertices with degree 1 are treated in a special way will be clear in the description of the algorithms.

The static algorithm we use consists of the following steps.

Degree-one Vertex Removal Vertices of degree one are temporarily removed.

Planarization A standard planarization [14] technique is applied.

Orthogonalization and Compaction We apply a variation of the technique presented in [13] for constructing orthogonal drawings (in the simplepodavsnef model) with vertices of prescribed size. The box representing a vertex v is a rectangle. Edges incident on v can incide the box only in the middle points of the sides. The length of the sides are chosen in such a way to have enough space to accommodate all the vertices that have been temporarily removed in the first step and that were adjacent to v.

Degree-one Vertex Re-insertion Each box representing a vertex v is partitioned into nine rectangles arranged into three rows and three columns. Denote them as $B_{i,j}$. Rectangle $B_{2,2}$ is used for drawing v. Rectangles $B_{1,1}$, $B_{1,3}$, $B_{3,1}$, and $B_{3,3}$ are used for drawing the degree-one vertices adjacent to v. Their incident edges are represented with straight-line segments, possibly overlapping other degree-one vertices. Actually, they are drawn on the back of the vertices. Rectangles $B_{1,2}$, $B_{2,1}$, $B_{3,2}$, and $B_{2,3}$ are used for hosting the connections of v to the other vertices. See Fig. 6.

The dynamic drawing algorithm allows to apply three primitives on the current map. **New-Edge(u,v)**: a new edge is added to the map between the two vertices u and v; vertices u and v must be already in the current map. **Attach-Vertex(u)**: a new vertex v is added to the map and connected to u with a new edge (u, v); vertex u must be already in the current map. **Insert-Vertex(u,v)**: a new vertex is added to the map by splitting edge (u, v); edge (u, v) must be already in the current map.

The algorithm computes the position of the new vertices and edges trying to optimize several aesthetic measures (number of crossings, number of bends, and edge length) at the same time, depending on the costs the user has chosen for each of them. The obtained drawing is guaranteed to have the same shape of the starting one, for the common parts.

In the following we give some details about the used data structure and how the three drawing primitives work. Vertices of degree one are temporarily absorbed into their incident vertices, with the techniques described above.

We maintain a copy of the map with the following main extra features:

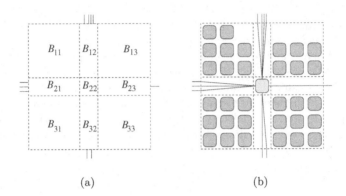

(a)   (b)

Fig. 6. Using a box to represent a vertex. (a) The nine regions. (b) Using the nine regions to reinsert degree-one vertices.

- All the crossings and bends of the map are replaced by dummy vertices, so that the topology of the drawing becomes planar.
- The map is simplified so that all vertices have degree less or equal than four. This is done with a standard technique adopted in the podevsnef model [3], where all the edges incident on the same vertex from the same side are collapsed into one single edge. See Fig. 7(a). A *thickness* is associated with each of the new edges, representing the number of edges it replaces. The transformation is recursively applied to all the vertices.
- New edges are added to the map for decomposing each face (including the external one) into rectangles, with the linear time algorithm described in [32]. We call *dashed* the new edges and *solid* the edges of the original map.
- An *incidence network* D is constructed. Such a network is defined as follows (see Fig. 7(b)): (1) The nodes of D are the (solid and dashed) edges of the map. (2) Nodes corresponding to solid edges have associated a cost equal to their thickness multiplied by a constant χ (see Fig. 7(a)). Nodes corresponding to dashed edges have cost equal to zero. Intuitively, the cost associated with an edge represents the cost of a crossing involving that edge. (3) An arc is added to D for each pair of edges of the map sharing a face. Fig. 7(b) shows three arcs of D. (4) An arc between two horizontal (vertical) edges that lie on different sides of the same face has a cost that is equal to their vertical (horizontal) distance multiplied by a constant λ. See Fig. 7(b). Intuitively, the cost associated with an arc of this type represents a lower bound on the length of a possible new edge that follows the arc. (5) An arc between a horizontal and a vertical edge has a cost that is equal to the orthogonal distance between the centers of the edges multiplied by a constant λ plus β. See Fig. 7(b). Intuitively, the cost associated with an arc of this type represents a lower bound on the length of a possible new edge that follows the arc plus a cost for a bend. (6) An arc between two horizontal

(vertical) edges that lie on the same side of the same face has a cost that is equal to the distance between the centers of the edges plus 2 (multiplied by a constant λ) plus 2β. See Fig. 7(b). Intuitively, the cost associated with an arc of this type represents a lower bound on the length of a possible new edge that follows the arc plus a cost for two bends. (7) Intuitively, χ, β, and λ, represent the costs for one cross, one bend, and one unit of length, respectively. Their values can be set-up by the user. The ratios between χ, β, and λ may determine different behaviors of the algorithm.

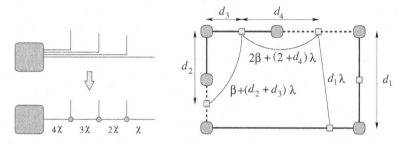

(a) Collapsing the edges incident on the same side of a vertex.

(b) Nodes (little squares) and arcs of the incidence network.

Fig. 7. Illustration of the dynamic algorithm.

Primitives New-Edge, Attach-Vertex, and Insert-Vertex are implemented as follows.

New-Edge(u,v) : In the case u and v are not on the same face two temporary nodes are added to D representing u and v. Also, temporary arcs are added to D between u and the nodes representing its incident edges. The same is done for v. The temporary arcs have zero cost.

A shortest path between u and v is computed. Such a path determines the route and the shape of the new edge. Namely, the new edge is inserted in the map following the arcs of the shortest path. The temporary nodes and arcs are removed an the new faces originated by the new edge are decomposed into rectangles.

In the case u and v are on the same face a simpler technique (not discussed here for brevity) is adopted.

Attach-Vertex(u) : A local evaluation of the edges incident on u is performed. The new edge is put preferably either on a direction around u where there is no incident edge or on a dashed edge.

Insert-Vertex(u,v) : Edge (u, v) is just split into two pieces. If the edge has a bend we put the new vertex preferably on that bend.

Once a primitive has been performed, the expansion technique described above is applied to make room for the vertices of degree one that were temporarily absorbed into their incident vertices.

Observe that D can have a number of arcs that is quadratic in the number of edges of the map. However, it is possible to see that D can be simplified to an equivalent net with a linear number of arcs.

Acknowledgements. We are grateful to Sandra Follaro and Antonio Leonforte for their fundamental contribution in the implementation of the dynamic algorithm. We are also grateful to Andrea Cecchetti for useful discussion on the repository.

References

1. Aprisma. Spectrum. On line. http://www.aprisma.com.
2. T. Bates, E. Gerich, L. Joncheray, J. M. Jouanigot, D. Karrenberg, M. Terpstra, and J. Yu. Representation of ip routing policies in a routing registry. On line, 1994. ripe-181, http://www.ripe.net, rfc 1786.
3. P. Bertolazzi, G. Di Battista, and W. Didimo. Computing orthogonal drawings with the minimum numbr of bends. *IEEE Transactions on Computers*, 49(8), 2000.
4. T. C. Biedl and M. Kaufmann. Area-efficient static and incremental graph darwings. In R. Burkard and G. Woeginger, editors, *Algorithms (Proc. ESA '97)*, volume 1284 of *Lecture Notes Comput. Sci.*, pages 37–52. Springer-Verlag, 1997.
5. U. Brandes and D. Wagner. Dynamic grid embedding with few bends and changes. In K.-Y. Chwa and O. H. Ibarra, editors, *ISAAC'98*, volume 1533 of *Lecture Notes Comput. Sci.*, pages 89–98. Springer-Verlag, 1998.
6. S. Bridgeman and R. Tamassia. Difference metrics for interactive orthogonal graph drawing algorithms. In S. H. Withesides, editor, *Graph Drawing (Proc. GD '98)*, volume 1547 of *Lecture Notes Comput. Sci.*, pages 57–71. Springer-Verlag, 1998.
7. S. S. Bridgeman, J. Fanto, A. Garg, R. Tamassia, and L. Vismara. Interactive-Giotto: An algorithm for interactive orthogonal graph drawing. In G. Di Battista, editor, *Graph Drawing (Proc. GD '97)*, volume 1353 of *Lecture Notes Comput. Sci.*, pages 303–308. Springer-Verlag, 1998.
8. M. Bukowy and J. Snabb. RIPE NCC database documentation update to support RIPE DB ver. 2.2.1. On line, 1999. ripe-189, http://www.ripe.net.
9. CAIDA. Otter: Tool for topology display. On line. http://www.caida.org.
10. CAIDA. Plankton: Visualizing nlanr's web cache hierarchy. On line. http://www.caida.org.
11. R. F. Cohen, G. Di Battista, R. Tamassia, and I. G. Tollis. Dynamic graph drawings: Trees, series-parallel digraphs, and planar ST-digraphs. *SIAM J. Comput.*, 24(5):970–1001, 1995.
12. Cornell University. Argus. On line. http://www.cs.cornell.edu/cnrg/topology_aware/discovery/argus.html.
13. G. Di Battista, W. Didimo, M. Patrignani, and M. Pizzonia. Orthogonal and quasi-upward drawings with vertices of prescribed sizes. In J. Kratochvil, editor, *Graph Drawing (Proc. GD '99)*, volume 1731 of *Lecture Notes Comput. Sci.*, pages 297–310. Springer-Verlag, 1999.
14. G. Di Battista, P. Eades, R. Tamassia, and I. G. Tollis. *Graph Drawing*. Prentice Hall, Upper Saddle River, NJ, 1999.

15. G. Di Battista, A. Garg, G. Liotta, R. Tamassia, E. Tassinari, and F. Vargiu. An experimental comparison of four graph drawing algorithms. *Comput. Geom. Theory Appl.*, 7:303–325, 1997.

16. G. Di Battista, R. Lillo, and F. Vernacotola. Ptolomaeus: The web cartographer. In S. H. Withesides, editor, *Graph Drawing (Proc. GD '98)*, volume 1547 of *Lecture Notes Comput. Sci.*, pages 444–445. Springer-Verlag, 1998.

17. M. Dodge. An atlas of cyberspaces. On line. http://www.cybergeography.com/atlas/atlas.html.

18. P. Eades, R. F. Cohen, and M. L. Huang. Online animated graph drawing for web navigation. In G. Di Battista, editor, *Graph Drawing (Proc. GD '97)*, volume 1353 of *Lecture Notes Comput. Sci.*, pages 330–335. Springer-Verlag, 1997.

19. P. Eades, W. Lai, K. Misue, and K. Sugiyama. Preserving the mental map of a diagram. In *Proceedings of Compugraphics 91*, pages 24–33, 1991.

20. U. Fößmeier and M. Kaufmann. Drawing high degree graphs with low bend numbers. In F. J. Brandenburg, editor, *Graph Drawing (Proc. GD '95)*, volume 1027 of *Lecture Notes Comput. Sci.*, pages 254–266. Springer-Verlag, 1996.

21. GDToolkit:. Graph drawing toolkit. On line. http://www.gdtoolkit.com.

22. B. Huffaker. Tools to visualize the internet multicast backbone. On line. http://www.caida.org.

23. IPMA. Internet performance measurement and analysis project. On line. http://www.merit.edu/ipma.

24. Merit Network, Inc. Radb database services. On line. http://www.radb.net.

25. D. Meyer. University of oregon route views project. On line. http://www.antc.uoregon.edu/route-views.

26. K. Miriyala, S. W. Hornick, and R. Tamassia. An incremental approach to aesthetic graph layout. In *Proc. Internat. Workshop on Computer-Aided Software Engineering*, 1993.

27. K. Misue, P. Eades, W. Lai, and K. Sugiyama. Layout adjustment and the mental map. *J. Visual Lang. Comput.*, 6(2):183–210, 1995.

28. S. North. Incremental layout in DynaDAG. In F. J. Brandenburg, editor, *Graph Drawing (Proc. GD '95)*, volume 1027 of *Lecture Notes Comput. Sci.*, pages 409–418. Springer-Verlag, 1996.

29. A. Papakostas and I. G. Tollis. Interactive orthogonal graph drawing. *IEEE Transactions on Computers*, 47(11):1297–1309, 1998.

30. C. Rachit. Octopus: Backbone topology discovery. On line. http://www.cs.cornell.edu/cnrg/topology_aware/topology/Default.html.

31. Y. Rekhter. A border gateway protocol 4 (bgp-4). IETF, rfc 1771.

32. R. Tamassia. On embedding a graph in the grid with the minimum number of bends. *SIAM J. Comput.*, 16(3):421–444, 1987.

33. R. Tamassia, G. Di Battista, and C. Batini. Automatic graph drawing and readability of diagrams. *IEEE Trans. Syst. Man Cybern.*, SMC-18(1):61–79, 1988.

Drawing Hypergraphs in the Subset Standard
(Short Demo Paper)

François Bertault and Peter Eades

[1] Tom Sawyer Software, 804 Hearst Avenue, Berkeley, CA 94710, USA.
bertault@tomsawyer.com
[2] The University of Sydney, F09 - Madsen, NSW 2006, Australia.
eades@cs.usyd.edu.au

Abstract. We report an experience on a practical system for drawing hypergraphs in the subset standard. The PATATE system is based on the application of a classical force directed method to a dynamic graph, which is deduced, at a given iteration time, from the hypergraph structure and particular vertex locations. Different strategies to define the dynamic underlying graph are presented. We illustrate in particular the method when the graph is obtained by computing an Euclidean Steiner tree.

1 Introduction

Hypergraphs can be viewed as an extension of classical graphs in which edges can represent relationships between more than two vertices. Figure 1 is an example of a possible graphical representation of an hypergraph, where inclusion relationships are represented by curves that enclose node. An interpretation of the drawing could be that Paul and John are familiar with GUIs, Paul and Chris with OSs, Alex, Mary and Bob with SEs, while only Mary and Chris know about DBase. Sadly, only a few papers in the literature studies the drawing of hypergraphs [4,7,6].

Johnson and Pollak introduced two notions of planarity of hypergraphs, inspired by the Venn diagram representations of sets [9], and have given NP-completeness results in [4]. In the *hyperedge-based Venn diagrams* and *vertex-based Venn diagrams*, hyperedges are represented by faces of a planar graph that satisfy some connectivity property.

Mäkinen introduced two kinds of hypergraph drawings in [7]. In both cases, vertices are represented by points in the plane. In the *edge standard*, an hyperedge e is represented by connecting the points that represent the vertices that define e by smooth curve lines. Two vertices belong to the same hyperedge if there is a smooth curve between the points that represent these vertices. In the *subset standard*, an hyperedge is represented by a closed curve that contains exclusively the points that represent the vertices that define the hyperedge. Figure (2,left) is an example of drawing of an hypergraph using these two representations. A method for drawing hypergraphs in the edge standard is given in [6].

J. Marks (Ed.): GD 2000, LNCS 1984, pp. 164–169, 2001.

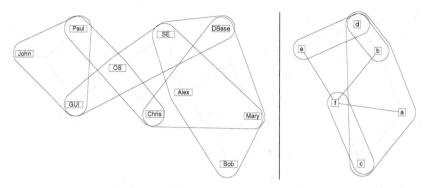

Fig. 1. A simple hypergraph (left) and a higraph (right), drawn with PATATE .

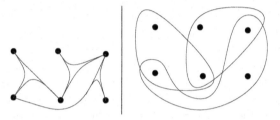

Fig. 2. Drawing of a hypergraph in the edge standard (left) and in the subset standard (right).

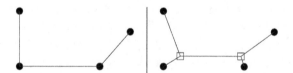

Fig. 3. Example of minimum Euclidean spanning tree (left) and Euclidean Steiner tree (right).

Our PATATE system focuses on the representation of hypergraphs in the subset standard. In this representation, the emphasis is on the representation of hypergraphs as set intersections. This representation is also a step toward the drawing of higraphs, structures introduced by Harel in [3]. The higraph model can be viewed as an hypergraph structure in which edges between nodes or hyperedges can be added. PATATE can already handle drawings for a subset of higraphs; the edges can be defined between nodes only, and are then represented by straight lines. Figure (2,right) is an example of such a higraph drawing obtained with PATATE.

2   The Method

A *graph* $G = (V, E)$ is defined by a finite set V of *vertices* and a finite set E of *edges*, that is, unordered pairs of vertices. A *hypergraph* $H = (V, E)$ is defined by a finite set V of *vertices* and a finite set E of *hyperedges*, that is, unordered non-empty finite sets of vertices. In this paper, we consider *simple* hypergraphs, that is hypergraphs that contain hyperedges between at least two elements.

The method implemented in PATATE is as follows:

1) Assign random locations to vertices of H
2) For a given number of iterations
   a) Construct a graph G from the current positions of the nodes of hypergraph H, using one of the three methods described below. For every vertex v in H, there will be a vertex $\nu(v)$ in G.
   b) Set the locations of the vertices in G to be the locations of the associated vertices in H according to ν.
   c) Apply a force directed drawing algorithm to compute new locations for the nodes in G.
   d) Set the locations of the vertices in H to be the locations of the associated vertices in G.
3) For each hyperedge $e = \{v_1, \ldots, v_k\}$ in H, build a curve that is obtained as the contour of the union of the edges in G that have both ends in e; the union of two edges is defined by the union of the drawing of the edge as thick round edges. In practice, we approximate the thick edges by polygons.
4) An optional step (convex) is to compute, for each hyperedge e, the convex hull $ch(e)$ of the curve associated with e. The curve of e is set to be the convex hull $ch(e)$ if only vertices that define e are included in $ch(e)$.

We assume that the reader is familiar with minimum Euclidean spanning trees and Euclidean Steiner trees (see [5]). Figure 3 is an example of such trees on a particular set of points. Given an hypergraph $H = (V, E)$, the underlying graph G, defined in the description of the PATATE method above, can be build using one of the three following methods, starting with an empty graph G:

(dummy)   For each vertex v in H, a vertex $\nu(v)$ is added in G and for each hyperedge e in H, a vertex $\nu(e)$ is added in G.
                Then, for each hyperedge $e = \{v_1, \ldots, v_k\}$, edges $\{\nu(e), \nu(v_1)\}$, $\ldots, \{\nu(e), \nu(v_k)\}$ are added in G. The location of $\nu(e)$ is set to be the barycenter of $v_1, \ldots, v_k$.

(span-tree)   For each vertex v in H, a vertex $\nu(v)$ is added in G. For each hyperedge e in H, a minimum Euclidean spanning tree that covers the vertices that define e is computed; the edges of the spanning tree are added to G.

(Steiner tree) For each vertex v in H, a vertex $\nu(v)$ is added in G. For each hyperedge e in H, a minimum Steiner tree whose leaves are the vertices that define e is computed. A vertex is added to G for every steiner point, and the edges of the tree are added to G.

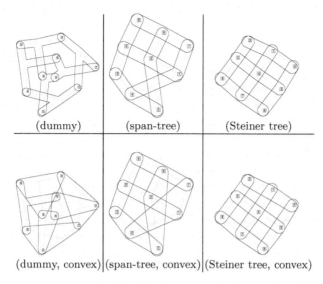

Fig. 4. Comparison between three different methods (dummy, span-tree, Steiner tree) to compute the underlying graph during the iteration steps, represented with or without the optional (convex) step.

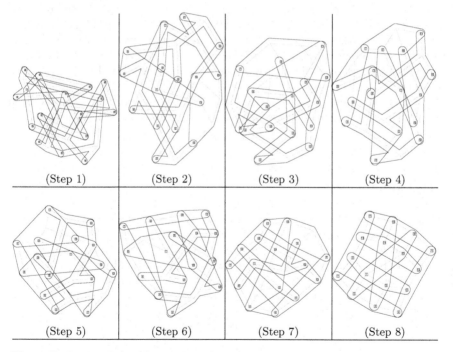

Fig. 5. Evolution of the solution at each iteration step of the PATATE method, when the Steiner and convex options are used.

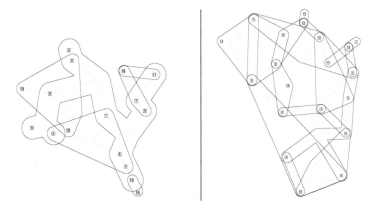

Fig. 6. Drawing of an hypergraph with a long hyperedge (left), and an hypergraph with 20 nodes, hyperedges of length at most 5, and vertices shared between at most 3 hyperedges (right).

3 Implementation

The PATATE method is implemented in C++ and uses the LEDA library [8] for the definition of the graph structures, computation of spanning trees and convex hulls. The geosteiner system [10] is used to compute Steiner trees. A modification of the force-directed algorithm by Fruchterman and Reingold [2,1] is used to compute the locations of the underlying graph G. The output of the program is a postscript file that represents the drawing of the hypergraph given as input.

Figure 4 represents drawings obtained using this three methods, and with or without the optional (convex) step. Figure 5 represents the evolution of the solution at each iteration step of the method, when the Steiner and convex options are used.

4 Conclusion

The method gives reasonable results for small hypergraphs, but can fail miserably when a vertex belongs to a high number of hyperedges (Fig. 7). The Steiner tree option can be viewed as a compromise between the (span-tree) and (dummy) options, and seems to produce the best results in practice. More experiments are needed: we did not investigate whether or not the use of Steiner trees reduces the number of iterations required to obtain a reasonable drawing. Other methods, such as the separation of two sets of points by curves with minimum perimeter, could be used instead of the simple (convex) option.

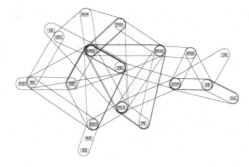

Fig. 7. Drawing of the Erdös-hypergraph from [4], using (span-tree) and (convex) options.

References

1. F. Bertault. A force-directed algorithm that preserves edge-crossing properties. *Information Processing Letters*, 74(1–2):7–13, 2000.
2. T. Fruchterman and E. Reingold. Graph drawing by force-directed placement. *Software-Practice and Experience*, 21(11):1129–1164, 1991.
3. D. Harel. On visual formalisms. *Communications of the ACM*, 31(5):514–530, 1988.
4. D. S. Johnson and H. O. Pollak. Hypergraph planarity and the complexity of drawing venn diagrams. *Journal of Graph Theory*, 1987.
5. R. M. Karp. On the computational complexity of combinatorial problems. *Networks*, 5:45–68, 1975.
6. H. Klemetti, I. Lapinleimu, E. Mäkinen, and M. Sieranta. A programming project: Trimming the spring algorithm for drawing hypergraphs. *SIGCSE Bulletin*, 27(3):34–38, 1995.
7. E. Mäkinen. How to draw a hypergraph. *International Journal of Computer Mathematics*, 1990.
8. K. Mehlhorn and S. Naher. LEDA, a Platform for Combinatorial and Geometric Computing. *Communications of the ACM*, 38(1):96–102, 1995.
9. John Venn. On the diagrammatic and mechanical representation of propositions and reasonings. *The London, Edinburgh, and Dublin Philosophical Magazine and Journal of Science*, 9:1–18, 1880.
10. P. Winter and M. Zachariasen. Euclidean steiner minimum trees: An improved exact algorithm. *Networks*, 30:149–166, 1997.

Knowledge Discovery from Graphs
(Invited Talk)

David Jensen[1]

Department of Computer Science
University of Massachusetts - Amherst, MA 01003
jensen@cs.umass.edu

Abstract. Knowledge discovery is the process of discovering useful and previously unknown knowledge by analyzing large databases. Knowledge discovery is also sometimes called "data mining" or "applied machine learning." A new generation of knowledge discovery tools are beginning to address data that can be expressed as large graphs. Example applications include fraud detection in telecommunication networks and classifying Web pages based on hyperlink structure. These new technologies for knowledge discovery are becoming increasingly relevant to graph drawing. Specifically, graph drawing can aid the process of knowledge discovery by providing visualizations that reveal useful patterns in the data. Conversely, knowledge discovery can provide guidance for graph drawing by identifying recurring substructures or by classifying nodes into distinct types. Attempts to exploit the synergy between the two fields raises interesting new research questions. How should knowledge about a domain affect the drawing of graphs about that domain? What types of knowledge are most easily discovered using visualization, as opposed to automated statistical algorithms? These questions were posed in the context of several examples of knowledge discovery applied to large graphical data sets.

J. Marks (Ed.): GD 2000, LNCS 1984, p. 170, 2001.

A Multilevel Algorithm
for Force-Directed Graph Drawing

C. Walshaw

School of Computing & Mathematical Sciences, University of Greenwich,
Park Row, Greenwich, London, SE10 9LS, UK. C.Walshaw@gre.ac.uk;
http://www.gre.ac.uk/~c.walshaw

Abstract. We describe a heuristic method for drawing graphs which
uses a multilevel technique combined with a force-directed placement
algorithm. The multilevel process groups vertices to form *clusters*, uses
the clusters to define a new graph and is repeated until the graph size
falls below some threshold. The coarsest graph is then given an initial
layout and the layout is successively refined on all the graphs starting
with the coarsest and ending with the original. In this way the multilevel
algorithm both accelerates and gives a more global quality to the force-
directed placement. The algorithm can compute both 2 & 3 dimensional
layouts and we demonstrate it on a number of examples ranging from 500
to 225,000 vertices. It is also very fast and can compute a 2D layout of
a sparse graph in around 30 seconds for a 10,000 vertex graph to around
10 minutes for the largest graph. This is an order of magnitude faster
than recent implementations of force-directed placement algorithms.

1 Introduction

Graph drawing is a basic enabling technology which can aid the understanding
of sets of inter-related data by producing 'nice' layouts (a comprehensive survey
can be found in [2]). Several layout algorithms are based on physical models and
the vertices are placed so as to minimise the 'energy' in the physical system.
Typically such algorithms are well able to display structures and symmetries in
small graphs but can have very high runtimes.

1.1 Motivation

The motivation behind our approach to the problem arises from our work in the
field of graph partitioning. In recent years it has been recognised that an effective
way of both accelerating graph partitioning algorithms and, perhaps more importantly, giving them a global perspective is to use multilevel techniques. The
idea is to match pairs of vertices to form *clusters*, use the clusters to define a
new graph and recursively iterate this procedure until the graph size falls below
some threshold. The coarsest graph is then partitioned (possibly with a crude algorithm) and the partition is successively refined on all the graphs starting with
the coarsest and ending with the original. This sequence of contraction followed

J. Marks (Ed.): GD 2000, LNCS 1984, pp. 171–182, 2001.

by repeated expansion/refinement loops is known as multilevel partitioning and has been successfully developed as a strategy for overcoming the localised nature of the Kernighan-Lin and other optimisation algorithms, e.g. [10]. The multilevel process has also recently been successfully applied to the travelling salesman problem and appears to work (for combinatorial optimisation problems at least) by sampling and smoothing the objective function, [15], thus imparting a more global perspective to the optimisation.

In this paper we apply multilevel ideas to force-directed placement (FDP) algorithms. In fact such ideas have been previously suggested in the graph drawing literature and for example, Fruchterman & Reingold, [7], suggest the possible use of 'a multigrid technique that allows whole portions of the graph to be moved' whilst Davidson & Harel, [1], suggest a multilevel approach to 'expedite the SA [simulated annealing] process'. More recently Hadany & Harel, [8], and in particular Harel & Koren, [9], have actually used multilevel ideas (or as they refer to them, *multiscale*) in combination with an FDP algorithm and are able to robustly handle graphs of 3,000 vertices (although their algorithm still contains an $O(N^2)$ component). Their approach, although derived independently (and using a different FDP algorithm), shares many features with the algorithm outlined here and in many ways confirms that the multilevel paradigm can be a powerful tool for force directed placement.

A related but somewhat different idea is that of multilevel drawings, e.g. [3,6]. Rather than using the multilevel process to create a good layout of the original graph, a multilevel graph is created, either by natural clustering which exists in the graph or by artificial means similar to those applied here. Each level is drawn on a plane at a different height and the entire structure can then be used to aid understanding of the graph at multiple abstraction levels, [5].

2   A Multilevel Algorithm for Graph Drawing

The multilevel FDP algorithm outlined here (and fully described in [14]) works by recursively coarsening the graph until its size falls below some threshold. The coarsest graph is then given an initial layout and the layout is successively refined on all the graphs starting with the coarsest and ending with the original. The algorithm does not actually operate simultaneously on multiple levels of the graph (as, for example, a multigrid algorithm might) but instead refines the layout at each level and then interpolates the result onto the next level down.

2.1   Contraction

Graph coarsening. Given a graph $G_l(V_l, E_l)$, there are many ways to create a coarser representation $G_{l+1}(V_{l+1}, E_{l+1})$ and clustering algorithms are an active area of research within the field of graph drawing, e.g. [3,12]. Often such clustering algorithms seek to retain the more important structural features of the graph in order that the visualisation of each level is meaningful in itself. However, here we are only interested in the drawing of the original graph. As such we seek a fast and efficient (i.e. not necessarily optimal) algorithm that coarsens gradually (aggressive clustering may depreciate the benefits of the multilevel paradigm) and uniformly (the coarsening should not change the inherent properties of the graph differently between different regions).

To suit these requirements we use a coarsening approach known as *matching* in which vertices are matched with at most 1 neighbour so that clusters are thus formed of at most 2 vertices. Computing a matching is equivalent to finding a maximal independent subset of graph edges which are then collapsed to create the coarser graph. The set is independent if no 2 edges in the set are incident on the same vertex (so no 2 edges in the set are adjacent), and maximal if no more edges can be added to the set without breaking the independence criterion. Having found such a set, each selected edge is collapsed and the vertices, $u_1, u_2 \in V_l$ say, at either end of it are merged to form a new vertex $v \in V_{l+1}$ with weight $|v| = |u_1| + |u_2|$.

The problem of computing a matching of the vertices is known as the maximum cardinality matching problem. Although there are optimal algorithms to solve this problem, they are of at least $O(N^{2.5})$, e.g. [11]. Unfortunately this is too slow for our purposes and, since it is not too important for the multilevel process to solve the problem optimally, we use a variant of the edge contraction heuristic proposed by Hendrickson & Leland, [10]. Their method of constructing a matching is to create a randomly ordered list of the vertices and visit them in turn, matching each unmatched vertex with an unmatched neighbouring vertex (or with itself if no unmatched neighbours exist). Matched vertices are removed from the list. If there are several unmatched neighbours the choice of which to match with can be random, but in order to keep the coarser graphs as uniform as possible, and after some experimentation, we choose to match with the neighbouring vertex with the smallest weight (note that even if the original graph G_0 is unweighted, G_l for $l > 0$ will be weighted).

The initial layout. Having constructed the series of graphs until the number of vertices in the coarsest graph is smaller than some threshold, the normal practice of the multilevel partitioning strategy is to carry out an initial partition. In terms of graph drawing the analogue is to compute the initial layout. However, if the graph is coarsened down to 2 vertices (which, because of the mechanisms of the coarsening, will be connected by a single weighted edge) we can simply place these vertices at random with no loss of generality. Note that contraction down to 2 vertices should always be possible provided the graph is connected, [14].

Layout interpolation. Having refined the layout on a graph G_l, it is interpolated onto its parent G_{l-1}. The interpolation itself is a trivial matter and matched pair of vertices, $v_1, v_2 \in V_{l-1}$, are placed at the same position as the cluster, $v \in V_l$, which represents them.

2.2   The Force-Directed Placement Algorithm

At each level we use a force-directed placement (FDP) or spring-embedder algorithm to draw the graph, G_l, and more importantly to provide initial positions for the parent graph G_{l-1}. The original FDP concept came from a paper by Eades, [4], and is based on the idea of replacing edges by springs. The vertices are given initial positions, usually random, and the system is released so that the springs move the vertices to a minimal energy state (i.e. so that the springs are compressed or extended as little as possible).

Unfortunately these local spring forces are insufficient to globally untangle a graph and so such algorithms also employ global repulsive forces, calculated between every pair of vertices in the graph, and thus the system resembles an n-body problem. Such repulsive forces between non-adjacent vertices do not have an analogue in the spring system but are a crucial part of spring-embedder algorithms to avoid minimal energy states in which the system is collapsed in on itself in some manner.

The particular variant of force-directed placement that we use is based on an algorithm by Fruchterman & Reingold (FR), [7], itself a variation of Eades' original algorithm. From the point of view of the multilevel approach it is attractive as it is an incremental scheme which iterates to convergence and which can reuse a previously calculated initial layout. We have made a number of modifications based on our experience with it and, in particular, because of the additional problems associated with drawing very large graphs. Space precludes a full description of the algorithm here but we describe the implementation in full in [14]. In principle however, it should be possible to use any iterative incremental algorithm for this part of the multilevel graph drawing, although in practice different algorithms can be somewhat sensitive and require tuning.

Natural spring length, k. A crucial part of the algorithm the choice of the natural spring length k_l (the length at which a spring is neither extended nor compressed). For the initial coarsest graph, G_L, the 2 vertices are placed at random and k_L set to be the distance between them. However at the start of the execution of the placement algorithm for graph G_l ($l < L$) the vertices will all be in positions determined by the layout calculated for graph G_{l+1}. We must therefore somehow set k_l relative to this existing layout in order not to destroy it. If k_l is too large, then the entire graph will have to expand from its current layout and potentially ruin any advantage gained via the multilevel process.

In fact we derive the new value for k by considering the coarsening a graph, G_l, with well placed vertices (i.e. all vertices are approximately at a distance k from each other) and in [14] justify our choice of $k_l = \sqrt{4/7} \times k_{l+1}$. Remarkably this simple formula works very robustly over all the examples that we have tested although we feel that this parameter could do with further investigation.

2.3   Reducing the Complexity

Unfortunately the complexity of the FR algorithm for each iteration on graph $G_l(V_l, E_l)$ is $O(|V_l|^2 + |E_l|)$. For the types of sparse graphs that we are interested in, the $|V_l|^2$ component heavily dominates this expression and we therefore use the FR grid variant for reducing the run-times. Their motivation was that long distance repulsive forces are sufficiently small enough to be neglected. If we set R to be the maximum distance over which repulsive forces will act we can then modify the algorithm by ignoring global forces between any pair of vertices further apart than R (and this can efficiently implemented using a superimposed grid to avoid calculating the distance between every pair of vertices).

In the original FR algorithm the value $R = 2k$ was used, but for the larger graphs that we are interested in this did not prove sufficient to 'untangle' them globally. Unfortunately the larger the value given to R the longer the algorithm takes to run and so although assigning $R = 20k$ gave better results, it did so with a huge time penalty. Fortunately, however, the power of the multilevel paradigm comes to our aid once again and we can make R a function of the level l. Thus for the initial coarse graphs we can set R_l to be relatively large and achieve some impressive untangling without too much cost (since $|V|$ is very small for these graphs). Meanwhile, for the final large graphs, when most of the global untangling has already been achieved we can make R_l relatively small without penalising the placement. In fact the first such schedule that we tried, with $R_l = 2(l+1)k_l$ for each graph G_l, worked so well that we have not experimented further. This also replicates the choice of $R = 2k$ for G_0 in the original FR algorithm.

2.4   Complexity Analysis

It is not easy to derive complexity results for the algorithm but we can state some bounds. Firstly the number of graph levels, L, is dependent on the rate of coarsening. At best the number of vertices will be reduced by a factor of 2 at every level (if the code succeeds in matching every vertex with another one) and in the worst case, the code may only succeed in matching 1 vertex at every level (e.g. if the graph is a star graph, a 'hub' vertex connected to every other vertex each of which is only connected to the hub). Thus we have $\log_2 |V| \leq L < |V|$. This indicates that the algorithm is not well suited to drawing star type graphs and in fact for the examples given in Section 3 the coarsening rate is close to 2.

The matching & coarsening parts of the algorithm are $O(|V_l| + |E_l|)$ for each level l but in fact the total runtime is heavily dominated by the FDP algorithm. Using the above simplification (§2.3) of neglecting long range repulsive forces we can see that each iteration of the FDP algorithm is bounded below by $O(|V_l| + |E_l|)$ although with a large coefficient. In fact if the graph is very dense, or in the worst case a complete graph, it may be that this is still $O(|V_l|^2 + |E_l|)$, dependent on the relative balance of attractive & repulsive forces. However, we suspect that no FDP algorithm is appropriate for very dense graphs (because the minimal energy state corresponds to a tightly packed 'blob').

The number of FDP iterations at every level is determined by the cooling schedule which sets t^i, the temperature at each iteration, to $t^i = (1 - \epsilon)t^{i-1}$. In the experiments below (and in [14]) we use an initial temperature $t^0 = k_l$ and $\epsilon = 0.1$ and the algorithm is deemed to have converged when all movement is less than $0.001k_l$ which means that all movement ceases at iteration i where $(1 - 0.1)^i < 0.001$ or in other words after 66 iterations.

In summary the total complexity at each level is close to $O(|V_l| + |E_l|)$ for sparse graphs and the runtime is heavily dominated by the FDP iterations.

Finally consider the FDP algorithm used, without coarsening, on a given sparse graph of size N (i.e. standard single-level placement (SLP)) and compare it with multilevel placement (MLP) used on the same graph. Let T_p be the time for the SLP algorithm to run on the graph and for MLP let T_c be the time to coarsen and contract it. If we suppose that the coarsening rate is close to 2 (which is true for the examples below) then for MLP this gives us a series of problems of size $N, N/2, \ldots, N/N$ whilst the (almost) linear complexity for FDP gives the total runtime for MLP as $T_c + T_p/N + \ldots + T_p/2 + T_p$. In all the example we have tested $T_c \ll T_p$ and so we can neglect it giving a total runtime of approximately $T_p/N + \ldots + T_p/2 + T_p = 2T_p$. In other words MLP should take *only* twice as long as SLP to run (and yet in the extended example below, §3.1, achieves far better results). In fact the final level of the MLP algorithm is likely to already have a very good initial layout which means that it should run even faster than SLP although this is neutralised somewhat by the fact that the coarsening rate is normally somewhat less than 2. Nonetheless this factor of 2 is a good 'rule of thumb' and note that if the chosen FDP algorithm were $O(N^2)$ or even $O(N^3)$ then a similar analysis suggests that the MLP runtime is substantially *less* than twice that of SLP.

3 Examples

We have implemented the algorithms described here within the framework of JOSTLE, a mesh partitioning software tool developed at Greenwich. The experiments were carried out on a Sun SPARC Ultra 10 with a 333 MHz CPU and 256 Mbytes of memory.

We have tested our multilevel algorithm on a number of example graphs, including some for which we already know a good layout (although interestingly, even the results on these graphs proved very illuminating). Most such graphs were drawn from genuine examples of meshes from various computational mechanics problems. Typically in such graphs the vertices either represent mesh nodes (the nodal graph) or mesh elements (the dual graph). However we have also considered graphs from other non mesh-based applications.

Table 1 gives a summary of some examples showing their sizes ($|V|$ & $|E|$), the maximum, minimum & average degree of the vertices, the MLP runtime and a short description. In all cases the MLP algorithm produced a good layout although sometimes with a certain amount of buckling (see [14]). However for none of these examples was the single-level FDP algorithm able to find a layout which untangled the graph in a global sense (despite adjusting the parameters and allowing considerably longer runtimes). For the shuttle, add32 & mesh100 graphs the layout was calculated in 3D and hence the runtime is a little higher than a 2D layout (e.g. compare the add32 & 4970 runtimes).

The second half of the table shows the results for a series of increasingly larger versions of the same mesh (created by adaptive mesh refinement) and gives an indication of the scalability of the algorithm (albeit on sparse & fairly

Table 1. A summary of the example graphs

graph	size		degree			runtime	graph type
	$|V|$	$|E|$	max	min	avg	(secs.)	
c-fat500-10	500	46627	188	185	186.51	12.53	max clique test
516	516	729	3	1	2.83	1.14	2D dual
shuttle	2851	15093	17	3	10.59	20.05	3D nodal
add32	4960	9462	31	1	3.82	48.09	electronic circuit
4970	4970	7400	3	2	2.98	13.93	2D dual
whitaker3	9800	28989	8	3	5.92	24.10	2D nodal
finan512	74752	261120	54	2	6.99	688.46	linear programming
sierpinski10	88575	177147	4	2	4.00	217.58	2D self-similar 'fractal'
mesh100	103081	200976	4	2	3.90	1245.43	3D dual
Laplace.0	23787	35281	3	2	2.97	64.16	2D dual
Laplace.2	40851	60753	3	2	2.97	114.84	2D dual
Laplace.5	88743	132329	3	2	2.98	247.52	2D dual
Laplace.8	185761	277510	3	2	2.99	477.36	2D dual
Laplace.9	224843	336024	3	2	2.99	598.87	2D dual

homogeneous graphs). This supports the complexity analysis in §2.4 that the runtime is approximately linear in $|V| + |E|$.

3.1   An Extended Example

In this section we demonstrate in a little more detail how the multilevel placement (MLP) algorithm works for the 516 graph. The algorithm first coarsens the problem (reducing the number of vertices by a factor of around 1.8 at each level) and constructs a series of 9 graphs of diminishing size. The initial layout is computed by placing the 2 vertices of G_9 at random and setting the natural spring length, k, to be the distance between them. Starting from $G_l = G_8$ the layout is interpolated from G_{l+1}, by simply placing vertices at the same position as the cluster representing them in the coarser graph, and then refined.

Figure 1(a) shows the final layout on G_4 and although over 10 times smaller than the original, the layout is already beginning to take shape. Figures 1(b)-1(d) meanwhile illustrate the placement algorithm on G_2. Figure 1(b) shows the initial layout as calculated on G_3 and with many of the vertices coincident whilst Figure 1(c) then shows the layout after the first iteration and where the coincident vertices have started to separate. Figure 1(d) finally shows the layout after the placement algorithm has converged for G_2. Notice an important feature of the multilevel process, common with partitioning, that on each level the final layout (partition), does not differ greatly from the initial one. Figure 1(e) shows the final layout on the original graph, G_0. The entire runtime of the MLP algorithm to compute this layout was just over 1 second.

For comparison, Figure 1(f) shows the single-level FDP algorithm used on a random initial layout. Possibly the algorithm is not well tuned for this pro-

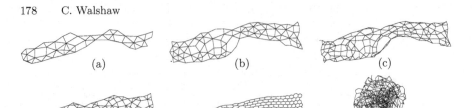

Fig. 1. The multilevel force directed placement illustrated for the mesh 516

blem (although the initial temperature was raised to $10k$), but it is clear that, although the micro structure has been reconstructed reasonably well, the single-level placement has not been able to 'untangle' the graph in a global sense.

3.2   Example Layouts

Space precludes any extended presentation of the examples in Table 1, but Figures 2-7 show some of the highlights (see also [14]). Despite the suggestion that the MLP algorithm is best suited to sparse graphs (§2.4), Figure 2 shows a dense regular graph, c-fat500-10 (generated to test algorithms for the maximum clique problem), and demonstrates that the layout nicely captures the symmetries. Meanwhile Figure 3 shows the layout calculated for the 4970 graph by the MLP algorithm where, in trying to equalise the edge lengths, the drawing has actually revealed far more of the graph than the original layout. Figure 4 shows the 3D layout of the shuttle graph which which reveals 3 weakly connected 'panels', the existence of which is not evident in the original layout. Meanwhile Figure 5 shows the layout calculated for a large Sierpinski graph, a self-similar 'fractal' type structure which has large holes. Finally, in the centre-piece of [14], Figure 6 shows the layout found by the MLP algorithm of finan512, a linear programming matrix with around 75,000 vertices. Once again this layout is highly illuminating; the graph is revealed to have a fairly regular structure and consists of a ring with 32 'handles' each of which has a number of fronds protruding. Figure 7 shows a detailed view of one of these handles.

4   Summary and Further Research

We have described a multilevel algorithm for force-directed graph drawing which coarsens the graph, refines the layout at each level and then interpolates the result onto the next level down. The algorithm is fast, e.g. about 1 second for 2D layout of a 500 vertex sparse graph and about 10 minutes for 225,000 vertices. This is an order of magnitude faster than recent implementations of force-directed placement (e.g. around 70 seconds for a 1,000 vertex graph in [13])

Fig. 2. The layout of c-fat500-10 computed with the multilevel placement algorithm

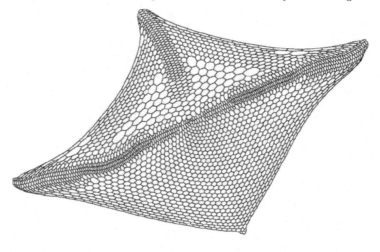

Fig. 3. The layout of 4970 computed with the multilevel placement algorithm

and indeed it is not even clear whether current single-level algorithms can produce reasonable layout for such large graphs. As such the multilevel paradigm broadens the scope of force-directed algorithms.

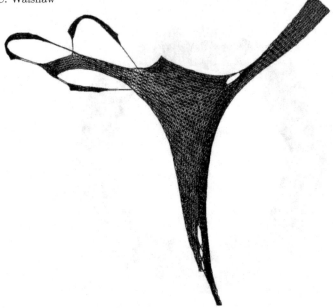

Fig. 4. The layout of shuttle computed with the multilevel placement algorithm

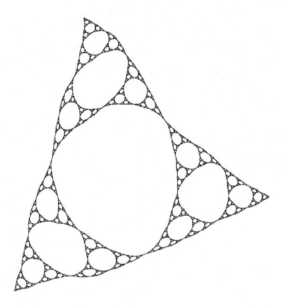

Fig. 5. The layout of sierpinski10 computed with the multilevel placement algorithm

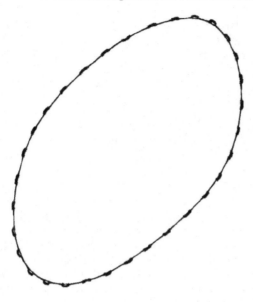

Fig. 6. The layout of finan512 computed with the multilevel placement algorithm

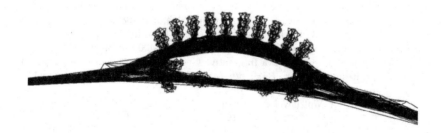

Fig. 7. finan512 computed with the multilevel placement algorithm: detail of the micro structure

We have not particularly tried to address graphs for which the technique might not work. It is likely that very dense graphs or even those which have a dense substructure are never going to be good candidates for any FDP algorithm and ours is no exception. It is also likely that graphs containing vertices of very high degree may not particularly suit the coarsening process (see §2.4). However we believe that the multilevel process can accelerate & enhance FDP algorithms for a range of useful graphs and further testing on different types of graph is an important subject for further research. We have not addressed disconnected graphs but feel that this requires only minor modifications.

References

1. R. Davidson and D. Harel. Drawing Graphs Nicely using Simulated Annealing. *ACM Trans. Graphics*, 15(4):301–331, 1996.
2. G. Di Battista, P. Eades, R. Tamassia, and I. G. Tollis. *Graph Drawing: Algorithms for the Visualization of Graphs*. Prentice-Hall, New Jersey, U.S.A., 1998.
3. C. A. Duncan, M. T. Goodrich, and S. G. Kobourov. Planarity-Preserving Clustering and Embedding for Large Planar Graphs. In J. Kratochvíl, editor, *Proc. 7th Int. Symp. Graph Drawing*, volume 1731 of *LNCS*. Springer, 1999.
4. P. Eades. A Heuristic for Graph Drawing. *Congressus Numerantium*, 42:149–160, 1984.
5. P. Eades and Q. Feng. Multilevel Visualization of Clustered Graphs. In *Proc. 6th Int. Symp. Graph Drawing*, volume 1190 of *LNCS*, pages 101–112. Springer, 1996.
6. P. Eades, Q. Feng, X. Lin, and H. Nagamochi. Straight-Line Drawing Algorithms for Hierarchical Graphs and Clustered Graphs. Tech. rep. 98-03, Dept. Comp. Sci., Univ. Newcastle, Callaghan 2308, Australia, 1998.
7. T. M. J. Fruchterman and E. M. Reingold. Graph Drawing by Force-Directed Placement. *Software — Practice & Experience*, 21(11):1129–1164, 1991.
8. R. Hadany and D. Harel. A Multi-Scale Algorithm for Drawing Graphs Nicely. Tech. Rep. CS99-01, Weizmann Inst. Sci., Faculty Maths. Comp. Sci., Jan, 1999.
9. D. Harel and Y. Koren. A Fast Multi-Scale Algorithm for Drawing Large Graphs. Tech. Rep. CS99-21, Weizmann Inst. Sci., Faculty Maths. Comp. Sci., Nov, 1999.
10. B. Hendrickson and R. Leland. A Multilevel Algorithm for Partitioning Graphs. In S. Karin, editor, *Proc. Supercomputing '95*. ACM Press, 1995.
11. C. H. Papadimitriou and K. Stieglitz. *Combinatorial Optimization: Algorithms and Complexity*. Prentice Hall, Englewood Cliffs, NJ, 1982.
12. R. Sablowski and A. Frick. Automatic Graph Clustering. In *Proc. 6th Int. Symp. Graph Drawing*, volume 1190 of *LNCS*, pages 395–400. Springer, 1996.
13. D. Tunkelang. JIGGLE: Java Interactive General Graph Layout Environment. In S. H. Whitesides, editor, *Proc. 6th Int. Symp. Graph Drawing*, volume 1547 of *LNCS*, pages 413–422. Springer, 1998.
14. C. Walshaw. A Multilevel Algorithm for Force-Directed Graph Drawing. Tech. Rep. 00/IM/60, Univ. Greenwich, London SE10 9LS, UK, April 2000.
15. C. Walshaw. A Multilevel Approach to the Travelling Salesman Problem. Tech. Rep. 00/IM/63, Univ. Greenwich, London SE10 9LS, UK, Aug. 2000.

A Fast Multi-scale Method for Drawing Large Graphs

David Harel and Yehuda Koren

Dept. of Applied Mathematics and Computer Science
The Weizmann Institute of Science, Rehovot, Israel
{harel,yehuda}@wisdom.weizmann.ac.il

Early version appeared as Weizmann Institute Tech. Report MCS99-21, 1999.
An abridged version appeared in *Proc. Working Conf. on Advanced Visual Interfaces (AVI'2000)*, ACM Press, May 2000, pp. 282-285

Abstract. We present a multi-scale layout algorithm for the aesthetic drawing of undirected graphs with straight-line edges. The algorithm is extremely fast, and is capable of drawing graphs of substantially larger size than any other algorithm we are aware of. For example, the algorithm achieves optimal drawings of 1000 vertex graphs in about 2 seconds. The paper contains graphs with over 6000 nodes. The proposed algorithm embodies a new multi-scale scheme for drawing graphs, which was motivated by the recently published multi-scale algorithm of Hadany and Harel [7]. It can significantly improve the speed of essentially any force-directed method (regardless of that method's ability of drawing weighted graphs or the continuity of its cost-function).

1 Introduction

A graph $G(V, E)$ is an abstract structure that is used to model a relation E over a set V of entities. Graph drawing is a conventional tool for the visualization of relational information, and its usefulness depends on its readability, that is, the capability of conveying the meaning of the diagram quickly and clearly. In recent years, many algorithms for drawing graphs automatically were proposed (the state of the art is surveyed comprehensively in [2]).

We concentrate on the problem of drawing an undirected graph with straight-line edges. In this case the problem reduces to that of positioning the vertices by determining a mapping $L : V \longrightarrow \mathbb{R}^2$. A popular generic approach to this problem is the *force-directed* technique, which introduces a heuristic cost function (an *energy*) of the mapping L, which (hopefully) achieves its minimum when the layout is nice. Various variants of this approach differ in the definition of the energy, and in the optimization method that finds its minimum. Some known algorithms are those of [4], [10], [3] and [5]. Major advantages of force-directed methods are their relatively simple implementation and their flexibility (heuristic improvements are easily added), but there are some problems with them too. One severe problem is the difficulty of minimizing the energy when dealing with

J. Marks (Ed.): GD 2000, LNCS 1984, pp. 183–196, 2001.
© Springer-Verlag Berlin Heidelberg 2001

large graphs. The above methods focus on graphs of up to 100 vertices. For larger graphs the convergence to a minimum, if possible at all, is very slow.

We propose a new method for drawing graphs that can significantly improve the speed of every force-directed method. We build our algorithm around the Kamada-Kawai method, and the resulting algorithm, which is extremely fast, is capable of drawing graphs of substantially larger size than any other algorithm we are aware of. The algorithm, which was motivated by the recent multi-scale algorithm of Hadany and Harel [7], works by producing a sequence of improved approximations of the final layout. Each approximation allows vertices to deviate from their final place by an extent limited by a decreasing constant r. As a result, the layout can be computed using increasingly coarse representations of the graph, where closely drawn vertices are collapsed into a single vertex. Each layout in the sequence is generated very rapidly, by performing a local beautification on the previously generated layout.

2   Multi-scale Graph Drawing

The intuition of [7] for beauty in graph layout is that the graph should be nice on all scales. In other words the drawing should be nice at both the micro level and the macro level. Relying only on this intuition, we will formalize the notion of *scale* relevant to the graph drawing problem. The crucial observation is that global aesthetics refer to phenomena that are related to large areas of the picture, disregarding its micro structure, which has only a minor impact on the global issue of beauty. On the other hand, local aesthetics refer to phenomena that are limited to small areas of the drawing. Following this line of thinking, we will construct a *coarse scale* of a drawing by shrinking nodes that are drawn close to each other, into a single node, obtaining a new drawing that eliminates many local details but preserves the global structure of the original drawing.

An alternative view of our notion of coarsening is as an approximation of a nice layout. This approximation allows vertices to deviate from their final position by an amount limited to some constant r. As a consequence, we can unify all the vertices whose final location lies within a circle of radius r, and thus obtain the coarse scale representation.

Our presentation of the drawing scheme is preceded by some definitions:

Definition 21
A layout of a graph $G(V, E)$ is a mapping of the vertices to the Euclidean space: $L_G : V \longrightarrow \mathbb{R}^2$. We often omit the subscript $_G$.
For simplicity, we assume that there is a single optimal layout with respect to fixed set of aesthetic criteria accepted in force-directed algorithms. We term this layout nice. *The nice layout of G is denoted by L_G^*, or simply L^*.*

Definition 22
L is a locally nice *layout of $G(V, E)$ with respect to r, if the intersection of $L(V)$ with every circle of radius r induces a nice layout of the appropriate subgraph of G.*

Definition 23
L is a globally nice *layout of* $G(V, E)$ *with respect to* r *if*

$$\max_{v \in V}\{|L(v) - L^*(v)|\} < r$$

Definition 24
A locality preserving k-clustering *(a k-lpc for short) of* $G(V, E)$ *with respect to* r *is the weighted graph* $G(\{V_1, V_2, \ldots, V_k\}, E', w)$, *where:*

$$V = V_1 \cup V_2 \ldots V_k, \quad \forall i \ne j : V_i \cap V_j = \emptyset$$
$$E' = \{ (V_i, V_j) \mid \exists (v_i, v_j) \in E \wedge v_i \in V_i \wedge v_j \in V_j \}$$
$$w(V_i, V_j) = \frac{1}{|V_i||V_j|} \sum_{v \in V_i, u \in V_j} d_{vu}, \quad \forall_{(V_i, V_j) \in E'}$$

(d_{vu} is the shortest distance between u and v in G)

and for every i:

$$\max_{v, u \in V_i} \{|L^*(v) - L^*(u)|\} < r$$

i.e., all the vertices in one cluster are drawn relatively close in the nice layout. We sometimes call the vertices of a k-lpc clusters.

At first glance, this definition seems to be of a little practical value, as it refers to the unknown nice layout L^*. We will discuss this important point in the next section.

Definition 25
A multi-scale representation *of a graph* $G(V, E)$ *is a sequence of graphs* G^{k_1}, $G^{k_2}, \ldots, G^{k_l}$, *where* $k_1 < k_2 < \cdots < k_l = |V|$, *and for all* $1 \le i \le l$: G^{k_i} *is a locality preserving* k_i-*clustering of* $G(V, E)$ *with respect to* r_i, *where* $r_1 > r_2 > \cdots > r_l = 0$.

Remark: We naturally assume that in a nice layout of a weighted graph, the lengths of the edges have to reflect their weights.

Our method relies on the ease of drawing graphs with a small number of vertices and on the following two assumptions, which formalize what we think to be amenability to multi-scale aesthetics — independence between global and local aesthetics.

Assumption 21
Let G_r *be k-lpc of a graph* G *with respect to* r, *and let* $\hat{r} \ge r$. *L is a globally nice layout of* G_r *with respect to* $\hat{r}$ *if and only if* L *is a globally nice layout of* G *with respect to* $\hat{r}$. *(In G, we take* $L(v) = L(V_i)$ *for each* $v \in V_i$.)*

Corollary: If L is a nice layout of a k-lpc of a graph G with respect to r then L is a globally nice layout of G with respect to r.

The intuition of Assumption 21 is that global aesthetics is independent of the micro structure of the graph, so the differences between the layouts of G_r and of G are bounded with r.

Assumption 22

If L is both a locally and a globally nice layout of a graph G with respect to r, then it is a nice layout of G.

Now we present the multi-scale drawing scheme, which draws a graph by producing a sequence of improved approximations of the final layout.

The Multi-Scale Drawing Scheme.

1. Place the vertices of G randomly in the drawing area.
2. Choose an adequate decreasing sequence of radiuses $\infty = r_0 > r_1 > r_2 > \cdots > r_l = 0$.
3. *for $i=1$ to l do*
   3.1 Choose an appropriate value of k_i, and construct G^{k_i}, a k_i-lpc of G w.r.t. r_i.
   3.2 Place each vertex of G^{k_i} at the (weighted) location of the vertices of G that constitute it.
   3.3 Locally beautify local neighborhoods of G^{k_i}.
   3.4 Place each vertex of G at the location of its cluster (i.e., the vertex in G^{k_i}).
4. *end*

The viability of the scheme stems from the following observations:

1. In the first iteration, after step 3.3, we should have a nice layout of G^{k_1}. To guarantee this, the value of r_1 has to be large enough so that the resulting G^{k_1} will be small and can be easily drawn nicely.
2. Step 3.3 should yield a locally nice layout of G^{k_i} w.r.t. r_{i-1}. To guarantee this, we have to choose large enough neighborhood. Not too large, however, because we want to draw it optimally by a standard method. A sufficiently dense choice of the sequence of r_i's will do.
3. At the beginning of iteration i, we have a globally nice layout of G^{k_i} w.r.t. r_{i-1}. Hence, by Assumption 22, after step 3.4 we have a nice layout of G^{k_i}. This layout is a globally nice layout of $G^{k_{i+1}}$ w.r.t. r_i, by Assumption 21.

We remark that the choice of the multi-scale representation can be based upon either the decreasing sequence $r_1, \ldots, r_l = 0$ (as described above) or the increasing sequence $k_1, \ldots, k_l = |V|$ (as we have done in our implementation).

In Definition 24 we added weights to the edges of the k-lpc in order to retain the size proportions of G, which is necessary for making Assumption 21 valid. However, in practice, we conjecture that our scheme works well even without weighting the edges of the k-lpc, when using some variants of the spring-embedder method (e.g., those of [4] and [5]) as the local beatification method. The reason for this is that such methods benefit significantly from the better initialization, which, while not capturing the final size of the graph, often solves many large scale conflicts correctly.

3   The New Algorithm

In order to implement a multi-scale graph drawing algorithm based on the scheme of Section 2, we have to further elaborate on two points: (1) how to find the multi-scale representation of a graph (line 3.1 of the scheme), and (2) how to devise a locally nice layout (line 3.3 of the scheme).

3.1   Finding a Multi-scale Representation

In order to construct a multi-scale representation of a graph G based on Definition 25, we must find a k-lpc of G, in it vertices that are drawn close in the nice layout should be grouped together. The important question is: *How can we know which vertices will be close in the final picture, if we still do not know what the final picture looks like?* [1] Luckily we do have a heuristic that can help decide which vertices will be drawn closely. Moreover, this key decision can be made very rapidly, and is a major reason for the fast running time of our algorithm.

The heuristic is based on the observation that a nice layout of the graph should convey visually the relational information that the graph represent, so *vertices that are closely related in the graph (i.e., the graph theoretic distance is small) should be drawn close together.* This heuristic is very conservative, and all the force directed drawing algorithms use it heavily.

Employing this heuristic, we can approximate a k-lpc of G by using an algorithm for the well known *k-clustering* problem. In this problem we wish to partition V into k clusters so that the longest graph-theoretic distance between two vertices in the same cluster is minimized. In reality, we would like to identify every vertex in the cluster with a single vertex that approximates the barycenter of the cluster. Hence we use a solution to the closely related *k-center* problem, where we want to choose k vertices of V, such that the longest distance from V to these k centers is minimized. These fundamental problems arise in many areas and have been widely investigated in several papers (see e.g., [6] and [9]). Unfortunately, both problems are NP-hard, and it has been shown in [6] and [9] that unless P=NP there does not exist a $(2 - \epsilon)$-approximation algorithm for any fixed $\epsilon > 0$. [2] Nevertheless, there are various fast and simple 2-approximation algorithms for these problems.

---

[1] What is needed is only a sufficient (even if not necessary) condition that vertices are close.

[2] A δ-approximation algorithm delivers an approximate solution guaranteed to be within a constant factor δ of the optimal solution.

We will approximate a k-lpc as the solution to the k-center problem, adopting a 2-approximation method mentioned in [8]:

K-Centers $(G(V, E), k)$
Goal: Find a set $S \subseteq V$ of size k, such that $\max_{v \in V} \min_{s \in S} \{d_{sv}\}$ is minimized.

1. $S \leftarrow \{v\}$ for some arbitrary $v \in V$
2. *for* $i = 2$ *to* k *do*
   2.1 Find the vertex u farthest away from S
       (i.e., such that $\min_{s \in S} \{d_{us}\} \geqslant \min_{s \in S} \{d_{ws}\}$, $\forall w \in V$)
   2.2 $S \leftarrow S \cup \{u\}$
3. *return* S
4. *end*

Complexity: Line 2.1 can be carried out in time $\Theta(|E|)$ by BFS. In our case, it can be done faster, since, as we shall see, we have the all-pairs shortest distance at our hands (it is needed for the local beautification). Utilizing this fact and memorizing the current distance of every vertex from S, we can implement line 2.1 in time $\Theta(|V|)$, yielding a total time complexity of $\Theta(k|V|)$.

3.2   Local Beautification

We have chosen to use a variant of the Kamada and Kawai method [10] as our local drawing method. We found it to be very appropriate, because it relates every pair of vertices, so, when constructing a new coarse representation of the graph, we do not have to define which pairs of vertices are connected by an edge. Notice that this property of the Kamada and Kawai method has a price: it forces us to waste $\Theta(|V|^2)$ memory, even when the graph is sparse. Another advantage of the Kamada and Kawai method is that it can deal directly with weighted graphs, which is convenient in our case since the multi-scale representation of a graph contains weighted graphs.

The Energy Functional: We consider the graph $G(V, E)$, where each vertex v is mapped by the layout L into a point in the plane $L(v)$ with coordinates (x_v, y_v). The distance d_{uv} is defined as the length of the shortest path in G between u and v. We define the k-*neighborhood* of v to be: $N^k(v) = \{u \in V \mid 0 \leqslant d_{uv} < k\}$. In order to find a layout with aestheticaly pleasing k-neighborhoods, we use an energy functional that relates the graph theoretic distance between vertices in the graph to the Euclidean distance between them in the drawing, and is defined as follows:

$$E_k = \sum_{v \in V} \sum_{u \in N^k(v)} k_{uv} (\|L(u) - L(v)\| - l d_{uv})^2$$

where l is the length of a single edge, $\|L(u) - L(v)\|$ is the Euclidean distance between $L(u)$ and $L(v)$, and k_{uv} is a weighting constant that can be either $\frac{1}{d_{uv}}$ or $\frac{1}{d_{uv}^2}$.

The energy represents the normalized mean squared error between the Euclidean distance of vertices in the picture and the graph-theoretic distance. Only pairs in the same k-neighborhood are considered.

Local Minimization of the Energy: Our purpose is to find a layout that brings the energy E_k to a local minimum. The necessary condition of a local minimum is as follows:

$$\frac{\partial E_k}{\partial x_v} = \frac{\partial E_k}{\partial y_v} = 0, \quad \forall v \in V$$

To achieve this condition we iteratively choose the vertex that has the largest value of Δ_v, which is defined as:

$$\Delta_v = \sqrt{(\frac{\partial E_k}{\partial x_v})^2 + (\frac{\partial E_k}{\partial y_v})^2}$$

and move this vertex, v, by the amount of (δ_x^v, δ_y^v). The computation of (δ_x^v, δ_y^v) is carried out by viewing E_k as a function of only $L(v) = (x_v, y_v)$, and the use of a two-dimensional Newton-Raphson method. As a result, the unknowns δ_x^v and δ_y^v are found by solving the following pair of linear equations:

$$\frac{\partial^2 E_k}{\partial x_v^2}\delta_x^v + \frac{\partial^2 E_k}{\partial x_v \partial y_v}\delta_y^v = -\frac{\partial E_k}{\partial x_v}$$

$$\frac{\partial^2 E_k}{\partial y_v \partial x_v}\delta_x^v + \frac{\partial^2 E_k}{\partial y_v^2}\delta_y^v = -\frac{\partial E_k}{\partial y_v}$$

The interested reader can find further details in [10].

Algorithms for Locally Nice Layout
The following algorithm, which is from [11] and as mentioned in [1], is a modification of [10]. The algorithm computes a nice layout of every k-neighborhood of a graph:

LocalLayout $(d_{V \times V}$ (all-pairs shortest dist.), L (initialized layout))
Goal: Find a locally nice layout L by beautifying k-neighborhoods

1. *for $i=1$ to Iterations $\cdot |V|$ do*
    1.1 Choose the vertex v with the maximal Δ_v
    1.2 Compute δ_x^v and δ_y^v by solving the above mentioned equations
    1.3 $L(v) \leftarrow L(v) + (\delta_x^v, \delta_y^v)$
2. *end*

A typical value of the parameter *Iterations* is 4. (When running this algorithm as a stand alone (not as a part of the multi-scale algorithm) the value should be 10, at least.)

Complexity: The computation time of step 1.1 is $\Theta(|V|)$, since we memorize the first derivatives of E_k, and update them in time $\Theta(|N^k(v)|)$ after the movement of each vertex v. The computation time of δ_x^v and δ_y^v is $\Theta(|N^k(v)|)$. These computations are carried out *Iterations* $\cdot |V|$ times, so the overall time complexity is $\Theta(|V|^2)$. We can select the vertex v with the maximal Δ_v in constant time, by making this selection through sampling constant sized subsets of V, without serious harm to the quality of the results. Moreover, if the degree of G is bounded, $\Theta(|N^k(v)|)$ is constant. As a result, the overall time complexity in the bounded degree case is $\Theta(|V|)$.

In the full version of the paper we describe an alternative beautification method, which eliminates the choice of the vertex with the maximal Δ_v.

3.3   The Multi-scale Drawing Algorithm

We now describe the full algorithm:

Layout $(G(V, E))$
Goal: Find L, a nice layout of G

1. Compute the all-pairs shortest distance $(d_{V \times V})$
2. Set up a random layout L
3. $k \leftarrow Threshold$
4. *while* $k \leqslant |V|$ *do*
       4.1 $Centers \leftarrow$ **K-Centers**$(G(V, E), k)$
       4.2 **LocalLayout** $(d_{Centers \times Centers}, L(Centers))$
       4.3 *for* every $v \in V$ *do*
           4.3.1 $L(v) \leftarrow L(center(v)) + \xi$
       4.4 $k \leftarrow k \cdot Ratio$
5. *end*

Comments: In line 4.3.1, we assume that the call *center(v)* returns the center that is closest to v. We add a small random noise $(0, 0) < \xi < (1, 1)$, because our local beautification algorithm performs badly when the vertices are initialized to the same point. *Threshold* and *Ratio* are constants, with typical values of 10 and 3, respectively. One might try to improve the global aesthetics by iteratively repeating lines 4.1–4.3 a number of times that decreases with k (e.g., $\frac{|V|}{|k|}$ times).

Complexity: The overall asymptotical complexity is determined by the computation of the all-pairs shortest distance (line 1), which we implemented by initiating a BFS from every vertex. It thus takes time $\Theta(|V||E|)$. For 1000-vertex sparse graphs, the local beautification (line 4.2) consumes 80-90% of the running time, and the computation of the all-pairs shortest distance consumes the remaining 10-20%. All the other parts of the algorithm consume only negligible time. As the size of the graphs becomes larger, the computation of the all-pairs shortest distance becomes more dominant. The space complexity is $\Theta(|V|^2)$, since we have to memorize the all-pairs shortest distance matrix.

4   Examples

This section contains examples of the results of our algorithm. The implementation is in C++, and runs on a Pentium III 533Mhz PC. We set the **LocalLayout** procedure to beautify neighborhoods of radius 7 × *appr*, where *appr* is the approximate distance between neighbor vertices in the current representation. The constant *iterations* was set to be 4. Typical execution time with these parameters is about 2sec for 1000-vertex graphs, under 12sec for 3000-vertex graphs, and 60sec for 6000-vertex graphs. Well optimized code will probably do better in the 6000-vertex case.

As can be seen the layouts are extremely natural looking; almost "optimal" in aesthetics. The virtually perfect layouts shown in Figures 1– 5 are typical of the power of the algorithm. The graphs in Figures 6 and 7 are particularly impressive as the "correct" (grid or torus) appearance is retained despite the partiality of information available to the algorithms.

Drawing the full binary trees in Figure 8 took more than the normal time. Force-directed methods behave better on bi-connected graphs, since there are forces that work in many ways and directions. They perform quite badly on trees. Our method succeeded in finding a nice layout of the trees only when it considered the entire graph as one in its "local" beautification stages. The problem is that the local beautification scheme is based only on neighborhoods in the graph-theoretic sense, but in deep kinds of trees, leaves should show up close to each other even if they are far apart in the graph-theoretic sense.

We mention that even in laying out full binary trees our method has two advantages when compared to Kamada and Kawai's method [10]. First, its running time is still much faster (about 10sec for the 1023-vertex tree), on account of the relatively few beautification iterations. Second, the resulting picture is almost planar and has no global distortions, in contrast to Kamada-Kawai's method, which will be trapped in many local minima, and will always result in many edge crossings.

Figure 9, which is a grid with some of the horizontal edges removed, illustrates a similar problem. In Figure 9(a), we set things up so that the local beautification procedure considered larger than regular neighborhoods, as this is the only way to take into account together the vertices of two consecutive "lines" to make the resulting picture planar. On the other hand, since in Figure 9(b) the local beautification procedure considered neighborhoods that were too small, we get the overlapping.

It is interesting to mention that the fact that Figure 9(a) involves more global considerations by taking into account larger neighborhoods is not only an advantage. In fact, these global considerations sometimes come at the expense of important local aesthetics, resulting in overcrowded clusters of vertices. In general, multi-scale aesthetics have a fundamental advantage over regular aesthetics in that they separate global considerations from local ones. Hence, a multi-scale layout algorithm serves not only as a way to minimize a complex energy-function rapidly, but often also as a better general approach to graph drawing. Mixing the global and the local aesthetics, is a *compromise* that we can adopt when we cannot distinguish the primitive clusters of the graph and are willing to pay dearly in time for considering larger clusters. For example, consider the cylinder of Figure 5. In Figure 5(a), the beautification procedure considered neighborhoods of radius 4, and in Figure 5(b) larger neighborhoods were considered. Because the correct multi-scale structure of the graph was identified, Figure 5(a) is superior to Figure 5(b), in which the structure of the small circles is not apparent.

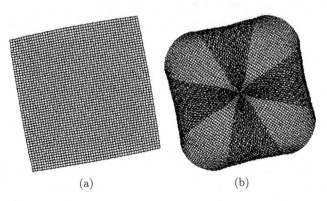

(a)   (b)

Fig. 1. (a) 55x55 (3025-vertex) square grid; (b) 80x80 (6400-vertex) square grid with each two opposite corners connected

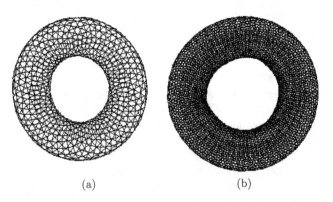

(a)   (b)

Fig. 2. Toruses: (a) 64x16 (1024-vertex) (b) 160x40 (6400-vertex)

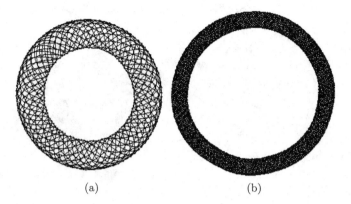

(a)   (b)

Fig. 3. Cayley graphs of the rings: (a) Z_{1000} with generators ± 9 and ± 11 (b) Z_{6000} with generators ± 13 and ± 17

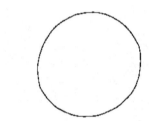

Fig. 4. 1000-vertex circle

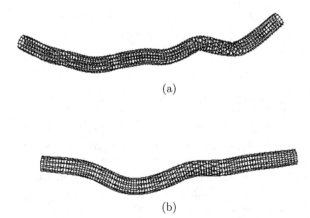

(a)

(b)

Fig. 5. 100x10 (1000-vertex) cylinder; beautification considered neighborhoods of radius 4 in (a) and 10 in (b)

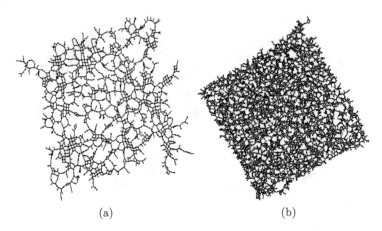

(a)   (b)

Fig. 6. (a) 40x40 (1600-vertex) grid with $\frac{1}{3}$ of the edges omitted at random; (b) 80x80 (6400-vertex) grid with $\frac{1}{4}$ of the edges omitted at random

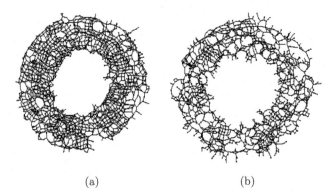

(a)   (b)

Fig. 7. 80x20 (1600-vertex) torus with $\frac{1}{5}$ and $\frac{1}{3}$ of the edges omitted at random

5   Conclusions and Future Work

We have presented a new multi-scale approach for drawing graphs nicely, and have suggested a useful formulation for the desired properties of the coarsenings. Our algorithm is able to deal extremely well and extremely fast with large graphs.

The algorithm was designed for speed and simplicity and does not require explicit representations of coarse graphs.

A more powerful and general implementations of the multi-scale drawing scheme will overcome some limitations of our algorithm. We can change the algorithm to require only linear time and space, by discarding the all-pairs shortest distance computation (and not relying any more on the Kamada-Kawai method). This will enable the algorithm to deal with graphs of over 10,000 vertices. Ano-

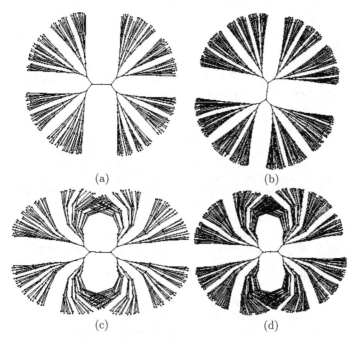

Fig. 8. Full binary trees: (a,c) 511-vertices, depth 8; (b,d) 1023 vertices, depth 9; beautification considered neighborhoods of radius 16, 18, 14 and 16, respectively; in (a) and (b) this is the whole graph.

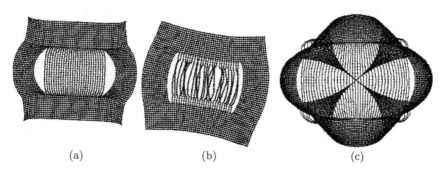

Fig. 9. (a,b) 55x55 (3025-vertex) sparse grid: (a) beautification considered larger than normal neighborhoods, of radius 35 (b) beautification considered usual neighborhoods of radius 7; (c) 80x80 (6400-vertex) sparse grid with each two opposite corners connected

ther improvement lies in the construction of the coarse scale representation, for which we used a simple heuristic. For many graphs this heuristic is fine, but for some graphs it may not be enough. For example consider graphs with a tiny diameter. Our heuristic may fail on such graphs, since the distances between each pair of vertices are roughly the same, and it is thus unable to distinguish between different clusters. We are aware of better heuristics that are more complicated, and a new implementation of the multiscale scheme is underway.

References

1. Brandenburg, F.J., Himsolt, M., and Rohrer, C., "An Experimental Comparison of Force-Directed and Randomized Graph Drawing Algorithms", *Proceedings of Graph Drawing '95*, Lecture Notes in Computer Science, Vol. 1027, pp. 76–87, Springer Verlag, 1995.
2. Di Battista, G., Eades, P., Tamassia, R. and Tollis, I.G., *Algorithms for the Visualization of Graphs*, Prentice-Hall, 1999.
3. Davidson, R., and Harel, D., "Drawing Graphs Nicely Using Simulated Annealing", *ACM Trans. on Graphics* **15** (1996), 301–331.
4. Eades, P., "A Heuristic for Graph Drawing", *Congressus Numerantium* **42** (1984), 149–160.
5. Fruchterman, T.M.G., and Reingold, E., "Graph Drawing by Force-Directed Placement", *Software-Practice and Experience* **21** (1991), 1129–1164.
6. Gonzalez, T., "Clustering to Minimize the Maximum Inter-Cluster Distance", *Theoretical Computer Science* **38** (1985), 293–306.
7. Hadany, R., and Harel, D., "A Multi-Scale Method for Drawing Graphs Nicely", *Discrete Applied Mathematics*, in press, 2000. (Also, *Proc. 25th Inter. Workshop on Graph-Theoretic Concepts in Computer Science* (WG '99), Lecture Notes in Computer Science, Vol. 1665, pp. 262–277, Springer Verlag, 1999.)
8. Hochbaum, D. S. (ed.), *Approximation Algorithms for NP-Hard Problems*, PWS Publishing Company, 1996.
9. Hochbaum, D.S., and Shmoys, D. B, "A Unified Approach to Approximation Algorithms for Bottleneck Problems", *J. Assoc. Comput. Mach.* **33** (1986), 533–550.
10. Kamada, T., and Kawai, S., "An Algorithm for Drawing General Undirected Graphs", *Information Processing Letters* **31** (1989), 7–15.
11. The LSD library, available from the Graphlet website at http://www.fmi.unipassau.de/Graphlet/download.html

FADE: Graph Drawing, Clustering, and Visual Abstraction

Aaron Quigley and Peter Eades

http://www.cs.newcastle.edu.au/~aquigley
Department of Computer Science and Software Engineering,
Univ. of Newcastle, Callaghan, NSW 2308, Australia

Abstract. A fast algorithm(FADE) for the 2D drawing, geometric clustering and multilevel viewing of large undirected graphs is presented. The
algorithm is an extension of the Barnes-Hut hierarchical space decomposition method, which includes edges and multilevel visual abstraction.
Compared to the original force directed algorithm, the time overhead is
$O(e + n \log n)$ where n and e are the numbers of nodes and edges. The
improvement is possible since the decomposition tree provides a systematic way to determine the degree of *closeness* between nodes without
explicitly calculating the distance between each node. Different types of
regular decomposition trees are introduced. The decomposition tree also
represents a hierarchical clustering of the nodes, which improves in a
graph theoretic sense as the graph drawing approaches a lower energy
state. Finally, the decomposition tree provides a mechanism to view the
hierarchical clustering on various levels of abstraction. Larger graphs can
be represented more concisely, on a higher level of abstraction, with fewer
graphics on screen.

1 Introduction

Force-directed algorithms are often used for graph drawing due to their flexibility,
ease of implementation and their often pleasant resultant drawings. Numerous
graph drawing systems have been developed based on such force directed algorithms [9,36,8,5]. The flexibility of the force directed approach allows for many
kinds of constraints [16,3,31,13]. However, one common feature of most force
directed methods, is their inability to scale to handle large graphs. The problem
stems from the high computational cost of computing a force between every pair
of nodes.

This paper introduces a fast algorithm for the drawing of large undirected
graphs based on the force directed approach. This algorithm hierarchical *groups*
nodes. The nonedge force on an individual node from other nodes close by is,
on average, evaluated by direct node-to node interaction, whereas the force due
to more distant nodes is included as a *group* contribution. The reduction in the
computational cost of the force directed approach allows larger graphs to be
drawn in real-time.

J. Marks (Ed.): GD 2000, LNCS 1984, pp. 197–210, 2001.

The fact that larger graphs can be drawn using this method is not enough for visualization. It is easy to draw a great deal of graphical information but the limits on both the available screen space and more importantly human cognition are quickly reached [35,34,33,20].

The rest of this paper is organized as follows. In section 2, we review the concepts of *Graph Theoretic Clustering* and *Geometric Clustering* and how they relate to the visualization of large graphs. In section 3 we outline *hierarchical space decomposition* and *tree codes*. Our algorithm FADE is described in section 4. Section 5 describes the *progressive cycle* of clustering, based on a step-wise improvement in the drawing. We discuss the generation of the *visual abstraction* in section 6. And finally, in section 7 our results for performance, error rates and clustering are presented.

2   Clustering

2.1   Graph Theoretic Clustering

A *clustering* of a graph, $G = (V, E)$ consists of a partition $V = V_1 \cup V_2 \cup .... \cup V_k$ of the node set of G. *Graph theoretic clustering* is the process of forming clusters based on the structure of the graph [22,29,23,6,24,30]. The usual aim is to form clusters that exhibit a high *cohesiveness* and a low *coupling*; these may be measured, for example, by the number of *intra*-cluster edges and the number of *inter*-cluster edges respectively. For example, the graph with the adjacency table in Figure 1 has been divided into two clusters with only one edge coupling them. Methods for graph theoretic clustering, motivated by applications in parallel computing, data mining, and VLSI design are well developed [6,24,30].

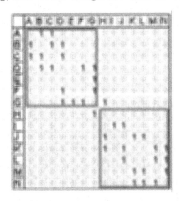

Fig. 1. Graph Theoretic, single linkage clustering example

One of the main problems for graph theoretic clustering methods is selecting a threshold of coupling, without some *a priori* knowledge of the data. A low threshold results in too many clusters with too few elements. A high threshold results in large tree-like clusters (*dendograms*), with the degenerate case of all the nodes collapsing into a single cluster.

2.2   Clustered Graphs

A clustered graph is a graph with a recursive clustering, or partitioning, of the node set of G [26]. Graph drawing algorithms have been developed to draw clustered graphs [37,19,27,7,26]. Recently, methods have been developed that exploit clusters to improve the performance of the underlying drawing algorithm. Harel et al. [11,10] show how a force directed method can use the graph theoretic structure to approximate long range force interactions; here *node to cluster* forces are computed rather than *node to node* forces. However, approximating forces based on the graph theoretic structure has some drawbacks:

1. The computational cost in discovering the recursive clustering can quickly dominate as the graphs become large.
2. If the graph theoretic clustering is poor, then there is no time saving from such an approximate force scheme.
3. The resultant graph drawing can be distorted, based on the clustering scheme chosen; this may adversely affect its usefulness in visualization.

The partitioning scheme chosen in [11] may be improved by the use of multi-level partitioning schemes [21,22] where the graph is coarsened, then the smaller graph is partitioned, and then uncoarsened to construct a partitioning of the original graph. It should be noted, that the scheme introduced in [11], produces nice drawings of large graphs that exhibit good graph theoretic clusters. For graphs, such as trees, it is noted that a post-processing *beautification* step, on the entire graph, must be performed.

2.3   Geometric Clustering

A *geometric clustering* of a set S of points in k-dimensional space is a partition of $S = S_1 \cup S_2 \cup .... \cup S_p$ into subsets. The usual aim of a geometric clustering method is to have points which are close to each other sharing a cluster and points that are far apart in different clusters. Geometric clustering is used in a wide variety of data mining applications; the k attributes of an entity are mapped to a point in k-dimensional space, and an appropriate distance function is used [32,6,24,30]. One such function is the *K-means algorithm*, which is based on a distance metric where the attributes are scaled so that the Euclidean distance between sites is appropriate [12].

In many circumstances, a geometric clustering is defined by a *space decomposition*, such as the Voronoi diagram in Figure 2. Geometric clustering may be applied to the location of nodes in a graph drawing; however, unlike graph theoretic clustering, the edges are not used to determine the geometric clustering.

2.4   Graph Drawing and Clustering

A Geometric clustering method [32,6,24] applied to a graph drawing induces a graph theoretic clustering, but it may have low quality in terms of coupling and cohesion [21]. To ensure that the graph theoretic clustering has high quality, the

Fig. 2. Voronoi diagram based on pseudnodes, real nodes shown as points

graph drawing must exhibit a strong relationship between geometric and graph theoretic distance between nodes.

Our graph drawing method (FADE) uses a fast recursive space decomposition, which induces a geometric clustering of the locations of the nodes; this in turn induces a graph theoretic clustering. This graph theoretic clustering is then used in a force directed algorithm that improves the relation between Euclidean and graph theoretic distances in the drawing, and this in turn improves the graph theoretic clustering. Iterating this process improves both the drawing and the clustering. Viewing the graph drawing on different abstraction levels, of the cluster tree, gives a set $A=A_1,A_2,A_3...A_x$ of abstractions, where x is the height of the decomposition tree.

3   Hierarchical Space Decomposition

Modeling large numbers of interacting particles has interested physicists for centuries. Typical examples include celestial mechanics, plasma simulations and the vortex method in fluid dynamics [25,14,18,15]. Such numerical simulations are referred to as the *N-body problem* and typically involve 10^{12} - 10^{23} particles. In classical systems, the *potential* (or force) has the form:

$$\phi = \phi_{short} + \phi_{long} + \phi_{external}, \tag{1}$$

where ϕ_{short} is a rapidly decaying function of distance (such are the Van der Waals potential in chemical physics), $\phi_{external}$ is a function which is independent of the number and relative positions of the particles (such as a magnetic field); these forces can be computed in time O(N). However, the computation of ϕ_{long}, the particle to particle interactions, takes time O(N^2). The ϕ_{long} force can be thought of as an *electrical repulsion* force in the *classical force-directed* methods (Chapter 10 [9]) for graph drawing. The cost of O(N^2) is excessive in most cases, and prohibitive in others. When using the force-directed approach for graph drawing the same limitations apply. In Physics different approaches have been developed to reduce the computational effort in calculating the long range ϕ_{long} forces by means of *approximation* or *estimation*.

One such method is called *particle-in-cell (PIC)*: a regular grid is laid over the simulation area and the particles contribute their masses to create a source density. These source densities are used in the estimation of the forces on a particular particle. PIC codes have been popular in particle simulations but have difficulties dealing with non-uniform particle distributions, that is, where the particle distribution is clustered [25]. The modified force directed algorithm of Fruchterman and Reingold [8] follows the PIC approach. PIC based methods prove unsuitable for approximating the forces in graph drawings which do not exhibit a uniform node distribution.

3.1   Tree Codes

A systematic and recursive division of space into a series of *cells*, called a *tree code* in the astrophysics literature [2,25,18,4,1], can be used to speed up n-body force calculations. The force on an individual particle from other particles close by is, on average, evaluated by direct particle-to-particle interaction, whereas the force due to more distant particles is included as a group contribution. Tree codes were first developed in the context of galactic simulations involving hundreds of billions of bodies. Tree codes may be based on regular (*Quad-tree, Oct-tree*) and irregular (*Voronoi*) space decompositions. A *nono-tree* (a nine-way recursive decomposition), is illustrated in Figure 3. Regular trees such as this can be built in linear time.

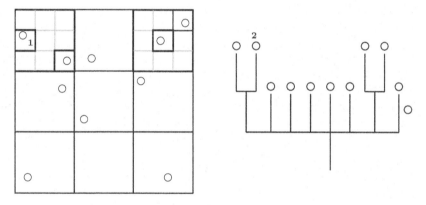

Fig. 3. Nono-tree space decomposition and data structure

Unlike PIC codes building a decomposition tree does not suffer from a non-uniform distribution of particles.

3.2   Tree-Codes and Clustering

The decomposition of the nodes of a graph drawing based on a tree-code method generates a recursive geometric clustering of the vertices of the graph. The recursive clustering, represented as sub-trees, does not directly produce a high *quality* geometric clustering of the node positions, nor is it necessarily a high quality graph theoretic clustering. However the clustering does facilitate an improvement in the performance of force directed algorithms, as in shown in Section 4.1 below. Further, it allows for multi-level viewing of huge graphs at various levels of abstraction. Finally, as the quality of the drawing improves (that is, it reaches a lower energy state of the force system), the quality of suggested clustering, exhibited by the decomposition tree, improves.

4   FADE Algorithm

The main idea of the FADE algorithm is the following. Using the decomposition tree the *nonedge force* on an individual node from other nodes close by is, on average, evaluated by direct node-to-node interaction, whereas the force due to more distant nodes is included as a node cluster contribution. The main difference with the graph theoretic cluster approach of Harel et al. [11,10] is that we approximate forces based on actual geometric information, rather than preprocessed approximate forces. This difference allows our force directed method to be applied to large scale graphs, without distorting the resultant drawing due to the underlying graph structure.

4.1   Implementation

We have implemented FADE, a drawing algorithm for undirected graphs, in two dimensions, using a simple force model and a quad-tree space decomposition. Force directed algorithms, such as those in [16,3,13,9,36,8] often have two kinds of forces: *edge-forces* such as Hooke's law spring forces, and *non-edge forces*, such as magnetic repulsion. Typical graphs in graph drawing applications have $O(n)$ edges and $\Omega(n^2)$ non-edges, where n is the number of nodes. The FADE algorithm computes node-node forces for edges, and approximates non-edges forces using a tree-code approach. Where the tree is reconstructed *ab initio* for every new iteration of the force directed algorithm FADE.

Roughly speaking, FADE works as follows:

```
REPEAT:
  Construct geometric clustering using a space decomposition
  Compute edge forces
  Compute nonedge forces (node-node and node-pseduonode)
  Move nodes
UNTIL convergence
```

The tree structure, provided by a recursive space decomposition such as a quad-tree, provides a systematic way to determine the degree of *closeness* between nodes without explicitly calculating the distance between each node, see for example Figure 8. The nonedge force between near nodes is calculated directly whereas more distant nodes are grouped together into super-nodes or pseudonodes.

The most basic means for determining closeness is a *tolerance criterion* [25, 2]. We test $s/d \leq \theta$, where s is the width of the cell, d is the distance between the current node and the center of mass of the cell and θ is the fixed-tolerance parameter. If $s/d \leq \theta$, then the internal nodes are ignored and the force contribution of the pseudonode is added to the cumulative force for that node. Otherwise, the pseudonode is resolved into its daughter pseudonodes (sub-trees), each of which is recursively examined. Pseudonodes are resolved by continuing the decent through the tree until either the tolerance criterion is satisfied or a leaf node is reached.

Using a quad-tree space decomposition and a value of 1.0 for θ the fixed-tolerance parameter, some examples are given for the node set shown on the left of Figure 4.

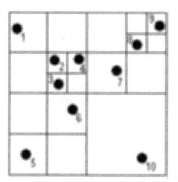 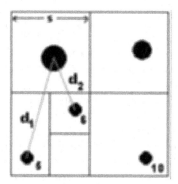

Fig. 4. Comparing node 5 and 6 with the North-West pseudonode, where s/d=0.80 and s/d=1.46 respectively

In Figure 4, the leaf node 5 is compared with the pseudo particle in the North-West quad. The weight of the pseudonode is the cumulative weight of the leaves in its sub-tree, in this case 4. The s value is the width of the cell the pseudonode covers and d_1 is the distance between the node and the pseudonode. $\frac{s}{d} = 0.80$ in this case. Based on the value 1.0 for θ the FADE algorithm computes the nonedge force between this pseudonode and node 5, it then adds this force to the cumulative nonedge force for node 5.

Leaf node 6, is compared with the pseudo particle and d_2 is the distance between the node and the pseudonode in the North-West quad to get, $\frac{s}{d} = 1.46$. Here the value does not fulfill the criterion $s/d \leq \theta$, hence the pseudonode is resolved into its sub-nodes (in this case leaf-node 1 and the pseudonode representing 2,3,4). The algorithm continues by testing $s/d \leq \theta$ for the pseudonode(2,3,4) and if it fulfills the criterion then a node to pseudonode force calculation is performed. The *interaction list* for a given node, is the list of pseudonodes (twig nodes) and nodes (leaf nodes) involved in the calculation of its nonedge force.

For nonzero θ the time required to compute the force in a single iteration of FADE scales as $O(n \log n)$ [4,14]. The actual interaction list for a given node depends strongly on the choice of the tolerance parameter θ. The case $\theta = 0$

is equivalent to computing all node-node nonedge forces, which is the classical approach in $O(N^2)$ force-directed methods. However a very large θ results in the calculation of only forces between nodes and very large pseudonodes. It would be very quick, approaching $O(N)$ but very inaccurate. The high level of inaccuracy exhibits itself as a poor drawing of the graph, based on a force computation.

There is a cross-over point in the number of nodes where hierarchical tree codes becomes more efficient than the direct method. Optimizing the tree construction and the force calculations by vectorisation or parallelization lowers this cross-over point. When the number of dimensions in the force law is low this can be in the order of a few hundred to several thousand nodes [25].

5   Progressive Cycle

The space decomposition tree generated at each timestep in the FADE algorithm represents a recursive geometric clustering of the nodes of the graph. The drawing typically moves to a lower energy state from timestep to timestep. This is as a result of the edge and nonedge forces. Overall, the drawing improves as nodes with an edge between them, are on average drawn closer than nodes without. The geometric clustering generated also improves as the drawing improves; this process in referred to as the *progressive cycle*. Various geometric measures of clustering can be applied to each recursive clustering, and this allows for a measure of how the clustering improves as the forces are applied.

Results from such a measure can be seen in figure 5, the measure here is the number of *inter-cluster edges* vs *intra-cluster* edges. Here we show this measure applied to different levels in the hierarchical geometric clustering over 100 iterations of *fade* for a graph of 1020 nodes. This measure shows the result of the drawing reaching a stable state, which induces an improvement in the quality of the clustering over time, measured in terms of the coupling.

6   Visual Abstraction

If the drawing of the graph cannot easily be viewed on screen, for example due to poor resolution, or the sheer volume of graphical information, then the visualization can no longer be considered useful to the user. Numerous viewing schemes have been proposed to overcome this problem [17,9,28,16,7]. FADE follows the clustered graph drawing approach. It has been shown, that the tree built by the FADE algorithm produces a hierarchical clustering of the vertices of the graph. Similar to other cluster graph drawing techniques, a particular level in the cluster tree (an *abridgement*) can be drawn on screen. The choice of level allows the graph to be concisely (i.e. more abstractly) represented with fewer graphics on screen.

The abridgement of the graph drawing results in a loss of precision. This has to be measured against the fact that a greater understanding of the overall graph is now possible. A visualization scheme based on a multi-level viewing still has the ability to mix levels of abstraction, allowing greater detail in certain areas of interest, while maintaining an abstract view of the rest of the graph drawing. The amount of graphical elements presented to the user in Figure 6(left side) (16 pseudonodes and 33 edges) is approximately 0.67% of that displayed in drawing of the underlying graph on the right side in Figure 7 (2500 nodes and 4900 edges). The first drawing can be easily be drawn in realtime and it still provides a visual approximation to the overall shape of the graph, and if varying pseudonode sizes were used, to the node distribution. Several examples of multi-level viewing, based on a regular recursive decomposition of space, are shown in the Appendix.

7   Experimental Results

We ran our experiment based on a Parlance implementation running on a Pentium III 500 Mhz processor. The performance results were as expected, from the physics literature [4,25], with a large improvement over the classical direct approach per iteration. The error rate is a vector measure computed from the direct non-edge forces and the approximate non-edge forces computed in the FADE algorithm. It is important to note the error rates, which are less than 1%, do not grow as the graphs get larger. The error rate is further damped when the edge forces are considered.

Table 1. Experimental Comparison of tree-code Vs direct force calculation

Nodes	Direct (sec)	FADE (sec)	% Error
512	0.455	0.04	0.513
1020	1.82	0.088	0.592
1442	3.61	0.168	0.675
2500	10.88	0.202	0.622
6000	62.66	0.676	0.673
10510	192	1.704	0.449
22800	920	3.36	0.561
30000	1593	3.546	0.517
40960	2979	5.592	0.567
49284	4316	6.730	0.628
105233	19604	13.371	0.481

8   Conclusions and Future Work

FADE is an algorithm for the drawing and multi-level visualization of large undirected graphs. It is based on an extension of tree-codes for the N-body problem in physics. The algorithm is effective in drawing large graphs and the use of the space decomposition tree provides a new and novel method for the multi-level viewing of large graphs and provides a geometric clustering of the nodes of the graph. Future work will investigate the use of other quality measures for graph

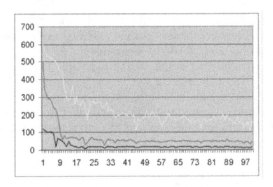

Fig. 5. Cluster measure for levels 4, 6, 8 in the clustering

clustering and what affects other regular and irregular space decompositions have on these measures and on the visual abstraction.

Finally, approaches to data mining based on the generation of clusters from a hierarchical space decomposition have been developed [39,38]. These methods analyze weights of sub-trees to discover candidate clusters and then compose adjacent sub-trees, based on various measures, to form the clusters. The decisions are based on the weights of the sub-trees, which are inherently localized. For geometric graph cluster analysis, it should be possible, to integrate the force model into the decision process, thereby taking global information into account. Not only the size of the sub-tree but also how often this sub-tree is considered to be a pseudonode in the force computation.

References

1. Richard J. Anderson, *Tree data structures and n-body simulation*, SIAM J. Comput. **28** (1999), no. 6, 1923–1940.
2. J. Barnes and P. Hut, *A hierarchical o(n log n) force-calculation algorithm*, Nature **324** (1986), no. 4, 446–449.
3. Francois Bertault, *A force-directed algorithm that preserves edge crossing properties*, Seventh International Symposium on Graph Drawing (Prague, Chezh Republic) (Jan Kratochvíl, ed.), Springer, September 1999, pp. 351–358.
4. G. Blellcoh and G. Narlikar, *A practical comparison of n-body algorithms*, Technical report, Wright Laboratory, 1991.
5. Peter Eades, *A heuristic for graph drawing*, Congresses Numerantium **42** (1984), 149–160.
6. G. Erhard, *Advances in system analysis vol. 4, graphs as structural models: The application of graphs and multigraphs in cluster analysis*, Vieweg, 1988.
7. Q.W. Feng., *Algorithms for drawing clustered graphs*, Ph.D. thesis, The University of Newcastle, Australia, 1997.
8. T. Fruchterman and E. Reingold, *Graph drawing by force-directed placement*, Software-Practice and Experience **21** (1991), no. 11, 1129–1164.
9. Roberto Tamassia G. Di Battista, P. Eades and I. G. Tollis, *Graph drawing, algorithms for the visualization of graphs*, Prentice-Hall Inc., 1999.

10. R. Hadany and David Harel, *A multi-scale method for drawing graphs nicely*, 25th International Workshop on Graph-Theoretic Concepts in Computer Science, 1999.

11. David Harel and Yehuda Koren, *A fast multi-scale method for drawing large graphs*, Technical report, Dept. of Applied Mathematics and Computer Science, Weizmann Institute, Rehovot, Israel, November 1999.

12. J. Hartigan, *Clustering algorithms*, Wiley, 1975.

13. W. He and K. Marriott, *Constrained graph layout*, Symposium on Graph Drawing, GD '96 (Stephen North, ed.), vol. 1190 of Lecture notes in Computer Science, Springer, 1996, pp. 217–232.

14. L. Hernquist, *Hierarchical n-body methods*, Computational Physics Communications **48** (1988), 107–115.

15. R. W. Hockney and P. M. Sloot, *Computer simulations using particles*, McGraw-Hill, 1981.

16. Mao Huang, *On-line animated visualization of huge graphs*, Ph.D. thesis, The University of Newcastle, Australia, 1999.

17. Maurice M. de Ruiter Ivan Herman, Guy Melançon and Maylis Delest, *Latour - a tree visualisation system*, Seventh International Symposium on Graph Drawing (Prague, Chezh Republic) (Jan Kratochvíl, ed.), Springer, September 1999, pp. 392–399.

18. L. Greengard J. Carriert and V. Rokhlin, *A fast adaptive multipole algorithm for particle simulations*, SIAM Journal on Scientific Computing **9** (1988), no. 4, 669–686.

19. Arunabha Sen Jubin Edachery and Franz J. Brandenburg, *Graph clustering using distance-k cliques*, 7th International Symposium on Graph Drawing, GD '99 (Jan Kratochvíl, ed.), vol. 1731 of Lecture notes in Computer Science, Springer, 1999, pp. 98–106.

20. P. Eades et al K. Misue, *Layout adjustment and the mental map*, Journal of Visual Languages and Computing (1995), 183–210.

21. George Karypis and Vipin Kumar, *A fast and high quality multilevel scheme for partitioning irregular graphs*, SIAM Journal on Scientific Computing (1998).

22. _____, *A parallel algorithm for multilevel graph partitioning and sparse matrix ordering*, Journal of Parallel and Distributed Computing (1998).

23. B. W. Kernighan and S. Lin, *An efficient heuristic procedure for partitioning graphs*, The Bell System Techinical Journal (1970).

24. M. Lorr, *Cluster analysis for social scientists*, Jossey-Bass Ltd., 1987.

25. Susanne Pfalzner and Paul Gibbon, *Many-body tree methods in physics*, Cambridge University Press, 1996.

26. R. Cohen Q. W. Feng and P. Eades, *How to draw a planer clustered graph*, COCOON, vol. 959, Lecture notes in Computer Science, Springer, 1995, pp. 21–31.

27. Reinhard Sablowski and Arne Frick, *Automatic graph clustering (system demonstration)*, Symposium on Graph Drawing, GD '96 (Stephen North, ed.), vol. 1190 of Lecture notes in Computer Science, Springer, 1996, pp. 395–400.

28. Manojit Sarkar and Marc Brown, *Graphical fisheye views of graphs*, ACM SIGCHI '92 Conference on Human Factors in Computing Systems, March 1992.

29. Horst Simon and Shang-Hua Teng, *How good is recursive bisection*, Nasa - publication, Ames Research Center NASA, June 1993.

30. H. Spath, *Cluster analysis algorithms for data reduction and classification of objects*, Horwood, 1980.

31. K. Sugiyama and K. Misue, *A simple and unified method for drawing graphs: Magnetic-spring algorithm*, Proceedings of Graph Drawing, GD '94 (Roberto Tamassia and Ioannis Tollis, eds.), vol. 894 of Lecture Notes in Computer Science, Springer, 1995, pp. 364–375.

32. S. Teng, *Points, spheres, and separators: A unified geometric approach to graph partitioning*, Ph.D. thesis, School of Computer Science, Carnegir Mellon University, 1991.

33. Edward R. Tufte, *The visual display of quatitative information*, Graphics Press, 1983.

34. _____, *Envisioning information*, Graphics Press, 1990.

35. _____, *Visual explanations*, Graphics Press, 1997.

36. Daniel Tunkelang, *Jiggle: Java interactive general graph layout environment*, Sixth International Symposium on Graph Drawing (McGill University, Canada) (Sue Whitesides, ed.), Springer, August 1998.

37. Andrej Mrvar Vladimir Batagelj and Matjaž Zaveršnik, *Partitioning approach to visualization of large graphs*, 7th International Symposium on Graph Drawing, GD '99 (Jan Kratochvíl, ed.), vol. 1731 of Lecture notes in Computer Science, Springer, 1999, pp. 90–97.

38. Jiong Yang Wei Wang and Richard Muntz, *Sting: A statistical information grid approach to spatial data mining*, 23rd International Conference on Very Large Data Bases(VLDB), IEEE, 1997, pp. 186–195.

39. _____, *Sting+: An approach to active spatial data mining*, IEEE, 1999, pp. 116–125.

9 Appendix

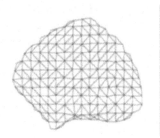

Fig. 6. Graph of 2500 nodes shown on levels 3 and 5 of the decomposition tree

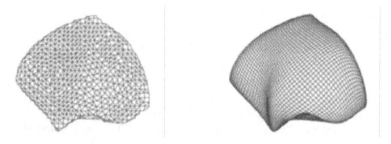

Fig. 7. Graph of 2500 nodes shown on level 6 and the lowest level of the decomposition tree

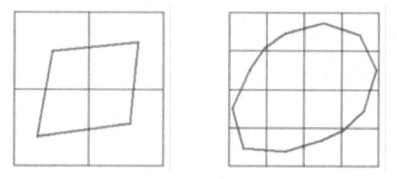

Fig. 8. Cycle of 400 nodes shown on levels 2 and 3 with the space decomposition overlaid

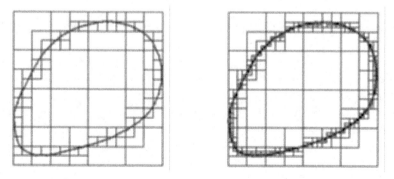

Fig. 9. Cycle of 400 nodes shown on level 5 and the lowest level, with the space decomposition overlaid

Fig. 10. Graph of 4700 nodes shown on levels 3 and 5 of the decomposition tree

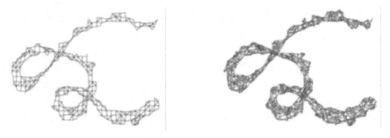

Fig. 11. Graph of 4700 nodes shown on level 8 and the lowest level of the decomposition tree

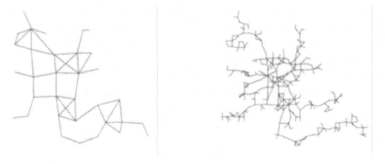

Fig. 12. Graph of 1400 nodes shown on levels 4 and 6 of the decomposition tree

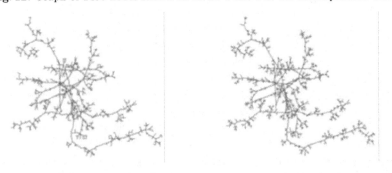

Fig. 13. Graph of 1400 nodes shown on level 8 and the lowest level of the decomposition tree

A Multi-dimensional Approach to Force-Directed Layouts of Large Graphs[*]

Pawel Gajer[1], Michael T. Goodrich[1], and Stephen G. Kobourov[2]

[1] Department of Computer Science
Johns Hopkins University
Baltimore, MD 21218
[2] Department of Computer Science
University of Arizona
Tucson, AZ 85721

Abstract. We present a novel hierarchical force-directed method for drawing large graphs. The algorithm produces a graph embedding in an Euclidean space $\mathbb{E}$ of any dimension. A two or three dimensional drawing of the graph is then obtained by projecting a higher-dimensional embedding into a two or three dimensional subspace of $\mathbb{E}$. Projecting high-dimensional drawings onto two or three dimensions often results in drawings that are "smoother" and more symmetric. Among the other notable features of our approach are the utilization of a maximal independent set filtration of the set of vertices of a graph, a fast energy function minimization strategy, efficient memory management, and an intelligent initial placement of vertices. Our implementation of the algorithm can draw graphs with tens of thousands of vertices using a negligible amount of memory in less than one minute on a mid-range PC.

1 Introduction

Graphs are common in many applications, from data structures to networks, from software engineering to databases. Typically, small graphs are drawn manually so that the resulting picture best shows the underlying relationships. The task of drawing graphs by hand becomes more challenging as the complexity of the graphs increases. Graph drawing tools have been the focus of the graph drawing community for at least the last two decades, see [5,6] for a comprehensive reviews of the field. Numerous algorithms have been developed for drawing special classes of graphs such as trees and planar graphs. There are fewer general purpose graph drawing algorithms, however. Force-directed methods are the methods of choice for drawing general graphs. Substantial interest in force-directed methods stems from their conceptual simplicity, applicability to general graphs, and aesthetically pleasing drawings.

With few exceptions, most existing automated systems have trouble dealing with graphs of thousands of vertices. In this paper we present a new algorithm

---

[*] This research partially supported by NSF under Grant CCR-9625289, and ARO under grant DAAH04-96-1-0013.

J. Marks (Ed.): GD 2000, LNCS 1984, pp. 211–221, 2001.
© Springer-Verlag Berlin Heidelberg 2001

which allows for drawing simple undirected graphs with tens of thousands of vertices in under a minute. Even larger graphs can be displayed using this algorithm in conjunction with a fisheye view [14,19,22] or a multi-level display algorithm [7] which would allow us to accommodate graphs with more vertices than the number of pixels of the display device. However, the effectiveness of fisheye and multi-level views depends on a good recursive clustering, which in turn depends on a good initial embedding of the graph. Creating a good embedding for large graphs has been prohibitively expensive using existing algorithms. Our algorithm allows us to create excellent initial embeddings in very reasonable times; hence, it can be used either by itself or as a preprocessing step to the above large-graph layout methods. The key features of the algorithm are: (1) intelligent initial placement of vertices; (2) multi-dimensional drawing; (3) a simple recursive coarsening scheme; (4) fast energy function minimization; (5) space and time efficiency.

The rest of this paper is organized as follows: In Section 2 we review previous work in visualization of large graphs and force-directed algorithms for automated graph drawing. In Section 3 we describe our algorithm and in Section 4 we present some concluding remarks.

2   Previous Work

2.1   Visualization of Large Graphs

Visualizing large graphs presents unique problems which require non-orthodox solutions. Drawings that display the entire graph have the advantage of showing global graph structure. For large graphs such drawings become impractical as the limited resolution of display devices makes details hard to discern. Partially drawing graphs allows for display of larger graphs but fails to convey their global structure. Two other approaches to visualization of large graphs are of particular interest: fisheye views and multi-level displays. Fisheye views [14,19, 22] show an area of interest quite large and detailed while showing other areas successively smaller and in less detail. Multi-level views allow us to view large graphs at multiple abstraction levels. A natural realization of such multiple level representations is a 3D drawing with each level drawn on a plane at a different z-coordinate, and with the clustering structure drawn as a tree in 3D.

The multi-level display algorithms are introduced by Eades and Feng [9] in the context of visualization for clustered graphs. Compound and clustered graphs are studied in [10,11,20,23]. Creating a graph clustering based on binary space partitions and using it to display large graphs was introduced by Duncan, Goodrich, and Kobourov [7]. The quality of the resulting multi-level drawings depends on the initial embedding of the graph in the plane.

2.2   Force-Directed Algorithms

The force-directed placement algorithm of Quinn and Breur [21] and the spring embedder of Eades [8] are among the first practical algorithms for graph drawing.

In the latter algorithm the graph is modeled as a physical system of rings and springs. Classical force-directed methods start from a random embedding of a graph and utilize standard optimization methods to find a minimum of an energy function of their choice. The use of an energy function E is a characteristic feature of force-directed layout algorithms. It is used to assign to each embedding $\rho : G \to \mathbb{R}^n$ of a graph G in some Euclidean space $\mathbb{R}^n$ (typically $n = 2$ or $n = 3$) a non-negative number $E(\rho)$. Force-directed methods are based on the premise that minima of reasonably chosen energy functions produce aesthetically pleasing graph drawings. The main differences between force-directed algorithms are in the choice of energy function and the methods for its minimization. Examples of force-directed algorithm include the algorithms of Kamada and Kawai [18], Davidson and Harel [4], Fruchterman and Reingold [13], and Frick et al [3,12].

The main problem with most standard force-directed algorithms is their inability to draw large graphs. Even the best classical algorithms can draw graphs with a maximum of only several hundred vertices. When presented with a computationally expensive graph algorithm, a standard approach is to associate with the graph a hierarchy of graphs. The needed computation is done starting with the smallest graph in the hierarchy, then proceeding to larger and larger graphs and using at each stage the results of the previous computation. This strategy has been brought to the area of force-directed graph drawing from particle physics [1,2] in the multi-scale algorithm of Hadany and Harel [16]. In [17] Harel and Koren introduce several simplifications to the algorithm resulting in faster drawings and allowing for larger graphs.

However, as one of the underlying steps of the algorithm in [17], all-pairs shortest paths are computed, which is both time and space expensive. The quadratic space complexity incurred by the matrix of distances between vertices of the graph is another problem when we draw large graphs. Other computationally expensive procedures include the clustering procedure for a construction of a hierarchy of graphs and the Newton-Raphson optimization method for scaling the displacement vectors. Finally, the algorithm in [17] creates drawings in 2D and as it is based on the Newton-Raphson method, extending it to 3D considerably slows down the algorithm. The algorithm described in the next section addresses the above problems and introduces several new features.

3   The Algorithm

3.1   Algorithm Overview

The pseudo-code for the algorithm can be seen in Fig. 1. In the first stage we create a filtration of the set of vertices of the given graph and set up the scheduling function nbrs(), described in Sections 3.2 and 3.3, respectively. The main for-loop runs through all levels of the filtration, starting at V_k. At stage i for each vertex $v \in V_i - V_{i+1}$ we find sets $N_i(v), N_{i-1}(v), \dots, N_0(v)$ and find an initial position pos[v] of v. The vertex neighborhood $N_i(v)$ is a set of nbrs(i) closest to v elements of V_i. The method for determining $N_i(v), N_{i-1}(v), \dots, N_0(v)$ and for determining the initial positions are in Sections 3.3 and 3.4, respectively.

MAIN ALGORITHM
   create a filtration $\mathcal{V}: V_0 \supset V_1 \supset \ldots \supset V_k \supset \emptyset$
   set up scheduling function nbrs()
   for $i = k$ **to** 0 **do**
      for each $v \in V_i - V_{i+1}$ **do**
         find vertex neighborhood $N_i(v), N_{i-1}(v), \ldots, N_0(v)$
         find initial position pos$[v]$ of v
      repeat rounds times
      for each $v \in V_i$ **do**
         compute local temperature heat$[v]$
         disp$[v] \leftarrow$ heat$[v] \cdot \vec{F}_{N_i}(v)$
      for each $v \in V_i$ **do**
         pos$[v] \leftarrow$ pos$[v] +$ disp$[v]$
   add all edges $e \in E$

Fig. 1. After creating the vertex filtration and setting up the scheduling function the algorithm processes each filtration set, starting with the smallest one. Here pos$[v]$ is a point in $\mathbb{R}^n$ corresponding to vertex v and rounds is a small constant. In the refinement stage heat$[v]$ is scaling factor for the displacement vector disp$[v]$, which in turn is computed over a restriction $N_i(v)$ of the vertices of G.

The refinement stage is repeated **rounds** times, where **rounds** is a small constant. Within the refinement stage, the displacement vector disp$[v]$ of v is set to a local Kamada-Kawai force vector. Here local means that the force vector $\vec{F}_{N_i}(v)$ is computed over v's vertex neighborhood $N_i(v)$ rather than over all vertices in G. The displacement vector is scaled by a local temperature factor heat$[v]$. In Section 3.5 we describe the process of calculating heat$[v]$.

3.2   Vertex Set Filtrations

Faced with the problem of drawing a large graph, it is natural to associate with it a hierarchy of graphs and produce a drawing starting with the smallest graph in the hierarchy, and drawing larger and larger graphs using at each stage the previous drawing. Two important properties of such a hierarchy are its depth and the distribution of vertices. A constant depth hierarchy implies that as we go from one level to the next, more than a constant fraction of the vertices are added and this makes the drawing of the old level insufficient for placement of new vertices. On the other hand, a linear depth hierarchy is too time consuming to traverse. Thus, logarithmic depth is highly desirable. The effectiveness of this scheme is also dependent on the uniformity of the distribution of the vertices at all levels of the hierarchy. The hierarchy of graphs can be thought of as containing different levels of abstraction of the underlying graph. Uniform distribution of the vertices implies more accurate levels of abstraction which in turn implies better drawings on each level.

Hadany and Harel [16] create a hierarchy of graphs based on the cluster number, the degree number, and the homotopic number. Harel and Koren [17]

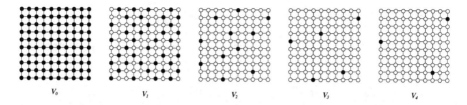

Fig. 2. An example of a MIS filtration. Here the underlying graph $G = (V, E)$ is a rectangular mesh of size 10×10. The dark vertices are included in the filtration. Here $V = V_0$, V_1 is a standard maximal independent set, V_2 is a maximal subset of V_1 so that the distances between its elements are at least $2^1 + 1 = 3$, and so on.

use a simpler method to create the hierarchy of graphs, which relies on a 2-approximation of the k-centers problem. The algorithm of [17] begins by producing a *graph centers (GC) filtration* $V = V_0 \supset V_1 \supset \ldots \supset V_k \supset \emptyset$ of the set V of vertices of the graph G, with $|V_i| = c \cdot x^{k-i}$, where $x > 1$ and $c = |V_k|$ is a constant. A cluster of vertices closest to each center is created for each center and on every level. A set of weighted edges is computed between elements of V_i, so that the weights correspond to the number of edges between the elements of the corresponding clusters. Thus the GC filtration together with the edges forms a hierarchy of graphs.

While having proper graphs on each level is necessary in many applications utilizing graph hierarchies, in the context of graph drawing we can save time and space by using just a filtration of the vertex set. Note that in a filtration there are no edges but only vertices. As we already pointed out, logarithmic depth and "uniform" filtrations are best for graph drawing purposes. We have developed and tested one specific such filtration that we call a *maximal independent set (MIS) filtration* and we use it in this algorithm.

Recall that $S \subset V$ is an *independent set* of a graph $G = (V, E)$ if no two elements of S are connected by an edge of G. Equivalently, S is an independent set of G if the graph distance between any two elements of S is at least two. The *graph distance* between two vertices is defined as the length of the shortest path between them in the graph. A maximal independent set filtration of G is a family of sets $V = V_0 \supset V_1 \supset \ldots \supset V_k \supset \emptyset$, such that each V_i is a maximal subset of V_{i-1} for which the graph distance between any pair of its elements is at least $2^{i-1} + 1$, see Fig. 2.

There are several advantages of MIS filtrations over GC filtrations. First, the number of vertices in the sets V_is in the case of MIS filtrations is controlled by the geometry of the graph, whereas in the graph centers filtration the sizes are arbitrarily set by the user. Moreover, we can build a MIS filtration using little time and space, whereas graph centers filtrations require knowledge about the distances between all pairs of vertices and the all-pairs shortest path is a serious time and space bottleneck when dealing with large graphs.

MIS filtrations can be constructed as follows. Suppose we constructed an order i independent set V_i of G. To construct V_{i+1} let $V^* = V_i$ be an auxiliary set of vertices from which we will draw elements of V_{i+1}. Take a random element $v_0 \in V^*$ out of V^*, and place it in V_{i+1}. Next remove all elements of V^* whose graph distance to v_0 is less than or equal to 2^i. This distance factor is important in ensuring that vertices are well distributed and in guaranteeing small depth of the filtration. Choose another element v_1 of V^*, and remove from V^* the chosen vertex and all vertices whose distance to v_1 is less than or equal to 2^i. Place v_1 in V_{i+1}. Repeat this procedure until V^* is empty. Note that the set V_1 produced by this procedure is an ordinary maximal independent set of G. An example of a maximal independent set filtration is shown in Fig. 2.

The construction of a MIS filtration stops at level k so that $2^k > \delta(G)$, where $\delta(G)$ is the diameter of G. Therefore, each MIS filtration has depth $O(\log \delta(G))$. MIS filtrations provide excellent distribution of the vertices by construction, a property needed for high quality filtrations.

3.3   Finding Vertex Neighborhoods $N_i(v)$

One of the key ideas of the hierarchical force-directed graph layout method is that at each stage of the construction a force-directed position refinement method is applied to a given layer V_i of a filtration only locally. More precisely, for a given energy function E and $v \in V_i$, the gradient of E at $\texttt{pos}[v]$ is computed not for E but for the restriction of E to some neighborhood $N_i(v)$ of v in V_i. Utilization of a good filtration of V and a local position refinement strategy are the key means of escaping a quadratic lower bound for space and time complexity of the classical force-directed methods.

This section describes a procedure of constructing $N_i(v)$ sets and the definition of the scheduling function $\texttt{nbrs}()$. Intuitively, at each stage of the hierarchical graph drawing strategy we should be getting a better and better approximation of the final drawing of the graph. Ideally, at the last stage, when we perform a force-directed local refinement of the position of each vertex v of the graph, it should be enough to take $N_0(v)$ to be the set of adjacent vertices of v. The time complexity of this last stage calculation is $c \cdot \sum_{v \in V} N_0(v) = c \cdot n \cdot \texttt{avgDeg}(G)$, where $\texttt{avgDeg}(G)$ is the average degree of G. We would like to make $c \cdot n \cdot \texttt{avgDeg}(G)$ an upper bound for the complexity of calculations at each stage of graph drawing construction. Therefore, we set $\texttt{nbrs}(i) = \Theta(\frac{\texttt{avgDeg}(G) \cdot n}{|V_i|})$.

Suppose $\mathcal{V}$ is a logarithmic depth filtration of the set V of vertices of G. The calculation of the sets $N_i(v), N_{i-1}(v), \ldots, N_0(v)$ is performed for each element $v \in V$ only once, when it is added to a set of already placed vertices, see Fig. 1. We require that $N_k(v)$ contains $\Theta(\texttt{nbrs}(k))$ elements for each $k = i, i-1, \ldots, 0$. Therefore, the space complexity of this strategy is bounded above by

$$\sum_{i=0}^{k} |V_i - V_{i+1}| \left(\texttt{nbrs}(1) + \texttt{nbrs}(2) + \cdots + \texttt{nbrs}(i) \right). \tag{1}$$

Since $V_{i+1} \subset V_i$, we have $|V_i - V_{i+1}| = |V_i| - |V_{i+1}|$, and after simplifications (1) takes the form

$$\sum_{i=0}^{k} |V_i| \text{nbrs}(i) \leq c_0 \sum_{i=0}^{k} |V_i| \frac{\text{avgDeg}(G) \cdot n}{|V_i|} = c_0 \sum_{i=0}^{k} \text{avgDeg}(G) \cdot n =$$

$$= c_0 \text{avgDeg}(G) \cdot (k+1)n. \qquad (2)$$

Similarly we can show that there exists a positive constant c_1 so that equation (1) is greater than $c_1 \text{avgDeg}(G) \cdot (k+1)n$. Thus, the storage complexity of the above strategy for finding $N_i(v), N_{i-1}(v), \ldots, N_0(v)$ for all $v \in V$ is $\Theta(\text{avgDeg}(G)kn)$. If G is of bounded degree, then $\Theta(\text{avgDeg}(G)kn) = \Theta(kn)$, where $k = \log n$ for a GC filtration, and $k = \log \delta(G)$ for a MIS filtration.

Let the *depth of a vertex*, $\text{depth}(v)$, with respect to $\mathcal{V}$ be the largest d, such that $v \in V_d$. The sets $N_i(v), N_{i-1}(v), \ldots, N_0(v)$ are created by repeated application of a breadth-first search algorithm. A new vertex with depth d is placed in each of $N_j(v)$, for $j \leq d$, if $N_j(v)$ is not full already. The process stops when all $N_j(v)$s are full. Note that the running time of this procedure is bounded above by

$$\sum_{i=1}^{k} |V_i|(1 \cdot \text{nbrs}(1) + 2 \cdot \text{nbrs}(2) + \cdots + i \cdot \text{nbrs}(i)). \qquad (3)$$

As in the case of the expression (1), (3) is equal to

$$\sum_{i=0}^{k} i|V_i|\text{nbrs}(i) \leq c_0 \sum_{i=0}^{k} i|V_i| \frac{\text{avgDeg}(G) \cdot n}{|V_i|} = c_0 \sum_{i=0}^{k} i\text{avgDeg}(G) \cdot n =$$

$$= c_0 \text{avgDeg}(G) \cdot \frac{(k+1)k}{2}n. \qquad (4)$$

Similarly we can show that there exists a positive constant c_1 so that equation (3) is greater than $c_1 \text{avgDeg}(G) \cdot \frac{(k+1)k}{2}n$. The time complexity of this strategy for finding $N_i(v), N_{i-1}(v), \ldots, N_0(v)$ for all $v \in V$ is $\Theta(\text{avgDeg}(G)k^2n)$. If G is of bounded degree, then $\Theta(\text{avgDeg}(G)k^2n) = \Theta(k^2n)$, where $k = \log n$ for a GC filtration, and $k = \log \delta(G)$ for a MIS filtration.

3.4 Initial Placement of Vertices

Most graph drawing algorithms begin by placing all the vertices of the graph randomly in the plane or in 3D. In this algorithm we have adopted a different approach in that we add vertices to the current drawing one at a time and only after we have found a suitable place for them. Here we describe the process in two dimensional space, but in practice it can be done in any Euclidean space $\mathbb{E}$. Recall that in the first step of the algorithm we compute a filtration $\mathcal{V} = V_0 \supset V_1 \supset \ldots \supset V_k \supset \emptyset$. If necessary, we modify the last one or two sets of the filtration so that the last one has exactly three elements, $V_k = \{u, v, w\}$.

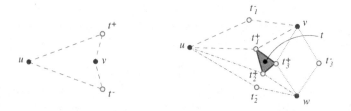

Fig. 3. Initial placement for a new vertex t; darkly shaded vertices have already been placed. (a) Given two vertices in $\mathbb{R}^2$, there are up to two possible places for t, based on its graph distance to u and v. (b) Using three vertices in $\mathbb{R}^2$ results in a better placement.

We start the process of drawing G by placing u, v, and w as follows: we find a triangle with endpoints given by $\text{pos}[u], \text{pos}[v], \text{pos}[w]$, so that $d_{\mathbb{R}^2}(u,v) = d_G(u,v)$, $d_{\mathbb{R}^2}(v,w) = d_G(v,w)$, $d_{\mathbb{R}^2}(w,u) = d_G(w,u)$, where $d_{\mathbb{R}^2}(u,v)$ is the Euclidean distance between $\text{pos}[u]$ and $\text{pos}[v]$, and $d_G(u,v)$ is the graph distance between u and v.

In general, after refining the positions of the vertices in V_i, we need to find initial positions for the vertices in $V_{i-1} - V_i$. Once all vertices in V_{i-1} are placed their positions are refined, and we proceed to the next level. This two-stage process continues until all vertices have been drawn. A natural way to place a new vertex given the placement of several others is to use the graph distance from the new vertex to several of its closest neighbors that have already been placed. We base our placement strategy on this simple idea.

Suppose that we are looking for a place for a new vertex $t \in V_{i-1} - V_i$. Furthermore, suppose that we know two vertices $u, v \in V_i$ which have already been placed. Then using their position vectors, $\text{pos}[u]$ and $\text{pos}[v]$, and the graph distances $d_G(u,t)$ and $d_G(v,t)$, it is straightforward to find a position $\text{pos}[t]$ of t in the plane so that $d_{\mathbb{R}^2}(u,t) = d_G(u,t)$, $d_{\mathbb{R}^2}(v,t) = d_G(v,t)$, as shown in Fig. 3(a). This idea can be generalized so that three or more already placed vertices are used to determine the location of new vertices. For each vertex $t \in V_{i-1} - V_i$ we find its three closest neighbors $u, v, w \in V_i$, see Fig. 3(b). Since u, v and w have already been placed we can obtain a suitable place for t by solving the following system of equations for u, v, w, and t

$$\begin{cases} (x - x_u)^2 + (y - y_u)^2 = d_G(u,t)^2 \\ (x - x_v)^2 + (y - y_v)^2 = d_G(v,t)^2 \\ (x - x_w)^2 + (y - y_w)^2 = d_G(w,t)^2, \end{cases}$$

where $\text{pos}[u] = (x_u, y_u)$, $\text{pos}[v] = (x_v, y_v)$, $\text{pos}[w] = (x_w, y_w)$, $\text{pos}[t] = (x, y)$. Since this system of equations is over-determined and may not have any solutions, we solve the following three pairs of equations instead:

$$\begin{cases} d_{\mathbb{R}^2}(u,t) = d_G(u,t) \\ d_{\mathbb{R}^2}(v,t) = d_G(v,t) \end{cases} \qquad \begin{cases} d_{\mathbb{R}^2}(v,t) = d_G(v,t) \\ d_{\mathbb{R}^2}(w,t) = d_G(w,t) \end{cases} \qquad \begin{cases} d_{\mathbb{R}^2}(u,t) = d_G(u,t) \\ d_{\mathbb{R}^2}(w,t) = d_G(w,t) \end{cases}$$

Solving these three systems of quadratic equations we obtain up to six different solutions. We choose the three closest to each other, call them t_1^+, t_2^+, t_3^+, and place t are their barycenter: $\mathtt{pos}[t] = (t_1^+ + t_2^+ + t_3^+)/3$, see Fig. 3(b).

3.5 Local Temperature Calculations

A common problem with most force-directed algorithms is determining the scaling factor of the displacement vector at each phase. Clearly, in the early iterations vertices should move farther than in the last iteration, but coming up with a schedule for scaling the displacement vector that works well for most graphs is generally difficult. One of the reasons for this difficulty is that initially the vertices are placed at random and as a result can be arbitrarily far from their final position. As a result of the intelligent placement of vertices in our algorithms, this is much less of a problem. The local temperature $\mathtt{heat}[v]$ of v is simply a scaling factor of the displacement vector $\mathtt{disp}[v]$ of v. One particular implementation is considered in detail in [15] but regardless of the specifics of the implementation, the time complexity for updating the local temperature for each v is constant and thus the total time complexity for local temperature calculations is linear.

3.6 Multi-dimensional Drawing

One of the major advantages of a simple local temperature calculation is that unlike the Newton-Raphson and the majority of other classical optimization methods, it works with minor changes in any dimension. In order to obtain an embedding of a graph in $\mathbb{R}^n$, we can simply make $\mathtt{pos}[v]$ an n dimensional vector. A problem with drawings in dimensions higher than three is that they cannot be trivially displayed. An obvious solution to this problem is to find a projection from $\mathbb{R}^n$ into $\mathbb{R}^3$ or $\mathbb{R}^2$.

Consider the case in which a four dimensional drawing is projected down to three dimensions. The projection method described below generalizes to higher dimensions as well. We begin by taking a random vector e_0' in $\mathbb{R}^4$ and normalizing it $e_0 = \frac{e_0'}{\|e_0'\|}$. Next we find three vectors $e_1', e_2', e_3' \in \mathbb{R}^4$ so that e_0, e_1', e_2', e_3' are linearly independent in $\mathbb{R}^4$. We find these vectors by repeatedly choosing a random vector and checking if it is independent from the previous ones until we have four vectors. We then use the Gram-Schmidt orthogonalization process to produce an orthonormal basis e_0, e_1, e_2, e_3 of $\mathbb{R}^4$ using e_0, e_1', e_2', e_3'. The three vectors e_1, e_2, e_3 span a 3 dimensional subspace S of $\mathbb{R}^4$ which is perpendicular to the vector e_0. The orthogonal projection $\rho : \mathbb{R}^4 \to S$ from $\mathbb{R}^4$ onto S in the direction of the vector e_0 is given by the formula $\rho(v) = v - (e_0, v) * e_0$, where (e_0, v) is the scalar product between e_0 and v. Yet to display v on the screen using OpenGL, we need the coordinates (v_1, v_2, v_3) of the projection $\rho(v)$ of v onto S with respect to the basis vectors e_1, e_2, e_3. We get these by a simple scalar product calculation $v_1 = (e_1, v)$, $v_2 = (e_2, v)$, $v_3 = (e_3, v)$.

The above procedure easily generalizes to higher dimensions. Our experiments with 4D drawings yield better results than regular three dimensional drawings. In particular, note the problems with the drawings of the Moebius bend

Fig. 4. (a-b) Moebius bands on 600 and 1500 vertices drawn in 3D. Note the rough twists.(c-d) The same graphs but drawn in 4D and projected in 3D.

directly in 3D in Fig. 4(a-b) and the improved drawings when the same graphs are drawn in 4D and projected to 3D in Fig. 4(c-d).

3.7   Space and Time Complexity

Main Theorem. *If G is a graph of bounded degree and $\mathcal{V}$ is a GC filtration or a MIS filtration of the set V of vertices of G, then the time complexity of our algorithm, after constructing $\mathcal{V}$, is $\Theta(n \cdot k^2)$ and the space required is $\Theta(n \cdot k)$, where $k = \log n$ if $\mathcal{V}$ is a GC filtration, and $k = \log \delta(G)$ if $\mathcal{V}$ is a MIS filtration.*

Proof. The proof of the theorem follows from the fact that after building a filtration $\mathcal{V}$, all parts of the algorithm take linear time and space, except the procedure for finding $N_i(v), N_{i-1}(v), \ldots, N_0(v)$ for each element v of V. Thus both time and space complexity of the algorithm is determined by the time and space complexity of the procedure for finding $N_i(v)$s. In Section 3.3, we showed that the time required for finding the sets $N_i(v)$ is $\Theta(n \cdot k^2)$ and the space required is $\Theta(n \cdot k)$, which concludes the proof.

4   Conclusion

We have presented a novel algorithm for drawing large graphs. The algorithm employs a vertex filtration together with intelligent placement of vertices and fast energy minimization. The algorithm produces drawings in two, three, and higher dimensions in sub-quadratic time and space. While the algorithm works very well for sparse graphs and graphs of low degree, it does not produce high quality drawings for all graphs. In particular, well-connected graphs pose significant challenges as the vertex filtrations become very shallow.

References

1. A. Brandt. Multilevel computations of integral transforms and particle interactions with oscillatory kernels. *Computer Physics Communications*, 65:24–38, 1991.
2. A. Brandt. Multigrid methods in lattice field computations. *Nucl. Phys. B*, 26:137–180, 1992. Proc. Suppl.
3. I. Bruß and A. Frick. Fast interactive 3-D graph visualization. In *Graph Drawing (Proc. GD '95)*, pages 99–110, 1995.

4. R. Davidson and D. Harel. Drawing graphics nicely using simulated annealing. *ACM Trans. Graph.*, 15(4):301–331, 1996.
5. G. Di Battista, P. Eades, R. Tamassia, and I. G. Tollis. Algorithms for drawing graphs: an annotated bibliography. *Computational Geometry: Theory and Applications*, 4:235–282, 1994.
6. G. Di Battista, P. Eades, R. Tamassia, and I. G. Tollis. *Graph Drawing: Algorithms for the Visualization of Graphs*. Prentice Hall, Englewood Cliffs, NJ, 1999.
7. C. A. Duncan, M. T. Goodrich, and S. G. Kobourov. Balanced aspect ratio trees and their use for drawing very large graphs. In *Proceedings of the 6th Symposium on Graph Drawing*, pages 111–124, 1998.
8. P. Eades. A heuristic for graph drawing. *Congressus Numerantium*, 42:149–160, 1984.
9. P. Eades and Q. Feng. Multilevel visualization of clustered graphs. In *Proceedings of the 4th Symposium on Graph Drawing (GD '96)*, pages 101–112, 1996.
10. P. Eades, Q. Feng, and X. Lin. Straight-line drawing algorithms for hierarchical graphs and clustered graphs. In *Proceedings of the 4th Symposium on Graph Drawing (GD '96)*, pages 113–128, 1996.
11. Q. Feng, R. F. Cohen, and P. Eades. How to draw a planar clustered graph. In *Procs. of the 1st Annual International Conference on Computing and Combinatorics (COCOON '95)*, pages 21–31, 1995.
12. A. Frick, A. Ludwig, and H. Mehldau. A fast adaptive layout algorithm for undirected graphs. In R. Tamassia and I. G. Tollis, editors, *Graph Drawing (Proc. GD '94)*, LNCS 894, pages 388–403, 1995.
13. T. Fruchterman and E. Reingold. Graph drawing by force-directed placement. *Softw. – Pract. Exp.*, 21(11):1129–1164, 1991.
14. G. W. Furnas. Generalized fisheye views. In *Proceedings of ACM Conference on Human Factors in Computing Systems (CHI '86)*, pages 16–23, 1986.
15. P. Gajer and S. G. Kobourov. GRIP: Graph dRawing with Intelligent Placement. In *To appear in Proceedings of the 8th Symposium on Graph Drawing*, 2000.
16. R. Hadany and D. Harel. A multi-scale algorithm for drawing graphs nicely. In *Proc. 25th International Workshop on Graph Teoretic Concepts in Computer Science (WG'99)*, 1999.
17. D. Harel and Y. Koren. A fast multi-scale method for drawing large graphs. Technical Report MCS99-21, The Weizmann Institute of Science, Rehovot, Israel, 1999.
18. T. Kamada and S. Kawai. Automatic display of network structures for human understanding. Technical Report 88-007, Department of Information Science, University of Tokyo, 1988.
19. K. Kaugars, J. Reinfelds, and A. Brazma. A simple algorithm for drawing large graphs on small screens. In *Graph Drawing (GD '94)*, pages 278–281, 1995.
20. S. C. North. Drawing ranked digraphs with recursive clusters. In *Graph Drawing 93, Proceedings of the First International Workshop on Graph Drawing*, Sept. 1993.
21. N. Quinn and M. Breur. A force directed component placement procedure for printed circuit boards. *IEEE Transactions on Circuits and Systems*, CAS-26(6):377–388, 1979.
22. M. Sarkar and M. H. Brown. Graphical fisheye views. *Communications of the ACM*, 37(12):73–84, 1994.
23. K. Sugiyama and K. Misue. Visualization of structural information: Automatic drawing of compound digraphs. *IEEE Transactions on Systems, Man, and Cybernetics*, 21(4):876–892, 1991.

GRIP: Graph dRawing with Intelligent Placement[*]

Pawel Gajer[1] and Stephen G. Kobourov[2]

[1] Department of Computer Science
Johns Hopkins University
Baltimore, MD 21218
[2] Department of Computer Science
University of Arizona
Tucson, AZ 85721

Abstract. This paper describes a system for Graph dRawing with Intelligent Placement, GRIP. The system is designed for drawing large graphs and uses a novel multi-dimensional force-directed method together with fast energy function minimization. The system allows for drawing graphs with tens of thousands of vertices in under a minute on a mid-range PC. To the best of the authors' knowledge GRIP surpasses the fastest previous algorithms. However, speed is not achieved at the expense of quality as the resulting drawings are quite aesthetically pleasing.

1 Introduction

The GRIP system is based on the algorithm of Gajer, Goodrich, and Kobourov [6]. It is written in C++ and OpenGL and uses an adaptive Tcl/Tk interface. Given an abstract graph, GRIP produces a drawings in two and three dimensions either directly or by projecting higher dimensional drawings into 2D or 3D space. GRIP follows a number of force-directed drawing tools [1,3,4,5,7,8] but uses a novel intelligent placement approach to drawing in higher dimensions. A fast energy minimization function combined with the use of a simple vertex filtration allows GRIP to draw graphs with tens of thousands of vertices in under one minute.

An overview of the system and its three main phases is given in Fig. 1. Starting with a graph $G = (V, E)$, we first create a maximal independent set (MIS) filtration $\mathcal{V} : V = V_0 \supset V_1 \supset \ldots \supset V_k \supset \emptyset$ of the set V of vertices of G, so that $k = O(\log n)$, and $|V_k| = 3$. A filtration $\mathcal{V}$ of V is called a *maximal independent set filtration* if V_1 is a maximal independent set of G, and each V_i is a maximal subset of V_{i-1} so that the graph distance between any pair of its elements is at least $2^{i-1} + 1$. The *graph distance* between a pair of vertices is defined as the length of the shortest path between them in the graph. Note that the size of a maximal independent set filtration is $\log \delta(G)$, where $\delta(G)$ is the diameter of the graph. We can ensure that the last set has exactly three elements by modifying the last one or two sets in the filtration.

[*] This research partially supported by NSF under Grant CCR-9625289, and ARO under grant DAAH04-96-1-0013.

J. Marks (Ed.): GD 2000, LNCS 1984, pp. 222–228, 2001.
© Springer-Verlag Berlin Heidelberg 2001

Fig. 1. An overview of the algorithm. Given a graph G, the algorithm proceeds phases. The first phase creates a MIS filtration. The second and third phases use the filtration sets $V_k, V_{k-1}, \ldots, V_0$ to repeatedly add more vertices and refine the drawing.

Given a graph G, GRIP produces its drawing in three distinct stages: filtration, initial placement, and refinement. First, we generate an initial embedding of V_k. Since $|V_k| = 3$ the three vertices can be placed in $\mathbb{R}^n$ using their graph distances. Then, we add the vertices of V_{k-1} that are not in V_k, placing them initially at the positions determined by their graph distances to a subset of the elements of V_k. The positions of the vertices in V_{k-1} are modified using a force-directed layout method. This process of adding new vertices and refining their positions is repeated for $V_{k-2}, \ldots, V_1, V_0$. The refined positions of the elements of V_0 constitute the final layout of the vertices of G. Note that we only draw vertices of G up to this point. When all the vertices have been placed we draw the edges of G as straight line segments connecting their endpoints.

2   Building MIS Filtrations

A maximal independent set filtration of a graph G can be obtained by computing the distances between all pairs of vertices of G. A problem with this strategy is that the running time of the all-pairs shortest path algorithm is $\Omega(nm)$ and the storage complexity is $\Omega(n^2)$, e.g., see [2]. When dealing with graphs with tens of thousand of vertices, both the running time and the space complexity of the all-pairs shortest path algorithms pose serious problems.

Our solution is based on the observation that to construct MIS filtration we do not need the distances between all the pairs of vertices. Moreover, the information that has been used to construct $|V_i|$ is not needed to construct $|V_{i+1}|$. Therefore, we use the following "create then destroy" strategy for construction of MIS filtrations. Suppose we have already constructed set V_i. To create V_{i+1}, we build for each vertex of V_i a breadth-first search (BFS) tree up to depth 2^i, but store in it only elements of V_i. Note that this is all we need to build V_{i+1}.

In the process of creating V_{i+1} we may need to build many BFS trees, but we destroy them immediately once they have been used, so that by the time we enter the next phase (of building V_{i+2}), all memory has been freed. Note that as i decreases, the number of vertices for which we have to perform a BFS calculation increases, but at the same time the depth to which we have to build these BFS trees decreases as well. The storage required for this strategy is $\max_i \sum_{v \in V_i} |\mathtt{bfs}_{2^i}(v, V_i)|$, where $|\mathtt{bfs}_{2^i}(v, V_i)|$ is the number of elements of V_i that belong to the BFS tree of v of depth 2^i. The time complexity for this strategy in the case of bounded degree graph G is $\Theta(\sum_{i=0}^{k} \sum_{v \in V_i} |\mathtt{bfs}_{2^i}(v)|)$, where $|\mathtt{bfs}_{2^i}(v)|$ is the number of the BFS tree of v of depth 2^i. Clearly, if we build a complete BFS tree for each vertex of G, then the running time and space complexity of this procedure even in the bounded degree case would be $O(n^2)$. Our tests indicate that the running time and storage cost of the above MIS filtration construction procedure (in the bounded degree case) is less than quadratic as we only construct partial BFS trees and destroy them right away. In all of our experiments, the time spent creating the MIS filtration was less than 3% of the total running time, see Fig 4(a) and Fig 4(b), respectively.

We store a MIS filtration of a graph of n vertices in an array `misFlt` of size n so that the first $|V_k|$ entries in the `misFlt` array are the elements of V_k. The next $|V_{k-1}|$ entries in the array are the elements of V_{k-1}, followed by $|V_{k-2}|$ entries in the `misFlt` array of the elements of V_{k-2}, all the way to V_1. To keep track of where one set ends and another one begins we store the indices indicating the borders of different level sets of the filtration in a separate array `misBorder` of size $\log \delta(G)$. Thus the space complexity for storing a MIS filtration is $n + \log \delta(G)$. The same method can be applied to any filtration.

3   Initial Placement and Refinement of V_i

The second and third phase of the algorithm are the placement and refinement stages, respectively. In the ith placement stage, the vertices of set V_i are intelligently placed in $\mathbb{R}^n$. In the ith refinement stage a local force-directed method is used to obtain better positions for the vertices of V_i. After the placement and refinement phases for V_i have been completed, the process is repeated for V_{i-1}, V_{i-2}, all the way to V_1.

Consider the general placement case. Suppose the refinement and placement phases for V_i have been completed and we want to start the placement phase for V_{i-1}. All the vertices in V_i are also in V_{i-1}, since $V_{i-1} \supset V_i$ as defined by the construction of the filtration. Thus we are only concerned with the placement of the vertices in V_{i-1} that are not in V_i. The idea behind the intelligent placement is that every vertex t is placed "close" to its optimal position as determined by the graph distances from t to several already placed vertices. The intuition is that if we can place the vertices close to their optimal positions from the very beginning, then the refinement phases need only a few iterations of a local force-directed method to reach minimal energy state.

For example, the following "three closest to t vertices" strategy starts by setting `pos[t]` to the barycenter $(\mathtt{pos}[u] + \mathtt{pos}[v] + \mathtt{pos}[w])/3$ of u, v, and w, the three vertices closest to t. This is followed by a force-directed modification of the position vector of t with the energy function E calculated only at the three points u, v, w. This makes the procedure very fast, and in our tests it produced good results, see Fig. 2. More details about the placement algorithm can be found in [6].

While the refinement is calculated using a force-directed method, it is important to note that the forces are calculated locally, see Fig. 3(a). For each level of the filtration V_i, we perform $\mathtt{rounds}(i)$ updates of the vertex positions, where $\mathtt{rounds}(i)$ is a scheduling function which can be specified at the beginning of the

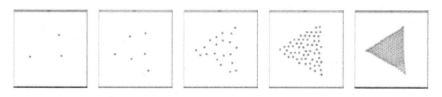

Fig. 2. Drawing of the vertices in the filtration sets. Here $\mathcal{V} : V_0 \supset V_1 \supset V_2 \supset V_3 \supset V_4$. The sizes of the sets are 231, 60, 21, 6, 3, respectively. The process begins with a placement for V_4, followed by V_3, etc. Note that edges are drawn only when all the vertices are placed.

execution. Typically, $5 \leq \text{rounds}(i) \leq 30$. At all levels of the filtration except the last one, the displacement vector $\text{disp}[v]$ of v is set to a local Kamada-Kawai force vector,

$$\overrightarrow{\mathbf{F}}_{\text{KK}}(v) = \sum_{u \in N_i(v)} \left(\frac{\text{dist}_{\mathbb{R}^n}(u, v)}{\text{dist}_G(u, v) \cdot \text{edgeLength}^2} - 1 \right) (\text{pos}[u] - \text{pos}[v]).$$

For the last level of the filtration, $V_0 = V$, when all the vertices have been placed, we set the displacement vector to a local Fruchterman-Reingold force vector,

$$\overrightarrow{\mathbf{F}}_{\text{FR}}(v) = \sum_{u \in \text{Adj}(v)} \frac{\text{dist}_{\mathbb{R}^n}(u, v)^2}{\text{edgeLength}^2} (\text{pos}[u] - \text{pos}[v]) +$$

$$+ \sum_{u \in N_i(v)} s \frac{\text{edgeLength}^2}{\text{dist}_{\mathbb{R}^n}(u, v)^2} (\text{pos}[v] - \text{pos}[u]),$$

Here, $\text{dist}_{\mathbb{R}^n}(u, v)$ is the Euclidean distance between $\text{pos}[u]$ and $\text{pos}[v]$, and $\text{dist}_G(u, v)$ is the graph distance between u and v. In the above equations, edgeLength is the unit edge length, $\text{Adj}(v)$ is the set of vertices adjacent to v, and s is a small scaling factor which is set to 0.05 in our program. Note that for a vertex $v \in G$ the force calculation is performed over a restriction $N_i(v)$ of the vertices of G. Each vertex neighborhood $N_i(v)$ contains a constant number of vertices closest to v which belong to V_i. Thus only a constant number of vertices which are near vertex v are used to refine v's position. This is why we call this type of force calculation local.

The local temperature $\text{heat}[v]$ of v is a scaling factor of the displacement vector for vertex v. The algorithm for determining the local temperature is in Fig. 3(b). To speed up the calculation, we maintain two auxiliary arrays oldDisp and oldCos, where $\text{oldDisp}[v]$ is the previous displacement vector for v , and $\text{oldCos}[v]$ is the previous value of the cosine of the angle between $\text{oldDisp}[v]$ and $\text{disp}[v]$. When a displacement vector of v is calculated for the first time, $\text{heat}[v]$ is set to a default value $\text{edgeLength}/6$. There are three cases for determining the local temperature: (i) if either $\text{oldDisp}[v]$ or $\text{disp}[v]$ is a zero vector, then the value of $\text{heat}[v]$ does not change; (ii) if v is oscillating around some stationary point or if it is moving in the same direction we add to it a factor $\text{heat}[v] * (1 + \cos * r * s)$; (iii) in all other cases we add a factor of $\text{heat}[v] * (1 + \cos * r)$.

$$\boxed{\begin{array}{ll}
\text{REFINEMENT OF } V_i & \texttt{updateLocalTemp}(v) \\
\quad \textbf{repeat rounds}(i) \textbf{ times} & \quad \textbf{if } \|\texttt{disp}[v]\| \neq 0 \textbf{ and } \|\texttt{oldDisp}[v]\| \neq 0 \\
\quad\quad \textbf{for each } v \in V_i \textbf{ do} & \\
\quad\quad\quad \textbf{if } i > 0 \textbf{ then} & \quad\quad \cos[v] = \dfrac{\texttt{disp}[v] * \texttt{oldDisp}[v]}{\|\texttt{disp}[v]\| * \|\texttt{oldDisp}[v]\|} \\
\quad\quad\quad\quad \texttt{disp}[v] = \overrightarrow{F}_{\text{KK}}(v) & \\
\quad\quad\quad \textbf{else} & \quad\quad r = 0.15, s = 3 \\
\quad\quad\quad\quad \texttt{disp}[v] = \overrightarrow{F}_{\text{FR}}(v) & \quad\quad \textbf{if } \texttt{oldCos}[v] * \cos[v] > 0 \textbf{ then} \\
\quad\quad\quad \texttt{heat }[v] = \texttt{updateLocalTemp}(v) & \quad\quad\quad \texttt{heat}[v]+ = (1 + \cos[v] * r * s) \\
\quad\quad\quad \texttt{disp}[v] = \texttt{heat}[v] \cdot \dfrac{\texttt{disp}[v]}{\|\texttt{disp}[v]\|} & \quad\quad \textbf{else} \\
\quad\quad \textbf{for each } v \in V_i \textbf{ do} & \quad\quad\quad \texttt{heat}[v]+ = (1 + \cos[v] * r) \\
\quad\quad\quad \texttt{pos}[v] = \texttt{pos}[v] + \texttt{disp}[v] & \quad\quad \texttt{oldCos}[v] = \cos[v]
\end{array}}$$

Fig. 3. Pseudocode for (a) the refinement phase and (b) the local temperature calculation.

4   Implementation

The GRIP system is written in C++ with OpenGL and uses a flexible Tcl/Tk interface. The system can generate several typical classes of graphs parametrized by their number of vertices, e.g. paths, cycles, square meshes, triangular meshes, and complete graphs. GRIP also contains generators for complete n-ary trees, random graphs with parametrized density, and knotted triangular and rectangular meshes. Different types of tori, as well as cylinders and Moebius bends can be generated with parametrized thickness and length. Finally, Sierpinski graphs in 2 and 3 dimensions (Sierpinski triangles and Sierpinski pyramids, respectively) are also available. In addition to the set of graphs that GRIP can generate, other graphs can be read from a file in several standard formats. The running times for the MIS creation and for the entire drawing process can be seen in Fig 4(a) and Fig. 4(b), respectively.

The parameters discussed in this paper can be changed via GRIP's flexible interface, thus allowing for experimentation with different scheduling functions, scaling parameters, filtrations, etc. There are controls for the drawing dimension and the drawing speed. The drawings produced by default are three dimensional, interactive, and use color and shading to aid three dimensional perception. For faster drawings the interactive display can be turned off and only the final drawing is shown. The size and colors of the vertices and edges can also be modified. Several drawings produced by GRIP are included in Fig. 5–10.

References

1. I. Bruß and A. Frick. Fast interactive 3-D graph visualization. In F. J. Brandenburg, editor, *Graph Drawing (Proc. GD '95)*, volume 1027 of *Lecture Notes Computer Science*, pages 99–110. Springer-Verlag, 1996.
2. T. H. Cormen, C. E. Leiserson, and R. L. Rivest. *Introduction to Algorithms.* MIT Press, Cambridge, MA, 1990.

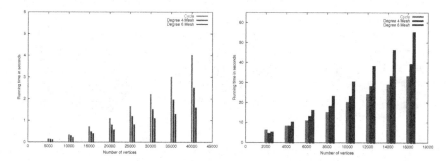

Fig. 4. (a) The left chart shows the running times for construction of a MIS filtration for cycles, and meshes of degree 4 and 6. (b) The right chart shows the total running time for the same graphs.

Fig. 5. Knotted rectangular (degree 4) meshes of 1600, 2500, and 10000 vertices.

Fig. 6. Cylinders of 1000, 4000, and 10000 vertices.

3. R. Davidson and D. Harel. Drawing graphics nicely using simulated annealing. *ACM Trans. Graph.*, 15(4):301–331, 1996.
4. A. Frick, A. Ludwig, and H. Mehldau. A fast adaptive layout algorithm for undirected graphs. In R. Tamassia and I. G. Tollis, editors, *Graph Drawing (Proc. GD '94)*, LNCS 894, pages 388–403, 1995.
5. T. Fruchterman and E. Reingold. Graph drawing by force-directed placement. *Softw. – Pract. Exp.*, 21(11):1129–1164, 1991.
6. P. Gajer, M. T. Goodrich, and S. G. Kobourov. A multi-dimensional approach to force-directed layouts of large graphs. In *To appear in Proceedings of the 8th Symposium on Graph Drawing*, 2000.
7. D. Harel and Y. Koren. A fast multi-scale method for drawing large graphs. Technical Report MCS99-21, The Weizmann Institute of Science, Rehovot, Israel, 1999.
8. T. Kamada and S. Kawai. Automatic display of network structures for human understanding. Technical Report 88-007, Department of Information Science, University of Tokyo, 1988.

Fig. 7. Tori of various length and thickness: 1000, 2500, and 10000 drawn in four dimensions and projected down to three dimensions.

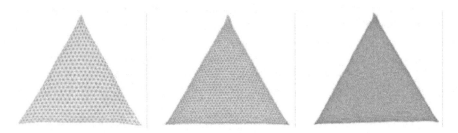

Fig. 8. Triangular (degree 6) meshes of 496, 1035, and 2016 vertices.

Fig. 9. Knotted triangular (degree 6) meshes of 496, 1035, and 2016 vertices.

Fig. 10. Sierpinski graphs in 2D and 3D (a) 2D Sierpinski of depth 6 (1095 vertices); (b) 3D Sierpinski of depth 5 (2050 vertices); (c) 3D Sierpinski of depth 6 (8194 vertices).

A Fast Layout Algorithm for k-Level Graphs

Christoph Buchheim, Michael Jünger, and Sebastian Leipert

Universität zu Köln, Institut für Informatik,
Pohligstraße 1, 50969 Köln, Germany
{buchheim,mjuenger,leipert}@informatik.uni-koeln.de

Abstract. We present a fast layout algorithm for k-level graphs with given permutations of the vertices on each level. The algorithm can be used in particular as a third phase of the Sugiyama algorithm [8]. In the generated layouts, every edge has at most two bends and is drawn vertically between these bends. The total length of short edges is minimized levelwise.

1 Introduction

When displaying hierarchical network structures, one usually has a partition of the vertices into k levels such that in the drawing all vertices of a common level are required to receive the same y-coordinate. This leads to the concept of k-level graphs, that is also used to draw arbitrary graphs in the algorithm of Sugiyama et al. [8]. This algorithm serves as a frame for many graph drawing algorithms, processing a graph in three phases. In a first phase, the vertices are assigned to levels $1, \ldots, k$, thus transforming the graph into a k-level graph. In the second phase, the number of edge crossings is reduced by permuting the vertices within the levels. Finally, a nice layout based on the results of the previous phases has to be determined, assigning y-coordinates to the levels and x-coordinates to the vertices and edge bends.

The first two phases of the Sugiyama algorithm have been examined intensively, see e.g. [5] for the crossing minimization, while the third phase has only been studied rarely, see e.g. [3] or [7]. In this paper, we present a new algorithm LEVEL_LAYOUT for the third phase. Every edge that traverses more than one level is drawn vertically except for its outermost segments. This improves readability (compare Fig. 1 and Fig. 2). Furthermore, the total length of short edges is minimized levelwise as described in Sect. 5.2. If the k-level graph is connected, LEVEL_LAYOUT performs in $O(\overline{m}(\log \overline{m})^2)$, where $\overline{m}$ is the number of edge segments in the k-level graph, i.e., the number of edges after introducing virtual vertices wherever an edge crosses a level. An implementation of the algorithm is contained in the AGD-Library [6].

2 Preliminaries

A *graph* G is a pair (V, E) where V is an arbitrary finite set and E is a subset of $\{\{v, w\} \mid v, w \in V, \ v \neq w\}$. Elements of V and E are called *vertices* and *edges*,

J. Marks (Ed.): GD 2000, LNCS 1984, pp. 229–240, 2001.

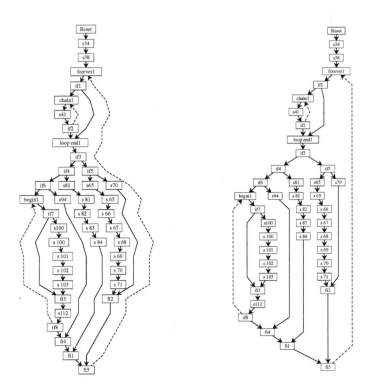

Fig. 1. Layout of an embedded 25-level graph generated by a straight-forward algorithm

Fig. 2. The same graph drawn by LEVEL_LAYOUT. Every edge has at most two bends

respectively. We usually denote an edge $\{v, w\}$ by (v, w). For a vertex $v \in V$, $\delta_G(v) = \{w \in V \mid (v, w) \in E\}$ is the set of its *neighbors*. For any nonnegative integer k, a *k-level graph* $G = (V, E, \lambda)$ is a graph $G = (V, E)$ equipped with a mapping $\lambda : V \to \{1, \ldots, k\}$ such that $\lambda(v) \neq \lambda(w)$ for every edge $(v, w) \in E$. If $v \in V$ is a vertex, $\lambda(v)$ is called the *level* of v.

An edge $e = (v, w)$ is called *short* if $|\lambda(v) - \lambda(w)| = 1$, otherwise *long*. Let e be a long edge and assume that $\lambda(w) > \lambda(v)$. We introduce a *virtual* vertex $\overline{v}_l$ for every level $l \in \{\lambda(v) + 1, \ldots, \lambda(w) - 1\}$ and set $\lambda(\overline{v}_l) = l$. We split up e into *edge segments* $(v, \overline{v}_{\lambda(v)+1}), (\overline{v}_{\lambda(v)+1}, \overline{v}_{\lambda(v)+2}), \ldots, (\overline{v}_{\lambda(w)-1}, w)$. Applying this to every long edge, we obtain a set of virtual vertices, disjoint from V, which is denoted by $\overline{V}$. Furthermore, we obtain a set $\overline{E}$ of edge segments. Obviously, this yields a new k-level graph $\overline{G} = (V \cup \overline{V}, \overline{E}, \lambda)$ without long edges. For the following, let $\delta = \delta_G$ and $\overline{\delta} = \delta_{\overline{G}}$. We will call the vertices $v \in V$ *original* vertices to distinguish them from virtual vertices. An edge segment is called *outer segment* if it is incident to an original vertex, otherwise it is called *inner segment*.

A *level embedding* of a k-level graph is a mapping that assigns to each $l \in \{1, \ldots, k\}$ a permutation of $\lambda^{-1}(l) = \{v \in V \cup \overline{V} \mid \lambda(v) = l\}$. For every vertex $v \in V \cup \overline{V}$, we define the *left direct sibling* of v to be the vertex preceding v in $\lambda^{-1}(\lambda(v))$, according to the given permutation. If the left direct sibling of v exists, it is denoted by $s_-(v)$, otherwise we set $s_-(v) = \star$. The *right direct sibling* $s_+(v)$ is defined analogously.

In a drawing of the graph, two direct siblings v and w must be separated by a minimal distance $m(v, w) > 0$ (which may be given by the user). The extension of m to arbitrary pairs of vertices on the same level is straightforward: Let $v_1, v_2, \ldots, v_r$ be a consecutive sequence of vertices on a common level, then we define $m(v_1, v_r) = \sum_{i=1}^{r-1} m(v_i, v_{i+1})$.

3   Outline of the Algorithm

For an arbitrary k-level graph with given level embedding, LEVEL_LAYOUT computes a layout having the following properties:

(1)   The minimal distance between direct siblings is respected and the given permutations on the levels are not changed.
(2)   Vertices belonging to the same level get the same y-coordinate.
(3)   The minimal distance between neighboring levels is respected.
(4)   All edge segments are drawn straight-line.
(5)   Inner segments of long edges are drawn vertically.

For simplicity, we treat the distances mentioned in (1) and (3) as distances between the centers of the vertices. To avoid overlapping, the vertex dimensions have to be included. In the following, we assume that long edges never intersect at inner segments. Otherwise, we cannot satisfy (1), (4), and (5) simultaneously. We propose to rule out this case by a preprocessing step, see [2] for details.

The algorithm LEVEL_LAYOUT performs in three phases. In the first phase, the x-coordinates of virtual vertices are determined. In the second phase, the x-coordinates of original vertices are computed, keeping the virtual vertices in fixed positions. Finally, the y-coordinates of the levels are determined.

> **LEVEL_LAYOUT**
>
> PLACE_VIRTUAL(x);
> PLACE_ORIGINAL(x);
> PLACE_LEVELS(x,y);

In the following sections, we demonstrate the proceeding of LEVEL_LAYOUT by a 10-level graph, see Fig. 3, Fig. 5 and Fig. 6.

4   Placement of the Virtual Vertices

The function PLACE_VIRTUAL places the virtual vertices as close to each other as possible in horizontal direction subject to properties (1) and (5) of Sect. 3.

First, all vertices are placed as far as possible to the left with respect to (1) and (5). This is done as explained by Sander [7], considering the segment ordering graph S of $\overline{G}$. While Sander needed this step only to get a preliminary placement with vertical inner segments, we compute final x-coordinates for virtual vertices. Since the placement is asymmetric, we determine a second placement of the vertices by placing them as far as possible to the right, analogously. Then we take the average positions of the two placements as final x-coordinates for the virtual vertices. See Fig. 3.

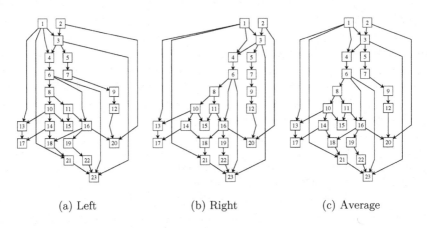

(a) Left   (b) Right   (c) Average

Fig. 3. Placement of the virtual vertices

If the segment ordering graph S is not connected, the placements of vertices belonging to parts of $\overline{G}$ induced by different components of S are not related to each other. This may lead to deformed drawings, since there can be short edges of $\overline{G}$ connecting these parts. To avoid this, we process each connected component of S separately and adjust the placements afterwards by minimizing the total length of these edges. We skip the technical details here, see [2].

5   Placement of the Original Vertices

The function PLACE_ORIGINAL places the original vertices. The positions of the virtual vertices computed by PLACE_VIRTUAL are denoted by $x \in \mathbf{R}^{\overline{V}}$ and regarded as fixed. We have a decomposition of V into maximal consecutive sequences of original vertices belonging to the same level. Let $S = v_1, \ldots, v_r$ be such a sequence. We define $b_- = s_-(v_1)$ and $b_+ = s_+(v_r)$, thus b_- is the virtual vertex bounding S to the left, or $\star$ if S has no siblings to the left, analogously for b_+. Observe that the positions of $v_1, \ldots, v_r$ are already fixed if $x(b_+) - x(b_-) = m(b_-, b_+)$, in this case S is called a *fixed sequence*. Usually, most sequences are not fixed. Our strategy is to process the sequences successively. When processing a sequence, we compute a placement that minimizes the total length of all edges connecting the vertices of the current sequence with their neighbors in the previously placed sequences, subject to the fixed positions of b_- and b_+. Obviously, the layout depends on the order of processing; the more neighbors have already been fixed, the more edges can be taken into account. We next discuss the order of processing the sequences and how to find the optimal placements.

5.1   The Order of Processing the Sequences

We use an array $D \in \{1, -1, 0\}^{\overline{V}}$ to encode this order and initialize it to zero. The array D is updated dynamically by a function ADJUST_DIRECTIONS discussed below.

The function PLACE_ORIGINAL first traverses the graph level by level downwards, and then, in a second step, it traverses the graph upwards. The direction of traversal is given by $d \in \{1, -1\}$, where a 1 is used to indicate the downward direction and a -1 to indicate the upward direction. For every level, the maximal original sequences are traversed from left to right. The currently examined sequence $S = v_1, \ldots, v_r$, bounded by b_- and b_+, is placed by PLACE_SEQUENCE if and only if $b_- = \star$ or $b_+ = \star$ or $D(b_-) = d$. When placing a sequence, we regard the positions of all neighbors that belong to the preceding level as fixed, i.e., the vertices in $\overline{\delta}(v_i, d) = \{v \in \overline{\delta}(v_i) \mid \lambda(v) = \lambda(v_i) - d\}$ for $i = 1, \ldots, r$ (see below for a justification).

Hence, if $b_- = \star$ or $b_+ = \star$, the positions for $v_1, \ldots, v_r$ are determined twice. The distances between the vertices $b_-, v_1, \ldots, v_r, b_+$ resulting from the first traversal are used as lower bounds for the distances computed in the second traversal. By this strategy, we take both neighboring levels into account for the final placement.

On the other hand, if b_- and b_+ are virtual vertices, the sequence is placed only once. It depends on $D(b_-)$ whether the sequence is placed while traversing upwards or while traversing downwards (for technical reasons, the sequence $v_1, \ldots, v_r$ is represented by its left virtual sibling b_-). It remains to determine D. This is done by ADJUST_DIRECTIONS dynamically, using an array $P \in \{\text{true}, \text{false}\}^{\overline{V}}$. For a virtual vertex v, $P(v)$ is true if and only if the original sequence to the right of v has been placed already. At the beginning, only fixed sequences are regarded as placed.

PLACE_ORIGINAL(x)

for all $b_- \in \overline{V}$
  let b_+ be the next virtual vertex to the right of b_-;
  if $b_+ \neq \star$
    set $D(b_-) = 0$;
    if $x(b_+) - x(b_-) = m(b_-, b_+)$ set $P(b_-) = $ true;
    else set $P(b_-) = $ false;
  to be continued...

for all $d = 1, -1$
    for all levels l traversed by direction d
        if level l contains a virtual vertex
            let b_- be the outermost left virtual vertex of level l;
            let $v_1, \ldots, v_r$ be the vertices to the left of b_-;
        else
            set $b_- = \star$;
            let $v_1, \ldots, v_r$ be all vertices of level l;
            PLACE_SEQUENCE(x,$\star$,b_-,d,v_1,$\ldots$,v_r);
            for $i = 1$ to $r - 1$ set $m(v_i, v_{i+1}) = x(v_{i+1}) - x(v_i)$;
            if $b_- \neq \star$ set $m(v_r, b_-) = x(b_-) - x(v_r)$;
            while $b_- \neq \star$
                let b_+ be the next virtual vertex to the right of b_-;
                if $b_+ = \star$
                     let $v_1, \ldots, v_r$ be the vertices to the right of b_-;
                     PLACE_SEQUENCE(x,b_-,$\star$,d,v_1,$\ldots$,v_r);
                     for $i = 1$ to $r - 1$ set $m(v_i, v_{i+1}) = x(v_{i+1}) - x(v_i)$;
                     set $m(b_-, v_1) = x(v_1) - x(b_-)$;
                else if $D(b_-) = d$
                     let $v_1, \ldots, v_r$ be the vertices between b_- and b_+;
                     PLACE_SEQUENCE(x,b_-,b_+,d,v_1,$\ldots$,v_r);
                     set $P(b_-)$=true;
                set $b_- = b_+$;
        ADJUST_DIRECTIONS(l,d,D,P);

After traversing the sequences of level l, the values of D for the next level $l + d$ are computed by ADJUST_DIRECTIONS. We first introduce the notion of a neighboring sequence. Let $S = v_1, \ldots, v_r$ be a maximal original sequence on level $l + d$. Let v_- be the next virtual vertex to the left of S that has a virtual neighbor w_- on level l. If no such vertex v_- exists, set $v_- = w_- = \star$. Analogously, we define v_+ and w_+. The *neighboring* sequences of S on level l are the maximal original sequences on level l between w_- and w_+. Furthermore, if $w_- \neq \star$ and $w_+ \neq \star$, S is said to be an *interior* sequence with respect to l. Otherwise, S is said to be an *exterior* sequence. Fig. 4 illustrates the definition of neighboring sequences.

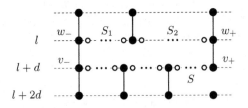

Fig. 4. Neighboring sequences of S. Filled and empty circles represent virtual and original vertices, respectively. Only inner edge segments are displayed. The neighboring sequences of S on level l are S_1 and S_2. The sequence S is interior with respect to l, but exterior with respect to $l + 2d$

We now give an informal description of ADJUST_DIRECTIONS. All interior maximal original sequences $S = v_1, \ldots, v_r$ on level $l + d$ are traversed. If all neighboring sequences of S on level l have already been placed (to check this we use the array P), we set $D(s_-(v_1)) = d$. This allows to place S in the next step of PLACE_ORIGINAL. To explain that this permission is justified, we have to show that every neighbor v on level l of a vertex v_i can be regarded as fixed. We distinguish three cases:

1. The neighbor v is virtual. Then its position is fixed by PLACE_VIRTUAL.
2. The neighbor v belongs to a neighboring sequence of S. Then v has been placed before, as checked by ADJUST_DIRECTIONS explicitly.
3. The neighbor v is original but does not belong to a neighboring sequence of S. In this case, the edge segment (v_i, v) crosses an inner edge segment that is drawn vertically by PLACE_VIRTUAL (like (v_-, w_-) or (v_+, w_+) in Fig. 4), so that the exact position of v does not affect the optimal placement of S.

Theorem 1. *PLACE_ORIGINAL applies PLACE_SEQUENCE to all maximal original sequences.*

Proof. Assume on the contrary that a sequence S on level l is not visited by PLACE_SEQUENCE. By construction of PLACE_ORIGINAL, there either exists a neighboring sequence of S on level $l-1$ that is unvisited as well, or S is exterior with respect to $l-1$. The same holds for level $l+1$. Applying this argument inductively, we get a chain of unvisited neighboring sequences, where the first and the last sequence is exterior. Now by construction of PLACE_VIRTUAL, every such chain must contain a fixed sequence, since otherwise the two parts of the graph divided by the chain could be placed closer to each other. Since fixed sequences are visited by construction, we have a contradiction.

In our example graph (see Fig. 5), the sequences are processed in the following order: The sequences (1,2), (13), (17), (21,22), and (23) are not bounded by virtual vertices, they are placed in both traversals. The sequences (9), (12), (14,15,16), and (20) are fixed. Traversing downwards, the sequence (18,19) is the only bounded sequence placed by PLACE_ORIGINAL, its neighboring sequence (14,15,16) is fixed. Traversing upwards, the first bounded sequence is (10,11), its only neighboring sequence (14,15,16) is fixed. The next one is (8), since its neighboring sequence (10,11) has just been placed. By the same reason, the sequences (6,7), (4,5), and finally (3) are placed.

5.2   The Computation of Optimal Placements

For a sequence of original vertices $S = v_1, \ldots, v_r$, PLACE_SEQUENCE finds an optimal placement $x(v_1), \ldots, x(v_r)$ in the following sense:

(*)   The placement $x(v_1), \ldots, x(v_r)$ minimizes $\sum_{i=1}^{r} \sum_{v \in \bar{\delta}(v_i, d)} |x(v) - x(v_i)|$ with respect to the minimal distances between $b_-, v_1, \ldots, v_r, b_+$.

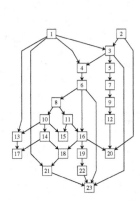

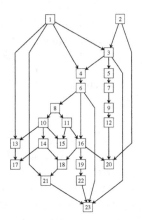

Fig. 5. Placement of the original vertices   **Fig. 6.** Final placement

Since PLACE_SEQUENCE uses a divide and conquer strategy, S is not necessarily maximal and b_- now denotes the next virtual vertex to the left of v_1 instead of $s_-(v_1)$, analogously for b_+.

> **PLACE_SEQUENCE**$(x,b_-,b_+,d,v_1,\ldots,v_r)$
>
> if $r = 1$
>   PLACE_SINGLE(x,b_-,b_+,d,v_1);
> if $r > 1$
>   set $t = \lfloor r/2 \rfloor$;
>   PLACE_SEQUENCE$(x,b_-,b_+,d,v_1,\ldots,v_t)$;
>   PLACE_SEQUENCE$(x,b_-,b_+,d,v_{t+1},\ldots,v_r)$;
>   COMBINE_SEQUENCES$(x,b_-,b_+,d,v_1,\ldots,v_r)$;

Finding a placement satisfying (*) for a single vertex is trivial, therefore we skip the description of PLACE_SINGLE. Next, we explain how to combine two optimal placements for $v_1,\ldots,v_t$ and $v_{t+1},\ldots,v_r$ to an optimal placement of the sequence $v_1,\ldots,v_r$. Let $m = m(v_t,v_{t+1})$. If $x(v_{t+1}) - x(v_t) \geq m$, then nothing has to be done. Otherwise, we transform the placement step by step, where in each step we increase the distance between v_t and v_{t+1} by either decreasing $x(v_t)$ or increasing $x(v_{t+1})$.

Let $p \in \mathbf{R}$ and $1 \leq i \leq t$. If $x(v_t)$ is decreased to position p, then $x(v_i)$ must be decreased to position $x_p(v_i) = \min\{x(v_i), p - m(v_i, v_t)\}$ in order to keep the two partial placements feasible. Let $j(p) \in \{1,\ldots,t\}$ be minimal such that $x_p(v_{j(p)}) < x(v_{j(p)})$; hence decreasing $x(v_t)$ implies decreasing $x(v_{j(p)}),\ldots,x(v_t)$. Let

$$r_-(p) = \sum_{i=j(p)}^{t} \left(\# \left\{ v \in \overline{\delta}(v_i,d) \mid x(v) \geq x_p(v_i) \right\} - \# \left\{ v \in \overline{\delta}(v_i,d) \mid x(v) < x_p(v_i) \right\} \right) .$$

Thus $r_-(p)$ is the number of edge segments getting longer when decreasing $x(v_t)$ past position p minus the number of edge segments getting shorter. This is called the *resistance* to decreasing $x(v_t)$ past p. Observe that $r_- : \mathbf{R} \to \mathbf{Z}$ is a piecewise constant monotone function with finitely many steps. Analogously, we define the resistance $r_+(p)$ to increasing $x(v_{t+1})$ past p. Now we proceed as follows. We decrease $x(v_t)$ if $r_-(x(v_t)) < r_+(x(v_{t+1}))$ or increase $x(v_{t+1})$ otherwise. If equality holds, we choose an arbitrary direction. Assume that $x(v_t)$ is decreased. Then we decrease $x(v_t)$ until either $x(v_{t+1}) - x(v_t) = m$ or $r_-(x(v_t))$ reaches a new step. In the latter case, we determine the new resistance and continue decreasing $x(v_t)$ or increasing $x(v_{t+1})$.

The function COMBINE_SEQUENCES computes the steps of r_- before starting to separate the vertices v_t and v_{t+1}. For every step of length c at position p, a pair (c,p) is stored on a heap R_- by CHANGES_LEFT. The heap R_- is sorted in a decreasing order with respect to the positions p. Analogously CHANGES_RIGHT stores the steps of r_+ on an increasing heap R_+. For runtime reasons, we only move v_t and v_{t+1}, and adjust the positions of $v_1, \ldots, v_{t-1}$ and $v_{t+2}, \ldots, v_r$ later.

> **COMBINE_SEQUENCES**$(x,b_-,b_+,d,v_1,\ldots,v_r)$
>
> let R_- and R_+ be heaps;
> CHANGES_LEFT(R_-);
> CHANGES_RIGHT(R_+);
> set $r_- = r_+ = 0$;
> while $x(v_{t+1}) - x(v_t) < m$
>   if $r_- < r_+$
>   if $R_- = \emptyset$ set $x(v_t) = x(v_{t+1}) - m$;
>   else
>   pop $(c_-, x(v_t))$ from R_-;
>   set $r_- = r_- + c_-$;
>   set $x(v_t) = \max\{x(v_t), x(v_{t+1}) - m\}$;
>   else
>   if $R_+ = \emptyset$ set $x(v_{t+1}) = x(v_t) + m$;
>   else
>   pop $(c_+, x(v_{t+1}))$ from R_+;
>   set $r_+ = r_+ + c_+$;
>   set $x(v_{t+1}) = \min\{x(v_{t+1}), x(v_t) + m\}$;
> for $i = t - 1$ down to 1
>   set $x(v_i) = \min\{x(v_i), x(v_t) - m(v_i, v_t)\}$;
> for $i = t + 2$ to r
>   set $x(v_i) = \max\{x(v_i), x(v_{t+1}) + m(v_{t+1}, v_i)\}$;

Assuming that $x(v_t)$ is decreased, we explain the computation of the steps of r_-. Three different situations lead to a step in the resistance function: The resistance changes by 2, if a vertex v_i passes a neighbor v (Fig. 7(a)). This coincides with $x(v_t)$ being decreased to $x(v)+m(v_i,v_t)$. Hence CHANGES_LEFT stores $(2, x(v) + m(v_i, v_t))$ on R_-. When the minimal distance between v_t and a vertex v_i is reached, the position of v_i is decreased as well (Fig. 7(b)). The resistance changes by

$$c_i = \#\{v \in \overline{\delta}(v_i, d) \mid x(v) \geq x(v_i)\} - \#\{v \in \overline{\delta}(v_i, d) \mid x(v) < x(v_i)\},$$

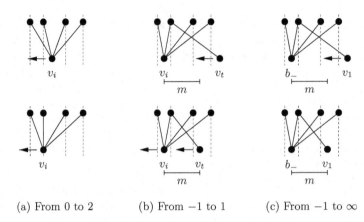

(a) From 0 to 2   (b) From −1 to 1   (c) From −1 to ∞

Fig. 7. Changes of the resistance to moving v_t to the left

and $(c_i, x(v_i) + m(v_i, v_t))$ is stored on R_-. Finally, we enforce the minimal distance between b_- and v_1 by adding $(\infty, x(b_-) + m(b_-, v_t))$ to the heap R_- (Fig. 7(c)).

CHANGES_LEFT(R_-)

for $i = 1$ to t
    set $c = 0$;
    for all $v \in \bar{\delta}(v_i, d)$
        if $x(v) \geq x(v_i)$ set $c = c + 1$;
        else
            set $c = c - 1$;
            push $(2, x(v) + m(v_i, v_t))$ to R_-;
        push $(c, x(v_i) + m(v_i, v_t))$ to R_-;
    if $b_- \neq \star$ push $(\infty, x(b_-) + m(b_-, v_t))$ to R_-;

From Theorem 1 we know that all maximal original sequences are visited. For the correctness of PLACE_ORIGINAL, it remains to show that placements computed by PLACE_SEQUENCE satisfy the minimality condition (*). Let $v_1, \ldots, v_r$ be the original sequence that has to be placed, and let x be a placement that satisfies (*) both for $v_1, \ldots, v_t$ and for $v_{t+1}, \ldots, v_r$. We show that COMBINE_SEQUENCES merges the two partial placements into a placement satisfying (*) for $v_1, \ldots, v_r$. We first give a lemma that allows us to restrict our attention to placements that are determined by the positions of v_t and v_{t+1}:

Lemma 1. *Let x be a placement satisfying (*) for $v_1, \ldots, v_t$ and for $v_{t+1}, \ldots, v_r$. Then there exists a placement x^* satisfying (*) for $v_1, \ldots, v_r$ such that the following conditions hold.*

(a) $x^*(v_i) = \min\{x(v_i), x^*(v_t) - m(v_i, v_t)\}$ for $i \leq t$
(b) $x^*(v_i) = \max\{x(v_i), x^*(v_{t+1}) + m(v_{t+1}, v_i)\}$ for $i \geq t + 1$.

Proof. Starting with a placement satisfying (*) for $v_1, \ldots, v_r$ but not necessarily (a) and (b), one can transform this placement by successively adjusting the position of v_j to condition (a), for $j = t, \ldots, 1$, such that condition (*) is not violated. For $j = t+1, \ldots, r$, one can proceed analogously to obtain (b). See [2] for a precise proof.

Theorem 2. *The placement $\widetilde{x}$ computed by COMBINE_SEQUENCES satisfies (*) for $v_1, \ldots, v_r$.*

Proof. Assume that $\widetilde{x}(v_{t+1}) - \widetilde{x}(v_t) < m(v_t, v_{t+1})$, otherwise there is nothing to show. For $p \in \mathbf{R}$ let

$$f_-(p) = \sum_{i=1}^{t} \sum_{v \in \overline{\delta}(v_i, d)} |x(v) - \min\{x(v_i), p - m(v_i, v_t)\}|,$$

and analogously

$$f_+(p) = \sum_{i=t+1}^{r} \sum_{v \in \overline{\delta}(v_i, d)} |x(v) - \max\{x(v_i), p + m(v_{t+1}, v_i)\}|.$$

By Lemma 1, we only need to consider placements satisfying (a) and (b) in order to check the minimality of $\widetilde{x}$. By construction, it is clear that $\widetilde{x}$ satisfies (a) and (b) and is feasible for $v_1, \ldots, v_t$. Hence $\widetilde{x}$ satisfies (*) if $\widetilde{x}(v_t)$ and $\widetilde{x}(v_{t+1})$ minimize $f_-(\widetilde{x}(v_t)) + f_+(\widetilde{x}(v_{t+1}))$ subject to $\widetilde{x}(v_{t+1}) - \widetilde{x}(v_t) \geq m(v_t, v_{t+1})$. However, the function f_+ is convex and piecewise linear, and the gradient to the left of a position p is the resistance to moving v_{t+1} to position p (analogously for f_- and v_t). Thus moving to the direction with lower resistance until $\widetilde{x}(v_{t+1}) - \widetilde{x}(v_t) = m(v_t, v_{t+1})$ yields a minimal placement.

6   Placement of the Levels

In most algorithms, neighboring levels get a fixed distance. Thus the length $|x(w) - x(v)|$ of an edge segment $(v, w) \in \overline{E}$ with $\lambda(v) = l$ and $\lambda(w) = l + 1$ has no influence on the distance between l and $l + 1$. It is however easy to see that long edge segments require a larger level distance than short ones in order to obtain good readability. In this section we propose a method for computing the y-coordinates of the vertices that considers this by adjusting the distance between l and $l + 1$ to the longest edge segment connecting neighboring levels.

Let the *gradient* of (v, w) be defined as $\nabla(v, w) = |x(w) - x(v)| / |y(w) - y(v)|$. We use a fixed maximal gradient GRADIENT. Then we determine the distance between l and $l + 1$ by

$$\max\{\nabla(v, w) \mid (v, w) \in \overline{E} \text{ and } \lambda(v) = l \text{ and } \lambda(w) = l + 1\} = \text{GRADIENT}.$$

Explicitly, the distance can be computed as

$$\max\{\text{GRADIENT} \cdot |x(w) - x(v)| \mid (v, w) \in \overline{E} \text{ and } \lambda(v) = l \text{ and } \lambda(w) = l + 1\}.$$

Fig. 6 shows the final layout with the new y-coordinates, including some local improvements that have been automated.

7   Runtime

Let $G = (V, E, \lambda)$ be a k-level graph with a given level embedding and let $\overline{G} = (V \cup \overline{V}, \overline{E}, \lambda)$ be the k-level graph resulting from G by introducing virtual vertices as explained in Sect. 2. For $\overline{m} = |\overline{E}|$ and $\overline{n} = |V \cup \overline{V}|$ we have

Theorem 3. *The algorithm LEVEL_LAYOUT computes a layout for the k-level graph G in $O((\overline{m} + \overline{n})(\log(\overline{m} + \overline{n}))^2)$ time.*

Proof. The left and right placement of virtual vertices can be computed in $O(\overline{n})$ time, if both segment ordering graphs are connected. Otherwise, the adjustment of different connected components can be done in $O(\overline{n} + \overline{m} \log \overline{m})$ (see [2]). The first loop of PLACE_ORIGINAL needs $O(\overline{n})$ time in total. The second loop applies PLACE_SEQUENCE to all maximal original sequences at most twice. The function COMBINE_SEQUENCES combines two sequences including r vertices and t incident edge segments in $O((r+t)\log(r+t))$, since at most $r+t+2$ changes of resistance are stored on the heap. By the logarithmic depth of the applied divide and conquer strategy we see that placing a sequence of r vertices with t incident edges can be performed by PLACE_SEQUENCE in $O((r+t)\log(r+t)\log t)$. In total, all calls of PLACE_SEQUENCE take $O((\overline{m}+\overline{n})\log(\overline{m} + \overline{n})\log \overline{m})$ time. Since ADJUST_DIRECTIONS needs $O(\overline{n})$, PLACE_ORIGINAL can be performed in $O((\overline{m} + \overline{n})\log(\overline{m} + \overline{n})\log \overline{m})$ time. The y-coordinates are computed in $O(\overline{m} + \overline{n})$, so we have the desired result.

References

1. F. J. Brandenburg, M. Jünger, and P. Mutzel. Algorithmen zum automatischen Zeichnen von Graphen. *Informatik-Spektrum 20*, pages 199–207, 1997.
2. C. Buchheim, M. Jünger, and S. Leipert. A fast layout algorithm for k-level graphs. Technical report, Institut für Informatik, Universität zu Köln, 1999.
3. E. R. Gansner, E. Koutsofios, S. C. North, and K.-P. Vo. A technique for drawing directed graphs. *IEEE Transactions on Software Engineering*, 19(3):214–230, 1993.
4. M. R. Garey and D. S. Johnson. Crossing number is NP-complete. *SIAM Journal on Algebraic and Discrete Methods*, 4(3):312–316, 1983.
5. M. Jünger and P. Mutzel. 2-layer straightline crossing minimization: Performance of exact and heuristic algorithms. *Journal of Graph Algorithms and Applications*, 1:1–25, 1997.
6. Petra Mutzel et al. A library of algorithms for graph drawing. In S. H. Whitesides, editor, *Graph Drawing '98*, volume 1547 of *Lecture Notes in Computer Science*, pages 456–457. Springer Verlag, 1998.
7. G. Sander. A fast heuristic for hierarchical Manhattan layout. In F. J. Brandenburg, editor, *Graph Drawing '95*, volume 1027 of *Lecture Notes in Computer Science*, pages 447–458. Springer Verlag, 1996.
8. K. Sugiyama, S. Tagawa, and M. Toda. Methods for visual understanding of hierarchical systems. *IEEE Transactions on Systems, Man, and Cybernetics*, 11(2):109–125, 1981.

Graph Layout for Displaying Data Structures

Vance Waddle

IBM Thomas J. Watson Research Center,
P.O.Box 704
Yorktown Heights, NY 10598
waddle@us.ibm.com

Abstract. Displaying a program's data structures as a graph is a valuable addition to debuggers, however, previous papers have not discussed the layout issues specific to displaying data structures. We find that the semantics of data structures may require constraining node and edge path orderings, and that nonhierarchical, leveled graphs are the preferred data structure display. We describe layout problems for data structures, and extend the Sugiyama algorithm to solve them.

1 Introduction

Displaying data structures graphically is a valuable addition to debuggers, especially for large collections of heap allocated data forming trees, graphs, etc. which are naturally displayed as directed graphs. Such a display gives a global view of the data and its structure; without it, the view is limited to a small portion of the data, like the seven blind men examining an elephant.

Previous efforts [12, 13, 16, 19, 24, 25] have explored the architectural and user interface facilities in tools displaying data structures, but have left layout as an architectural feature into which multiple, generic, layout algorithms could be substituted. This leads to the approach in [24] in which each layout algorithm requires a particular type of data structure, e.g., a tree, list, or DAG. However, as a program executes, its the data structures may evolve from one shape to another, and such a layout algorithm may produce bad layouts if given a different structure than it expects. Instead, we add constraints to a general purpose layout algorithm [20] to impose local "layout styles" on the graph.

We are developing a tool to display data structures, which we call DART (for "Data ARTist"). We originally intended to use our implementation of a Sugyiama-style algorithm [23] to lay out data structures with only some "minor tuning." However, when we built a prototype to determine the additional requirements, we discovered problems that were either different than those solved by published algorithms, or were more serious in this context:

1) **Ports - crossings inside nodes**: most layout algorithms were developed to lay out graphs in which the nodes are points. However, when data structures are displayed as in Figure 1 with structure members shown as nested fields, and edges due to pointers end inside the node, edge crossings can now occur inside (and near) the nodes, rather than just between them (Figure 2A, 2C).

J. Marks (Ed.): GD 2000, LNCS 1984, pp. 241–252, 2001.

The graph becomes unreadable much more quickly due to crossings inside the relatively smaller node area than it does for crossings between nodes. [6] calls these "ports", but only solves the problem when the fields are parallel to graph levels. This problem has similarities to the compound graphs of [21], but we are not free to rearrange the node's interior structure.

Fig. 1. Knuth-style Drawing of Tree

2) **User control over layout**: Layout algorithms assume more degrees of freedom in laying out the graph than may be allowable in displaying data structures. The two most serious examples uncovered by our prototype were:

i) *Assigning nodes to levels*: The common style of drawing data structures in [10] (Figure 1) uses edges within levels to show pointers to sibling nodes, visually organizing the graph by separating child and sibling links. Without this separation, sibling and child links can become tangled and hard to trace through the diagram. This style requires a consistent direction of edge flow and reduced edge crossings within a nonhierarchical, leveled graph. Orthogonal layout algorithms [3] produce nonhierarchical layouts, but do not preserve the edge flow. Sugiyama-style [20] algorithms reduce edge crossings and preserve edge flow, but their heuristics can not effectively handle edges within a level. Hierarchical layouts are also taller: A balanced M-ary tree ($M > 1$) of depth K (through *child* links - the tree in Figure 1 has depth of 3) requires max(K, K + (K-2) * (M-1)) levels. For K=10 and M=3, the tree takes 26 levels, or about 3 times as much zooming as a nonhierarchical graph, further degrading readability.

ii) *Reordering nodes and edges*: Layout algorithms reorder nodes and edges to reduce the number of edge crossings. However, the ordering of edges in a data structure may carry important semantic information. Consider a graph representing the expression "A - B", in which "-" is the node root, and "A" and "B" its children. A layout algorithm that reorders child nodes will eventually convert the expression "A - B" to "B - A". Few users appreciate this change.

We provide user control over the layout algorithm by adding constraints to specify the allowable relationships among nodes and edges. We extend the Sugiyama algorithm to process the supported constraints.

3) **Stability between successive drawings**: Debuggers display successive views of a data structure, where each view is slightly modified from its predecessor. Maintaining the stability of the unchanged parts of the display makes it easier to see the changes between views. This is the motivation for incremental layout algorithms [14]. However, it is difficult to create incremental layout algorithms, even without constraints. Although we have an incremental version of the Sugiyama algorithm [20], we considered adding constraints to its already complex processing to be too heroic an effort. Instead, the ordering constraints we required provided a simpler means of imposing stability on the graph.

2 Ports - Crossings Inside Nodes

Algorithms such as [20] are designed to lay out graphs whose nodes have no internal structure. The edges stop at the borders of their end nodes, and barycenters are computed as though the edges end at the node center. When data structures are represented as nodes with nested fields (as in Figures 1 and 2), edges ending at a fixed port inside the node generate crossings inside and around the node of which the algorithm is ignorant. These edge crossings are concentrated in a small area of the graph (Figures 2A and 2C), and quickly becomes crippling problem for data structures containing arrays of pointers like hash tables and B-trees. (We have omitted the field names, which are shown as gray fields in Figure 2.)

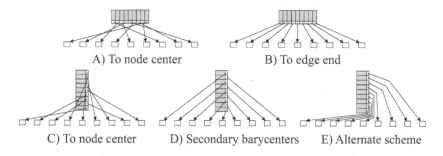

A) To node center   B) To edge end

C) To node center   D) Secondary barycenters   E) Alternate scheme

Fig. 2. Fields aligned parallel and perpendicular to levels

This problem divides into two cases: In the simpler case of Figure 2A, the pointer fields are parallel to the graph's levels. This layout could result from a naive application of [20]. This problem can be fixed by using coordinates for the actual size of the node (rather than the ordinal coordinates of [20]), and using the coordinate of the edge's endpoint within the node. It can also be solved by the auxiliary graph method [6], which minimizes total edge length in addition to positioning end nodes. Figure 2B shows the result of these revisions.

Fields stacked perpendicular to the levels have the same horizontal coordinate. Figure 2C shows the problem, and Figure 2D the desired result. ([12, 13, 19] avoid this problem by only displaying the fields parallel to levels.) We use the fact that [4, 20] are based on sorting, and a sort can have both primary and secondary sort keys, with the secondary sort key only significant for comparisons within the same primary key. We define a secondary barycenter, SBc, to produce the graph of Figure 2D. The secondary barycenter is assigned to the child nodes of the compound node. The lowest field goes to the center node to avoid crossing another edge, and edges above it go to nodes on alternate sides to maximize the distance between the edge paths. Given fields $F_0, \ldots, F_k$ in a column, where F_0 is the lowest field, and $i \in \{0, \ldots, k\}$ and $\delta > 0$, then $\mathrm{SBc}(F_i)$ is:

$$\mathrm{SBc}(F_0) = 0$$
if i is odd then $\mathrm{SBc}(F_i) = \frac{-(i+1)}{2}\delta$
if i is even and $i \neq 0$ then $\mathrm{SBc}(F_i) = \frac{i}{2}\delta$

This scheme causes edges to left of center child nodes to overlap the field names, which could be objectionable for large graphs. An alternate scheme (Figure 2E) avoids this problem by extra edge routing and longer edges. The parent node is centered over its children, although the graph would be visually improved by shifting it to the left. This is another instance in which traditional layout heuristics (center parent over children, minimize edge lengths) don't apply well. Where $m = \frac{k}{2}$, the secondary barycenters for this scheme are:

$$\mathrm{SBc}(F_m) = 0$$
if $i > m$ then $\mathrm{SBc}(F_i) = \mathrm{SBc}(F_{i-1}) + \delta$
if $i < m$ then $\mathrm{SBc}(F_i) = \mathrm{SBc}(F_{i+1}) - \delta$

3   User Control via Constrained Layout

Constrained layout schemes [1, 2, 3, 7, 8, 9, 17] typically use a constraint solver to perform layout by solving an optimization problem on some class of equations (linear in [1], linear or quadratic in [7, 8], [9, 17] are rule-based). Our constraints are different because they control the choices made by the layout algorithm within its existing layout style, providing a mechanism to override the default, global style with alternate, local criteria. From the results of our prototype, and surveying drawings of data structures, we determined the minimal set of constraints we needed to support were 1) constraining nodes to the same level, 2) ordering nodes within a level (and ordering edge paths), and 3) forcing edge reversals.

Abstractly, our constraints impose a total order on the layout algorithm's level assignment and crossing reduction phases. Although there are potentially a large variety of layout constraints, only a few are needed to impose a total order on these algorithm phases. The level assignment phase assigns nodes to levels, and reverses cycles. The constraints "node a is on the same level as node b;" "node a is on a level above node b;" and "reverse edge e" are enough to control these decisions. The crossing reduction phase orders nodes within their level, and is controlled by the constraint "node a precedes node b." However, the constraint systems of [1,7] also apply to the algorithm's final positioning phase, which we leave unconstrained. Supporting constraints on the positioning phase might require a more general framework for constraint solving.

In the following, we first give the rationale for each constraint, and then its processing. A constraint on a set of objects S is specified in the layout input by giving a list of the objects in S and a code for the constraint.

Constraining nodes to the same level: The *SameLevel(n, p)* constraint constrains nodes n and p to be on the same level. For convenience, we also provide a constraint that an edge is an in-level edge, which is converted internally into a *SameLevel* constraint. The edge constraint is the primary constraint that the user specifies.

Processing: The Sugiyama algorithm [20] provides crossing minimization on directed, leveled graphs, but is unable to handle the in-level edges in nonhierarchical, leveled graphs. [18] saw the same need for UML diagrams, and used the Sugiyama algorithm to layout subgraphs not containing in-level edges, then added in-level edges in a separate step. This kept the Sugiyama algorithm from seeing the in-level edges, with the result that crossing minimization was not performed on these edges. We have instead generalized the algorithm to handle in-level edges.

The problem with [20] handling in-level edges is that it sorts nodes in a level based on their barycenters. The barycenter for a node, n, is the average position of a set of nodes, S, related to n through edges. If the nodes in S are in a different level than n, their positions (and hence the barycenter) are constant during the sort. However, introducing in-level edges makes the value of the barycenters change during the sort, since they depend on the position of nodes that are being rearranged by the sort.

Note that this problem exists for the entire Sugiyama algorithm: If one attempted to use the barycenter heuristics without first partitioning the nodes into levels, the same circular dependency would exist of the barycenters on nodes whose positions are changing during the sort. The level structure is not an independent feature of the layout algorithm, but enables the barycenters to function properly. Within each level we create a secondary system of "virtual levels" by performing a topological sort on the level's in-level edges. These virtual levels do not show up directly in the display, but are used to enable barycenter sorting on in-level edges. Abstractly, the process of sorting a level with in-level edges now works by expanding the level's nodes into multiple levels whose structure is dictated by its virtual levels, sorting the resulting sequence of levels, and then re-embedding these temporary levels back to form the original level.

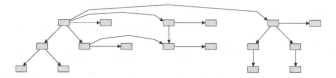

Fig. 3. Nonhierarchical graph showing routing scheme

A full description of the extensions to lay out nonhierarchical graphs is beyond the scope of this paper. For a full description, see [22]. Here, we describe processing the *SameLevel* constraint in the level assignment algorithm, after describing the routing scheme for in-level edges. Unless the end nodes of an in-level edge are neighbors in the level, the path of the edge must be routed around the intervening nodes. As shown in the first level of the graph in Figure 3, we route these edges by creating intermediate routing levels between the level containing in-level edges, and its predecessor level.

The algorithm to assign nodes to levels must also be extended to handle *SameLevel* constraints. These constraints are an example of equality constraints, and naturally give rise to equivalence classes. The extended level assignment algorithm first processes the constraints to produce the equivalence classes, i.e., sets of nodes that are on the same level. Each equivalence class is replaced by a single proxy node, and the normal level assignment algorithm is performed to create levels for the graph with proxy nodes. The proxy nodes are then expanded. Finally, for each level containing in-level edges, a variant of the level assignment algorithm is performed to assign the end nodes of in-level edges to the set of virtual levels associated with the level:

1) Create equivalence classes $N_1, \ldots, N_k$, where each N_i is the maximal set of nodes constrained to be on the same level.
2) Replace nodes in each N_i by a proxy node p for N_i. Mark edges between nodes in N_i as in-level edges. For edges e between a node $q \in N_i$ and $r \notin N_i$, replace q in e by p.
3) Perform level assignment on the revised graph, ignoring in-level edges.
4) Replace each proxy node p by nodes in its equivalence class N_i, and replace edges involving p by the original end node.
5) For each level, L, containing in-level edges, perform level assignment on its nodes that are end-nodes of in-level edges to assign them to L's virtual level structure.

Ordering nodes and edge paths: In graphs representing programming language structures like arithmetic expressions, the ordering of nodes and edges can carry important semantic information. Since layout algorithms often reposition the graph's components to achieve aesthetic heuristics (reduce edge-crossings, minimize edge length), we introduce ordering constraints to preserve this information. The constraints then preserve the semantic ordering at the possible cost of degrading the graph's aesthetics.

Consider the "sub-expression reuse graph" for the expression "A-B+A*C" shown in Figure 4A. (This type of graph shows the reuse of variables and sub-expressions in a calculation. Each variable occurs once as a leaf node, and arithmetic operations are non-leaf nodes. A node for a subexpression that is reused has multiple parent nodes.) Without constraints, the Sugiyama algorithm changes "A-B" into "B-A" in order to reduce edge crossings and edge lengths. (The semantically correct version in Figure 4B has an additional crossing and a longer edge.)

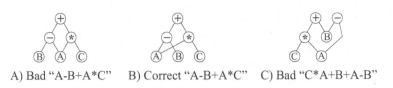

A) Bad "A-B+A*C"   B) Correct "A-B+A*C"   C) Bad "C*A+B+A-B"

Fig. 4. Subexpression reuse graphs

Previous efforts have provided constrained layout by incorporating a constraint solver into the layout process, e.g., [1] added linear constraints to the Sugiyama algorithm. This approach is attractive in that it adds a whole class of constraints via a single mechanism. However, when we examined how it would work in practice, we decided it did not solve our problem. [1] imposes an ordering constraint on the sub-expression "A-B" by the linear constraint "A.x<B.x". The graph in Figure 4C illustrates some problems with this scheme:

1) The positions of nodes A and B satisfy the constraint, but the expression A-B has still been reversed because node A has been moved to a lower level. The ordering could be enforced by constraining A and B to be in the same level, but the level assignment in the graph is the one we desire.

2) The constraint we really want is on the intermediate bend between "-" and "A", but the constraint mechanism specifies relationships on objects input to the layout algorithm. The only way to know that there will be a bend is to precompute the layout. This is particularly unattractive.

3) [1] uses the Sugiyama algorithm to generate an initial ordering that is input to the constraint solver as lower priority constraints than the explicit input constraints. This leaves crossing minimization as an artifact of the constraint solver's input, about which it is ignorant. A crossing reduction algorithm that solves constraints should be more effective in reducing edge crossings than separate algorithms.

These problems led us to implement "procedural constraints," in which each constraint has code dedicated to handling it in the layout algorithm. This allows our constraints to be expanded as necessary to apply to objects internal to the layout process like bend points. They may also contain implicit "guard predicates" such that the constraint condition only applies if the predicate is satisfied. (The constraints in pure constraint systems such as [1] always apply.) These properties allow our constraints to circumvent the problem of precomputing the layout. Unfortunately, each new constraint requires added programming.

We provide three constraints to order nodes and edge paths:

1) *NodePrecedes(x, y)* - node x precedes node y only when they are in same level. If they are in different levels, there is no constraint on their relative positions. The condition that the nodes are in the same level handles cases where the nodes have been shifted to different levels. For example, if x and y are children of node z, but y is involved in a cycle that was reversed, y may now be on the level above z. In this case, the ordering constraint may no longer make sense.

2) *PathPrecedes(e, f)* - this constraint orders the bend points and end nodes composing the paths of edges e and f with a common end node. Let e be an edge in a leveled graph between nodes x and y, whose k bend points are in the ordered set B_e. (If there are no bend points, $k = 0$.) Similarly, f is an edge between nodes x and z, with j bend points in the ordered set B_f. We define:

$Path(e) =$ the ordered set $\{B_e[1], \ldots, B_e[k], y\}$
$Path(e, i) = B_e[i]$ if $i \in \{1, \ldots, k\}$ else y if $i = k + 1$
$PathLength(e) = k + 1$

PathPrecedes(e, f) iff for all $i \in \{1, \ldots, \min(k + 1, j + 1)\}$ then
    NodePrecedes(Path(e, i), Path(f, i))
EdgeSuccessor(e, f) iff NodePrecedes(Path$(e, 1)$, Path$(f, 1)$) and
    Path$(e, 1)$ is the left neighbor of Path$(f, 1)$

The *PathPrecedes* constraint is converted into *NodePrecedes* constraints when edges spanning multiple levels are converted to a path of short (single-level) edges.

3) *PathPrefixPrecedes(length, e, f)* - this variant of the *PathPrecedes* constraint only applies for the first min(*length*, min(PathLength(e), PathLength(f)) nodes in the path. We use this constraint in our stabilization scheme when the initial part of e's path precedes f's path, but the edges cross.

Processing: After assigning nodes to levels, we convert *PathPrecedes* and *PathPrefixPrecedes* constraints into *NodePrecedes* constraints. Next, for the set of nodes N involved in ordering constraints, we create a precedence graph, $G = <N, E, L>$. For each constraint $NodePrecedes(x, y)$, there is an edge $e = <x, y>$ in E. G is a leveled graph, with the levels in L created by the same style of cycle-breaking topological sort used in the Sugiyama algorithm, except that we do not shorten long edges. The level assignment on G checks consistency by reversing unsatisfiable constraints. G also has the property a node's parents precede it in the constraint ordering, and its children follow it. This gives us a quick check for violations of ordering constraints.

We use the idea of a stable sort to incorporate the constraint ordering into the barycenter sort. Given a node n, let Bc(n) be its barycenter, and Pos(n) its level coordinate. The barycenter sort is stable [11] if given nodes x and y in a level, Bc$(x) = $ Bc(y), and Pos$(x) < $ Pos(y) before the sort, then Pos$(x) < $ Pos(y) after the sort, i.e., the relative ordering of nodes with the same sort key is unchanged. A stable sort places a node n in the proper order if, before the sort:

CO: For all nodes x such that $NodePrecedes(x, n)$ then either Bc$(x) < $ Bc(n) or Bc$(x) = $ Bc(n) and Pos$(x) < $ Pos(n)

Since we have implemented a stable sort, we only need to reorder nodes that fail **CO**. After computing the barycenters of nodes in a level, L, but before the barycenter sort, we perform a presort pass over L's order constrained nodes to find nodes failing condition **CO**. We give these nodes a new position and barycenter to make the sort place them in the proper order.

We check the constraint ordering by traversing the levels in the constraint graph G for the constrained nodes in L, from the root level downward. Any node, n, that violates **CO** with respect to its parent nodes in G will cause the sort to violate the constraint ordering by placing n in front a node it should follow. Let p be the rightmost node in L that is a parent of n in G. We force n into the proper order by placing it after p, and assigning n the barycenter and position of p. However, if several nodes are inserted after the same node, we then have a sequence of nodes with the same position. The comparison in **CO** requires using the position to determine the node order.

Recomputing the positions of all the nodes in the level with each insertion produces an $O(v^2)$ algorithm, where v is the number of nodes in the level. However, the duplicate positions only exist within a sequence of nodes inserted after a common ancestor node in G. We introduce a secondary positioning scheme which we call a collision sequence number. A node's collision sequence number is a secondary sort key, and is only used when two nodes have the same position. With the addition of the collision sequence number, Col, condition **CO** becomes

COS: For all nodes x such that $NodePrecedes(x, n)$ then either $Bc(x) < Bc(n)$ or $Bc(x) = Bc(n)$ and $Pos(x) < Pos(n)$ or $Bc(x) = Bc(n)$ and $Pos(x) = Pos(n)$ and $Col(x) < Col(n)$

Initially, all nodes have a collision sequence number of 0. When a node n repositioned to satisfy a constraint is inserted between nodes p and q, $Col(n) = Col(p) + 1$. We add 1 to the collision number of the nodes following n with non-zero collision numbers until reaching a node with a 0 collision number, which is not involved in a collision sequence.

This scheme is still $O(v^2)$ in the worst case, but now only within the sequence of nodes that have been repositioned after a common node, and only if nodes get inserted into the middle of the collision sequence. (If nodes are only added to the end of the sequence, there is no iteration.) If this is a problem in practice, we believe it is possible to develop a more sophisticated scheme that could avoid the need for renumbering. The systems of constraints we use to stabilize the layout do not produce this problem. Note that after the sort, the positions of nodes in the level are recomputed to their actual positions. Thus, the collision sequence scheme only exists temporarily, while sorting the level.

Forced edge reversal: Structures like doubly linked lists contain reverse pointers that we want the layout to reverse. However, the layout algorithm may choose to break cycles by reversing the edges representing forward pointers. The constraint $Reverse(e)$ prevents that by explicitly marking the edges to be reversed. This is mainly a convenience, since the tool could implement similar behavior by reversing the edges before giving them to the layout algorithm.

Level above: The constraint $LevelAbove(n, p)$ constrains node n to be placed on a level above the level assigned to node p. The combination of the $LevelAbove$ and $SameLevel$ constraints can impose a total ordering on node level assignments.

Processing: We form a precedence graph for nodes constrained by $LevelAbove$ constraints, whose level structure is input to the level assignment phase as the initial level assignment for the constrained nodes. This forces the resulting level assignments to obey the constraint.

4 Graph Stability Using Ordering Constraints

The minimum goal for an incremental layout algorithm is to preserve the stability [5] of the unchanged portions of a graph that is modified between successive views. Additionally, it may minimize processing costs by reusing the unchanged parts of the previous layout, thus providing fast lay out on a large graph to which small changes have been made. (See [14] for a fuller discussion.)

Alternatively, stability can be imposed [1] by adding constraints to the layout. This requires the entire graph to be laid out, even for small changes. However, it is easier to implement than an incremental algorithm, particularly since the required constraints were already implemented to preserve "semantic" ordering. Thus, we decided to provide stability via constraints. We add to the layout process the steps: 1) *constraint annotation* - constraints are added to the current layout, before any graph objects are modified, and 2) *modification phase* - nodes and edges are added to, and removed from, the graph to form the next view. The layout cycle is now: layout, constraint annotation, modification, next layout, etc.

1) *Constraint annotation*: We preserve the order of downward edges via *PathPrecedes* constraints when all of an edge's path precedes that of another edge, and *PathPrefixPrecedes* constraints when only a prefix of an edge's path precedes that of another edge. This preserves the inner ordering of subgraphs as they shift between levels. We order the root nodes on the graph's top level to preserve this order if they remain on the top level, even for unrelated subgraphs. Finally, for each edge $e = <r, n>$, where r is a root node, we constrain e to remain in order with its immediate predecessor and successor edges on n. This helps preserve the relative positioning on root nodes as they move in and out of the root level. Given a leveled graph, G, the annotation phase is given by **AddStabilityConstraints**:

CreateEdgePrecedes(e, f){
    find depth n to which nodes in e's path precede the nodes in fs path;
    if all nodes in e's path precede all nodes in fs path up an end node
        create constraint PathPrecedes(e, f);
    else
        create constraint PathPrefixPrecedes(n, e, f);
}
AddStabilityConstraints(G){
    for each node, n, on G's top level with a right neighbor p
        create constraint NodePrecedes(n, p);
    for each node n in G /* Constrain downward edges */
        for each edge e from n (in order of the EdgeSuccessor relation), with
                successor edge f, i.e., EdgeSuccess(e,f)
        CreateEdgePrecedes(e, f);
    for each root node r in G
        for each edge $e = <r, n>${
        if exists edge f such that EdgeSuccessor(f, e)
            CreateEdgePrecedes(f, e);
        if exists edge F such that EdgeSucessor(e, f)
            CreateEdgePrecedes(e, f);
    }
}

2) *Modification phase*: Nodes and edges added to the graph are unconstrained when first laid out. When a node y being removed has constraints
$NodePrecedes(x, y)$ and $NodePrecedes(y, z)$, the constraints are replaced by
$NodePrecedes(x, z)$, and similarly for edges and *PathPrecedes* constraints.
PathPrefixPrecedes constraints are similarly replaced, but the length of the
common path between the edges in the constraint is recalculated.

5   Conclusions

Previous papers on graphically displaying data structures have not examined
the layout problems specific to data structures. Data structure displays have
requirements and strong preferences that have not been recognized in the graph
drawing literature as significant problems. Among these are the "port" problem;
the need to display nonhierarchical graphs; and constraining the order of nodes
and edge paths in the layout. These are only a minimal set of requirements.
While other constraints could be provided, e.g., on node positioning, we do not
know which ones are practically important.

In addition to these requirements, many tools displaying data structures will
also benefit from stabilizing the unchanged parts of a graph between successive
layouts. We have given an algorithm to do this using ordering constraints. Although its processing is not incremental, it is simple to build using node and path
ordering constraints.

The algorithms we have described are implemented in the NARC graph tool
kit [23], which is DART's engine for graph display and layout. The next stage
of work will concentrate on the developing the other mechanisms required in a
working tool.

References

1. K.-F. Boehringer, and F. N. Paulisch Using Constraints to Achieve Stability in
   Automatic Graph Layout Algorithms, *ACM CHI '90 Proceedings*, pp. 43-51.
2. A. Borning, The Programming Language Aspects of ThingLab, a Constraint-
   Oriented Simulation Laboratory, *ACM Transactions on Programming Languages
   and Systems*, 3(4), pp. 252-387, 1981.
3. G. D. Battista, P. Eades, R. Tomassia, and I. G. Tollis, *Graph Drawing: Algorithms
   for the Visualization of Graphs*, Prentice Hall, 1999.
4. P. Eades and D. Kelly, Heuristics for Reducing Crossings in 2-Layered Networks,
   Ars Combin., 21.A, 89-98, 1986.
5. P. Eades, W. Lai, K. Misue, and K. Sugiyama, Preserving the Mental Map of a
   Diagram, *Proceedings Compugraphics '91*, pp. 24-33, 1991.
6. E.R. Gansner, E. Koutsofios, S.C. North and K.-P. Vo, A Technique for Drawing
   Directed Graphs, *IEEE Transactions on Software Engineering*, Vol. 19, No. 3. 1993.
7. W. He and K. Marriott, Constrained Graph Layout, *Proceedings of Graph Drawing
   GD'96*, pp. 217-232, Springer, 1996.
8. T. Kamps, J. Kleinz, and J. Read, Constraint-Based Sping-Model for Graph Layout, *Proceedings of Graph Drawing GD '95*, pp. 349-360, Springer.

9. C. Kosak, J. Marks, and S. Shieber, Automating the Layout of Network Diagrams with Specified Visual Organization, *IEEE Transactions on Systems, Man, and Cybernetics*, Vol. 24, No. 3, pp. 440-454.

10. D. E. Knuth, *The Art of Computer Programming, Vol 1: Fundamental Algorithms*, Second Edition, Addison-Wesley, 1973.

11. D. E. Knuth, *The Art of Computer Programming, Vol 3: Sorting and Searching*, Addison-Wesley, 1973.

12. J. Korn, A. W. Appel, Traversal-based Visualization of Data Structures, *IEEE Symposium on Information Visualization (InfoVis '98)*, pp 11-18.

13. B. Myers, INCENSE: A System for Displaying Data Structures, *Proc. SIGGRAPH 1983*, pp. 115-125.

14. S.C. North, Incremental Layout in DynaDAG, *Proc. of Graph Drawing GD '95*, pp. 409-418, Springer.

15. S.C. North and E. Koutsofios, Applications of Graph Visualization, *Graphics Interface '94*, pp. 235-245.

16. S.P. Reiss, *The Field Programming Environment: A Friendly Integrated Environment for Learning and Development*, Kluwer, 1995.

17. K. Ryall, J. Marks, and S. Shieber, An Interactive System for Drawing Graphs, *Proc. Graph Drawing GD '96*, pp. 387-393, Springer.

18. Jochem Seeman, Extending the Sugiyama Algorithm for Drawing UML Class Diagrams: Towards Automatic Layout of Object-Oriented Software Diagrams, pp. 415-427, *Proc. Graph Drawing '97*, Giuseppe DiBattista, ed. Springer.

19. T. Shimomura and S. Isoda, Linked-List Visualization for Debugging, *IEEE Software*, Vol. 8, No. 3, pp. 44-51, May 1991.

20. Sugiyama, K., Tagawa, S., and M. Toda, Methods for Visual Understanding of Hierarchical Structures, *IEEE Transactions on Systems, Man, and Cybernetics*, Vol. SMC-11, No. 2. Feb. 1981.

21. K. Sugiyama and K. Misue, Visualization of Structural Information: Automatic Drawing of Compound Digraphs, *IEEE Transactions on Systems, Man, and Cybernetics*, Vol 21, No. 4, pp. 876-892, July/August, 1991.

22. V. Waddle, A Sugiyama-Style Layout Algorithm for Nonhierarchical, Leveled Graphs, in preparation.

23. V. Waddle, and A. Malhotra, An E log E Line Crossing Algorithm for Leveled Graphs, *Proc. of Graph Draw GD '99*, pp. 59-71, Springer.

24. J. Yang, C.A. Shaffer, and L. S. Heath, SWAN: A Data Structure Visualization System, *Proc. of Graph Drawing GD '95*, pp 520-523.

25. A. Zeller and D. Luetkeaus, DDD - A Free Graphical Front-end for UNIX Debuggers, *SIGPLAN Notices*, 31(1):22-27, January 1996.

k-Layer Straightline Crossing Minimization by Speeding Up Sifting

Wolfgang Günther[1], Robby Schönfeld[2], Bernd Becker[1], and Paul Molitor[2]

[1] Institute for Computer Science
Albert-Ludwigs-University, Freiburg, Germany
{guenther,becker}@informatik.uni-freiburg.de
[2] Institute for Computer Science
Martin-Luther-University, Halle-Wittenberg, Germany
{schoenfe,molitor}@informatik.uni-halle.de

Abstract. Recently, a technique called sifting has been proposed for k-layer straightline crossing minimization. This approach outperforms the traditional layer by layer sweep based heuristics by far when applied to k-layered graphs with $k \geq 3$. In this paper, we present two methods to speed up sifting. First, it is shown how the crossing matrix can be computed and updated efficiently. Then, we study lower bounds which can be incorporated in the sifting algorithm, allowing to prune large parts of the search space. Experimental results show that it is possible to speed up sifting by more than a factor of 20 using the new methods.

1 Introduction

Graphs are commonly used to represent information in many fields such as software engineering, project management, and social sciences. A good visualization of this information is necessary to get general ideas of relations. Minimizing the number of crossings in a drawing is a key problem, since crossings make it difficult to interpret a graph.

The k-layer straightline crossing minimization problem is the problem to minimize the number of crossings if the vertices are placed in k *layers* (i.e. lines) and there are only edges between vertices of different layers. Vertices within each layer can be permuted in order to minimize the crossing number. Unfortunately, the problem is NP-hard even for two layers and a fixed order of the vertices of the first layer [3]. Many heuristics have been proposed in literature (for an overview, see [4,5]).

Recently, a new heuristic for k-layer straightline crossing minimization has been proposed [6]. This heuristic is based on a technique called sifting [10] which is usually applied to minimize *Binary Decision Diagrams* (BDDs) [1] that are widely used to model Boolean functions in formal logic verification and logic synthesis. The experiments made in [6] prove sifting to be very efficient, outperforming the heuristics known from literature for one-sided and k-layered straightline crossing minimization with $k \geq 3$. The drawback of this new heuristic is that its runtime is higher than the runtime of the best heuristics known from literature.

A similar problem is known for sifting applied to its original application. The size of BDDs largely depends on the underlying variable ordering [1]. Sifting dynamically optimizes the variable ordering by finding the optimal position for each variable while keeping the relative order of all other variables. Although

J. Marks (Ed.): GD 2000, LNCS 1984, pp. 253–258, 2001.
© Springer-Verlag Berlin Heidelberg 2001

sifting can be implemented efficiently, in many applications more than 90% of the CPU time is spent for variable reordering. Therefore, recently the use of lower bounds to speed up sifting has been proposed [2]: moving a variable is stopped if a lower bound on the resulting BDD size is larger than the best size found so far. In experiments it turned out that sifting can be sped up by 70% on average.

In this paper we propose the use of lower bounds for k-layer straightline crossing minimization using sifting. First, we show how the computation of the crossing matrix, which plays a key role in the algorithm, can be realized efficiently. Then, we formally prove lower bounds for the crossing number and we show how these lower bounds can be applied during sifting to prune parts of the search space which do not contain good solutions. We give experimental results showing that the sifting algorithm can be sped up by a factor of ten compared to the original algorithm of [6] by using the efficient computation of the crossing matrix. Using the lower bounds approach, sifting can be further sped up by a factor of 3. This clearly demonstrates the efficiency of the approach.

2   Preliminaries

2.1 k-Layer Crossing Minimization

A k-layered network is a directed graph $G = (V, E)$ whose vertex set V is partitioned into k subsets $V_1, \ldots, V_k$, i.e., $V_1 \cup \ldots \cup V_k = V$ and $(\forall i \neq j)\, V_i \cap V_j = \emptyset$. Set E only consists of edges which connect vertices of different levels V_i and V_j with $i < j$. V_i is called the i-th layer of the graph.

We assume that graph G is proper, i.e. that there are only edges between nodes of adjacent levels V_i and V_{i+1}. Thus, permuting the vertices in a layer V_i, $i \in \{1, 2, \ldots, k\}$, only affects crossings with adjacent layers.

The problem we have to solve is to find for all layers $i \in \{1, \ldots, k\}$ a permutation π_i of the vertices of layer V_i such that the number of edge crossings is minimized. Note that the number of edge crossings only depends on the permutations of the vertices and not on the exact positions of the vertices because edges are drawn as straight lines.

2.2   Sifting for Crossing Minimization

First, we only consider the one sided crossing minimization problem in order to introduce the basic sifting procedure. Assume that k equals 2 and the permutation π_1 of layer V_1 is fixed. Then, the sifting algorithm [6] successively considers all vertices of V_2. For a vertex v under consideration, the goal is to find a position for v on the second layer with a minimal number of crossings, assuming that the relative order of all the other vertices remains unchanged. It is shown in [6] that the order in which the vertices are considered does not have much influence on the quality of the result.

In order to find the best position of a vertex, it is moved to all positions of its layer. This is done in three steps:

1. The vertex is exchanged with its successors until it is the rightmost vertex.
2. The vertex is exchanged with its predecessors until it is the leftmost vertex.
3. The vertex is moved back to the closest position which has led to the minimal number of crossings.

Sifting can be extended to *k*-layer straightline crossing minimization using the following *global sifting* heuristic: In a first step, all vertices of the graph are sorted according to their degrees. We will denote this order by ϕ in the following. Now, each vertex is sifted within its layer, starting with the vertex with maximum degree. If no improvement is made, the Boolean variable *fail* is set to 1 and the vertices are sifted again, but starting with the vertex having minimum degree. Otherwise, the vertices are sifted again using the same order. As long as the value of the Boolean variable *fail* equals 0, the order ϕ is reversed and the last two steps are repeated.

2.3   Exchanging Neighboring Vertices

Let us assume that the order of all layers but layer *i* is fixed. In order to consider the effect of exchanging neighboring vertices in more detail, we have to introduce some new notations. We denote the set of edges incident with vertex *v* by $E(v)$. Furthermore, the variables

$$\delta^i_{uv} = \begin{cases} 1 \ \pi_i(u) < \pi_i(v) \\ 0 \ \text{otherwise,} \end{cases}$$

characterize the permutation π_i of the *i*th layer. δ^i_{uv} equals 1 if and only if vertex *u* is placed before vertex *v* in the ordering π_i. For each pair of nodes $u, v \in V_i$ with $u \neq v$, the crossing number c_{uv} is defined to be the number of crossings between $E(u)$ and $E(v)$ if $\pi_i(u) < \pi_i(v)$ holds. Using these notations, for a given permutation π_i the number of crossings between layers $i - 1$, i, and $i + 1$ is

$$C(\pi_i) = \sum_{u \in V_i} \sum_{v \in V_i} \delta^i_{uv} c_{uv}.$$

3   Crossing Matrix

In order to speed up the computation of the number of crossings in each step, we introduce a three-dimensional crossing matrix $c[i, p_1, p_2]$ which only requires update operations after interchanging two vertices. An entry $c[i, p_1, p_2]$ is set to $c_{\pi_i^{-1}(p_1), \pi_i^{-1}(p_2)}$ with respect to layer *i* and it contains the number of crossings that edges incident to vertex $\pi_i^{-1}(p_1)$ have with edges incident to vertex $\pi_i^{-1}(p_2)$ if $p_1 < p_2$ holds.

Now assume that we swap two vertices *u* and *v* of layer *i* with $p_1 = \pi_i(u)$, $p_2 = \pi_i(v)$. Thus, all entries of the crossing matrix for layer *i* and column p_1 have to be interchanged with the entries from column p_2. Analogously, all entries of row p_1 of layer *i* have to be interchanged with row p_2 of layer *i*. Note, that we only operate on the crossing matrix and do not need to consider the graph itself. This *swap operation* has to be called each time two vertices are swapped.

When the vertex under consideration is set to its locally optimal position, some entries of the crossing matrix which belong to the neighboring layers $i - 1$ and $i+1$ have to be updated, too. If vertex *v* is moved from its original position p_1 to position p_{opt}, some entries in the crossing matrix referring to the neighboring layers $i' \in \{i - 1, \ i + 1\}$ are also affected: entries of vertices adjacent to *v* or adjacent to one of the vertices $\pi_i^{-1}(p_1), \ldots, \pi_i^{-1}(p_{opt} - 1)$ for case $p_1 < p_{opt}$ or $\pi_i^{-1}(p_{opt} + 1), \ldots, \pi_i^{-1}(p_1)$ for case $p_1 > p_{opt}$ have to be modified. Figure 1 describes the *update* procedure for the case $p_1 < p_{opt}$ and layer $i' = i - 1$ in more detail. The case $p_1 > p_{opt}$ is handled analogously.

```
for x ← p₁ to p_opt − 1 do
    foreach vertex u_x adjacent π_i⁻¹(x)
        foreach vertex u_v adjacent π_i⁻¹(v)
            c[i', π_i'(u_x), π_i'(u_v)]−−;
            c[i', π_i'(u_v), π_i'(u_x)]++;
```

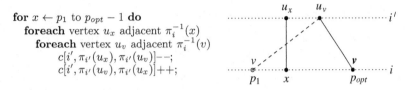

Fig. 1. The *update operation* for $p_1 < p_{opt}$ and $i' = i - 1$.

4   Lower Bound Sifting

When sifting one vertex, the order of one layer i is modified and the orders of the vertices of all the other layers remain unchanged. In this section, we give lower bounds on the resulting number of crossings and show how to use them during sifting.

In [4], a lower bound on the number of crossings for the one-sided crossing minimization is given:

$$LB := \sum_{u=1}^{|V_i|-1} \sum_{v=u+1}^{|V_i|} \min\{c_{uv}, c_{vu}\}$$

is a lower bound on the number of crossings if permutation π_i can be chosen arbitrarily and all other permutations are constant. In the following we show how this lower bound can be improved if only the position of one node of V_i can be modified and the relative order of all other nodes remains the same.

A lower bound for moving one vertex v in π_i to the right while keeping the relative order of all other vertices unchanged is the following:

$$LB^{\to}(v) := \sum_{\substack{u_1=1 \\ u_1 \neq v}}^{|V_i|} \sum_{\substack{u_2=1 \\ u_2 \neq v}}^{|V_i|} \delta^i_{u_1 u_2} c_{u_1 u_2} + \sum_{u=1}^{|V_i|} \delta^i_{uv} c_{uv} + \sum_{u=1}^{|V_i|} \delta^i_{vu} \min\{c_{vu}, c_{uv}\}$$

$$= C(\pi_i) - \sum_{u=1}^{|V_i|} \delta^i_{vu} \max\{0,\ c_{vu} - c_{uv}\}$$

Informally, this means that only those crossings are affected by moving vertex v to the right in which an edge incident to v and an edge incident to some node u on the right of v is involved. A similar lower bound can be used when moving a vertex to the left.

During sifting, moving a vertex v can be stopped if it cannot improve the order already computed. To decide this, these lower bounds can be used. A sketch of the resulting algorithm moving a vertex rightwards is the following:

```
ALGORITHM SiftingRight(v);
best = number_of_crossings();
while v is not the rightmost vertex and LB→(v) ≤ best do
    exchange_vertex_with_neighbor(v);
    if number_of_crossings() ≤ best then
        best = number_of_crossings();
```

If the initial number of crossings is given, the lower bound for a vertex v of the ith layer can be computed in time $O(|V_i|)$. It can be updated in constant time. Thus, computing lower bounds is inexpensive.

Table 1. Runtime requirements on sparse $4 \times n$ graphs.

| | runtime [sec] | | | | runtime [sec] | quality [%] | | |
n	siftingI	siftingII	siftingLB	barycenter	combined	combined	siftingLB	barycenter
50	21.7	1.7	0.6	0.01	0.4	100.0	110.1	123.2
60	29.6	3.0	1.1	0.01	0.8	100.0	112.3	123.5
70	50.5	5.2	1.9	0.02	1.2	100.0	110.9	122.5
80	80.2	8.6	3.1	0.02	2.0	100.0	110.3	122.9
90	126.1	12.9	4.6	0.03	3.0	100.0	110.3	122.9
100	200.8	20.3	7.2	0.03	4.8	100.0	111.3	123.1

Note that using lower bounds does not affect the quality of the procedure, which is measured in number of crossings, since the positions which are not considered cannot be locally optimal. However, the final result can be different, since the optimal position of one node is not unique. Using lower bounds, a different (optimal) position can be chosen for a node, which may affect the result of the following nodes. Practically, experimental runs on random generated graphs of different size and density result in an average deviation of quality by at most 0.3 percent.

5 Experimental Results

The implementations of sifting are using algorithms form the AGD library [8] which is based on LEDA [7]. The experiments have been carried out on a Sun UltraSPARC-IIi [1]. All runtimes are given in CPU seconds.

Since the quality of the global sifting algorithm of [6] is about 20% better than the layer by layer based heuristics known from literature [4], in the following we only focus on runtime. It is shown that the algorithm can be sped up tremendously while the quality does not change significantly.

The implementation of the global sifting that was presented in [6] is denoted by *sifting* I. The algorithm using the crossing matrix approach of Section 3 is named *sifting* II. Finally, algorithm *sifting* LB combines *sifting* II with the lower bound described in Section 4.

In order to get an impression of the time savings we compare the three sifting algorithms in terms of run time requirement. In these experiments, we consider sparse $k \times n$-graphs, i.e., k-layered graphs $G_k = (V_1 \cup \ldots \cup V_k, E)$ with $|V_i| = n$ for all $1 \le i \le k$ and $|E| = 2 \cdot (k-1) \cdot n$.

In the experimental run we consider in this paper, only the number of vertices is varied. Table 1 (left part) shows the results for the three sifting algorithms for sparse $k \times n$-graphs with $k = 4$ and $50 \le n \le 100$ taken over 100 samples. The first column gives the number n of vertices per layer, the second, third, and fourth column show the runtime of *sifting* I, *sifting* II, and *sifting* LB in seconds, respectively.

It can be seen that the improved implementation for the crossing matrix is much faster than the original method. In some cases, the runtime for *sifting* II is even up to 90% less than for *sifting* I. Using lower bounds, the runtime can be further reduced by more than 65%. Altogether, this is a speed-up of more than a factor of 20.

However when considering runtime requirement the *sifting*LB approach continues to be dominated by common layer by layer sweep based heuristics of which *barycenter* is the most efficient method. It can be seen (Column 5 of Table 1) that

---

[1] 440 MHz CPU, 256 MB RAM.

barycenter consumes practically constant computation time while the number of vertices per layer is being increased up to 100.

The performance of lower bound sifting can be further improved by taking advantage of the computations of the *barycenter* method. Using the permutation of vertices obtained by *barycenter* as initial ordering for our *sifting* LB heuristic has two major effects: the number of crossings is usually reduced further and the application of the lower bound is more efficient as most vertices are already located in their locally optimal positions when starting with a good initial order in each layer. Column 6 of Table 1 (right part) shows the runtime of the combined heuristics *barycenter* and *sifting* LB. The computation effort could be reduced by nearly 40% in comparison with the single *sifting* LB heuristic. The relative numbers of crossings using *sifting* LB and *barycenter* compared to the combined method are given in the last three columns. It can be seen that for the combined method also the quality improves by 10% compared to *sifting* LB and by 20% compared to *barycenter*.

6   Conclusions

In this paper, we presented two improvements to the global sifting approach of [6] for k-layer straightline crossing minimization. First, a three-dimensional crossing matrix was proposed. Furthermore, lower bounds have been used to speed up sifting by determining parts of the search space which do not have to be considered. Experimental results showed that it is possible to reduce the runtime by more than a factor of 20, without a loss of quality.

It is focus of current work to transfer other techniques known from BDD minimization like *group sifting* [9] to k-layer straightline crossing minimization.

References

1. R.E. Bryant. Graph - based algorithms for Boolean function manipulation. *IEEE Trans. on Comp.*, 35(8):677–691, 1986.
2. R. Drechsler and W. Günther. Using lower bounds during dynamic BDD minimization. In *Design Automation Conf.*, pages 29–32, 1999.
3. P. Eades and S. Whitesides. Drawing graphs in two layers. *Theoretical Computer Science*, 131:361–374, 1994.
4. M. Jünger and P. Mutzel. 2-layer straightline crossing minimization: Performance of exact and heuristic algorithms. *Journal of Graph Algorithms and Applications*, 1(1):1–25, 1997.
5. M. Laguna, R. Martí, , and V. Valls. Arc crossing minimization in hierarchical digraphs with tabu search. *Computers and Operations Research*, 24(12):1175–1186, 1997.
6. C. Matuszewski, R. Schönfeld, and P. Molitor. Using sifting for k-layer straightline crossing minimization. In *Graph Drawing Conference*, LNCS 1731, pages 217–224, 1999.
7. K. Mehlhorn and S. Näher. *The LEDA Platform of Combinatorial and Geometric Computing*. Cambridge University Press, 1999. Project home page at http://www.mpi-sb.mpg.de/LEDA/.
8. P. Mutzel, T. Ziegler, S. Näher, D. Alberts, D. Ambras, G. Koch, M. Jünger, C. Buchheim, and S. Leipert. A library of algorithms for graph drawing. In *International Symposium on Graph Drawing*, LNCS 1547, pages 456–457, 1998. Project home page at http://www.mpi-sb.mpg.de/AGD/.
9. S. Panda and F. Somenzi. Who are the variables in your neighborhood. In *Int'l Conf. on CAD*, pages 74–77, 1995.
10. R. Rudell. Dynamic variable ordering for ordered binary decision diagrams. In *Int'l Conf. on CAD*, pages 42–47, 1993.

Lower Bounds for the Number of Bends in Three-Dimensional Orthogonal Graph Drawings

David R. Wood*

Basser Department of Computer Science
The University of Sydney
Sydney NSW 2006, Australia
davidw@cs.usyd.edu.au

Abstract. In this paper we present the first non-trivial lower bounds for the total number of bends in 3-D orthogonal drawings of maximum degree six graphs. In particular, we prove lower bounds for the number of bends in 3-D orthogonal drawings of complete simple graphs and multigraphs, which are tight in most cases. These result are used as the basis for the construction of infinite classes of c-connected simple graphs and multigraphs ($2 \leq c \leq 6$) of maximum degree Δ ($3 \leq \Delta \leq 6$) with lower bounds on the total number of bends for all members of the class. We also present lower bounds for the number of bends in general position 3-D orthogonal graph drawings. These results have significant ramifications for the '2-bends' problem, which is one of the most important open problems in the field.

1 Introduction

The *3-D orthogonal grid* consists of *grid-points* in 3-space with integer coordinates, together with the axis-parallel *grid-lines* determined by these points. A *3-D orthogonal drawing* of a graph places the vertices at grid-points and routes the edges along sequences of contiguous segments of grid-lines. Edges are allowed to contain bends and can only intersect at a common vertex. 3-D orthogonal drawings have been studied in [2,3,4,5,6,7,8,12,13,14]. For brevity we say a 3-D orthogonal graph drawing is a *drawing*. A drawing with no more than b bends per edge is called a *b-bend drawing*. The graph-theoretic terms 'vertex' and 'edge' also refer to their representation in a drawing. The *ports* at a vertex v are the six directions, denoted by $\{X^+, X^-, Y^+, Y^-, Z^+, Z^-\}$, which the edges incident with v can use. For each dimension $I \in \{X, Y, Z\}$, the I^+ (respectively, I^-) port at a vertex v is said to be *extremal* if v has maximum (minimum) I-coordinate taken over all vertices. Clearly, orthogonal drawings can only exist for graphs with maximum degree six.

Drawings with many bends appear cluttered and are difficult to visualise. Therefore minimising the number of bends, along with minimising the bounding box volume, have been the most commonly proposed aesthetic criteria for

---

* Supported by the Australian Research Council Large Grant A49906214.

measuring the quality of a drawing. Using straightforward extensions of the corresponding 2-D NP-hardness results, optimising each of these criteria is NP-hard [4]. Kolmogorov and Barzdin [6] established a lower bound of $\Omega(n^{3/2})$ for the bounding box volume of a drawings of n-vertex graphs. In this paper we establish the first non-trivial lower bounds for the number of bends in 3-D orthogonal drawings. Lower bounds for the number of bends in 2-D orthogonal graph drawings have been established by Tamassia *et al.* [9] and Biedl [1].

Lower bounds for the maximum number of bends per edge: Obviously every drawing of K_3 has at least one bend. It follows from results in multi-dimensional orthogonal graph drawing by Wood [11] that every drawing of K_5 has an edge with at least two bends. It is well known that every drawing of $6K_2$ has an edge with at least three bends, and easily seen that $2K_2$ and $3K_2$ have at least one edge with at least one and two bends, respectively. Here kK_2 is the 2-vertex multigraph with k edges.

A natural candidate for the existence of a simple graph with a 3-bend edge in every drawing is K_7, as originally conjectured by Eades *et al.* [5]. A counterexample to this conjecture, namely a drawing of K_7 with at most two bends per edge, was first exhibited by Wood [11]. A more symmetric drawing of K_7 with at most two bends per edge is shown in Fig. 1(a). This drawing has the interesting feature of rotational symmetry about the line $X = Y = Z$. One may consider the other 6-regular complete multi-partite graphs $K_{6,6}$, $K_{3,3,3}$ and $K_{2,2,2,2}$ to be potential examples of simple graphs with a 3-bend edge in every drawing. 2-bend drawings of these graphs are presented in [14].

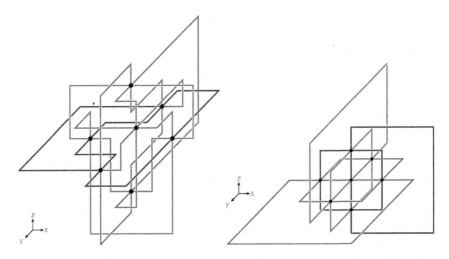

Fig. 1. (a) 2-bend drawing of K_7 (b) 4-bend drawing of K_7 with 24 bends.

Lower bounds for the total number of bends: In this paper we prove that drawings of the complete graphs K_4, K_5, K_6 and K_7 have at least 3, 7, 12 and 20 bends, respectively. For each of these graphs except K_7 there are well-known drawings with the corresponding number of bends. Figure 1(b) shows a drawing of K_7 with a total of 24 bends (compared with the total of 42 bends for the 2-bend drawing). We conjecture that there is no drawing of K_7 with fewer than 24 bends.

We use these lower bounds for the number of bends in complete graphs as the basis for the construction of infinite families of c-connected graphs of maximum degree Δ with lower bounds on the number of bends for each member of the class. Table 1 summarises these lower bounds.

Table 1. Lower bounds for the number of bends in drawings of m-edge c-connected graphs with maximum degree Δ.

Connectivity c	Simple Graphs				Multigraphs			
	$\Delta = 6$	$\Delta = 5$	$\Delta = 4$	$\Delta = 3$	$\Delta = 6$	$\Delta = 5$	$\Delta = 4$	$\Delta = 3$
0	$\frac{20}{21}m$	$\frac{4}{5}m$	$\frac{7}{10}m$	$\frac{1}{2}m$	$2m$	$\frac{8}{5}m$	$\frac{3}{2}m$	$\frac{4}{3}m$
2	$\frac{3}{4}m$	$\frac{7}{11}m$	$\frac{3}{7}m$	$\frac{1}{4}m$	$\frac{4}{3}m$	$\frac{6}{5}m$	m	$\frac{2}{3}m$
3	$\frac{8}{11}m$	$\frac{14}{23}m$	$\frac{2}{5}m$	$\frac{2}{9}m$	m	$\frac{4}{5}m$	$\frac{1}{2}m$	-
4	$\frac{12}{17}m$	$\frac{7}{12}m$	$\frac{3}{8}m$	-	$\frac{2}{3}m$	$\frac{2}{5}m$	-	-
5	$\frac{24}{35}m$	$\frac{14}{25}m$	-	-	$\frac{1}{3}m$	-	-	-
6	$\frac{2}{3}m$	-	-	-	-	-	-	-

Upper bounds: A number of algorithms have been proposed for 3-D orthogonal graph drawing [2,3,5,6,8,12,14] which explore the apparent tradeoff between the maximum number of bends per edge and the bounding box volume (see [14] for an overview). We now summarise the known upper bounds on the number of bends in the drawings produced by these algorithms. The 3-BENDS algorithm of Eades et al. [5] and the INCREMENTAL algorithm of Papakostas and Tollis [7] both produce 3-bend drawings[1] of multigraphs[2] with maximum degree six. As discussed above there exists simple graphs with at least one edge having at least two bends in every drawing. The following problem is therefore of considerable interest:

2-Bends Problem: Does every simple graph with maximum degree six admit a 2-bend drawing? [5]

---

[1] The 3-BENDS algorithm [5] produces drawings with $27n^3$ volume. By deleting grid-planes not containing a vertex or a bend the volume is reduced to $8n^3$. The INCREMENTAL algorithm [7] produces drawings with $4.63n^3$ volume. A modification of the 3-BENDS algorithm by Wood [14] produces drawings with $n^3 + o(n^3)$ volume.

[2] The 3-BENDS algorithm [5] explicitly works for multigraphs. The INCREMENTAL algorithm, as stated in [7], only works for simple graphs; with a suitable modification it also works for multigraphs [A. Papakostas, private communication, 1998].

The DIAGONAL LAYOUT AND MOVEMENT algorithm of Wood [12] solves the 2-bends problem in the affirmative for simple graphs with maximum degree five. For maximum degree six simple graphs, the same algorithm uses a total of at most $7m/3$ bends, which is the best known upper bound for the total number of bends in 3-D orthogonal drawings.

In this paper we provide a negative result related to the 2-bends problem. A 3-D orthogonal graph drawing is said to be in *general position* if no two vertices lie in a common grid-plane. The general position model is used in the 3-BENDS and DIAGONAL LAYOUT AND MOVEMENT algorithms. In this paper we show that the general position model, and the natural variation of this model where pairs of vertices share a common plane, cannot be used to solve the 2-bends problem, at least for 2-connected graphs.

The remainder of this paper is organised as follows. In Sect. 2 we establish a number of introductory results concerning 0-bend drawings of cycles. These results are used to prove our lower bounds on the total number of bends in drawings of complete graphs, established in Sect. 3. In Sect. 4 we use these lower bounds as the basis for lower bounds on the number of bends in infinite families of graphs of varying connectivity and maximum degree. In Sect. 5 we present lower bounds for the number of bends in general position drawings and their implications for the 2-bends problem.

Throughout this paper we consider n-vertex m-edge loop-free graphs with maximum degree six. By C_m we denote the m-edge cycle. A *chord* of a cycle C is an edge not in C whose end-vertices are both in C. We say two cycles are *chord-disjoint* if they do not have a chord in common. In this paper most proofs are outlined, and the proofs of the results in Table 1 concerning multigraphs are omitted; see [13] for details of all the proofs.

2   Drawings of Cycles

In this section we characterise the 0-bend drawings of the cycles C_k ($k \leq 7$). We then show that if a drawing of a complete graph contains such a 0-bend drawing of a cycle then there must be many edges with at least three bends in the drawing of the complete graph. These results are used in Sect. 3 in our lower bounds for the total number of bends in drawings of complete graphs.

A straight-line path in a 0-bend drawing of a cycle is called a *side*. A side parallel to the I-axis for some $I \in \{X, Y, Z\}$ is called an I-side, and I is called the *dimension* of the side. Clearly the dimension of adjacent sides is different, thus in a 2-dimensional drawing the dimension of the sides alternate around the cycle. Hence, there is no 2-dimensional 0-bend drawing of a cycle with an odd number of sides. If there is an I-side in a drawing of a cycle for some $I \in \{X, Y, Z\}$ then clearly there is at least two I-sides. Therefore a drawing of a cycle with X-, Y- and Z-sides, which we call *truly 3-dimensional*, must have at least six sides. Hence there is no truly 3-dimensional 3-, 4- or 5-sided 0-bend drawing of a cycle. There is also no two-dimensional 3- or 5-sided 0-bend drawing of a cycle. We therefore have the following observation.

Observation 1 *(a) There is no 3- or 5-sided 0-bend drawing of a cycle, (b) there is no 0-bend drawing of C_3, and (c) all 0-bend drawings of C_4 and C_5 have four sides.*   □

The next result forms an important component of the lower bounds to follow.

Lemma 1. *If a drawing of a complete graph contains a 0-bend 4-cycle (respectively, 5-cycle) then there are at least two (four) chords of the cycle each with at least three bends.*

Proof. By Obs. 1(c) all 0-bend drawings of C_4 and of C_5 have four sides. As illustrated in Fig. 2(a), the chord connecting each pair of diagonally opposite vertices in a 4-sided drawing of a cycle has at least three bends. Hence, if a drawing of a complete graph contains a 0-bend C_4, then the two chords each have at least three bends. Also, in the case of C_5, the two edges from the vertex not at the intersection of two sides to the diagonally opposite vertices both have at least three bends, as in Fig. 2(b). Hence, if a drawing of a complete graph contains a 0-bend C_5, then the four chords each have at least three bends.   □

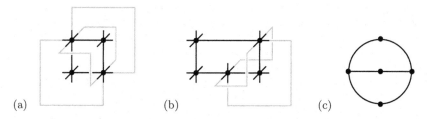

(a)   (b)   (c)

Fig. 2. 3-bend edges 'across' (a) C_4 and (b) C_5. (c) The graph H_1

Consider the graph shown in Fig. 2(c), which we call H_1.

Observation 2 H_1 *does not have a 0-bend drawing.*

Proof. H_1 contains C_4. As in Lemma 1, an edge between the non-adjacent vertices of a 4-sided cycle needs at least three bends. Hence the 2-path in H_1 between the non-adjacent vertices of the 4-cycle has at least one bend. Therefore H_1 does not have a 0-bend drawing.   □

We now consider 6-sided 0-bend drawings of a cycle.

Lemma 2. *If a drawing of a complete graph contains a 0-bend 6-cycle then there are at least four chords of the cycle each with at least three bends.*

Proof. We can assume without loss of generality that there is a 0-bend drawing of C_6 contained in a drawing of K_6. By Obs. 1(a), all 0-bend drawings of C_6 have four or six sides. In a 4-sided 0-bend drawing of C_6 the two vertices not at the intersection of adjacent sides can be (a) on the same side, (b) on adjacent sides, or (c) on opposite sides. In each case there are at least six chords from a vertex at the intersection of two sides to a vertex on neither of these sides, each with at least three bends.

Case analysis shows that the only 6-sided 0-bend drawings of C_6 (up to symmetry) are those in Fig. 3. For each such drawing, the chords of C_6 shown in Fig. 3 each require at least three bends. In the case of the drawing in Fig. 3(c) there are at least six chords each requiring at least three bends. For the drawing in Fig. 3(a) (respectively, Fig. 3(b)) it can be shown, by considering certain sets of chords for which all edge routes with at most two bends pass through a single grid point, that a further two (four) chords have at least three bends. Hence there are at least six chords of the cycle each with at least three bends.   □

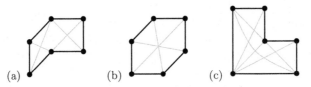

Fig. 3. Edges with at least 3 bends in a drawing of K_6 containing a 6-sided 0-bend C_6.

We now consider 7-sided 0-bend drawings of a cycle.

Lemma 3. *If a drawing of K_7 contains a 0-bend 7-cycle then there are at least four chords of the cycle each with at least three bends.*

Proof. By Obs. 1(a), a 0-bend drawing of C_7 has four, six or seven sides. In a 4-sided 0-bend drawing of C_7 the three vertices not at the intersection of two adjacent sides can be (a) all on the same side, (b) two on one side and one on an adjacent side, (c) two on one side and one on the opposite side, or (d) all on different sides. In each case there are at least eight chords from a vertex at the intersection of two sides to a vertex on neither of these sides, each with at least three bends. The 6-sided 0-bend drawings of C_7 can be obtained from the 6-sided 0-bend drawings of C_6 by placing one new vertex at each of the essentially different edges of each drawing. By Lemma 2, at least six of the chords of C_6, and therefore of C_7, have at least three bends. Case analysis shows that the only 7-sided 0-bend drawings of C_7 are those shown in Fig. 4. For each drawing there are at least four chords which need at least three bends.   □

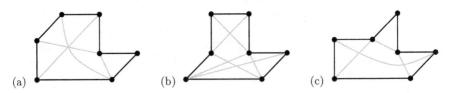

Fig. 4. Edges with at least three bends in a 7-sided 0-bend drawing of C_7.

3   Drawings of Complete Graphs

In this section we establish lower bounds for the total number of bends in 3-D orthogonal drawings of K_4, K_5, K_6 and K_7. We omit our proofs that the well-known drawings of K_4, K_5 and K_6 shown in Fig. 5 are bend-minimum. They are similar to the proof of the lower bound for K_7 which follows.

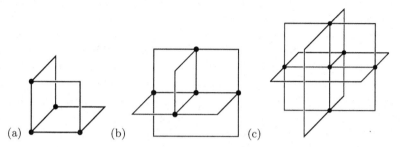

(a)   (b)   (c)

Fig. 5. (a) K_4 with 3 bends, (b) K_5 with 7 bends and (c) K_6 with 12 bends.

Theorem 1. *Every drawing of K_4 has at least three bends. Every drawing of K_5 has at least seven bends. Every drawing of K_6 has at least twelve bends.* □

Figure 1(b) shows a drawing of K_7 with a total of 24 bends.

Theorem 2. *Every drawing of K_7 has at least 20 bends.*

Proof. Suppose to the contrary, that there is a drawing of K_7 with at most 19 bends. The subgraph of K_7 consisting of the 0-bend edges is called the *0-bend subgraph*. Let k_i $(i \geq 0)$ be the number of i-bend edges. Hence

$$\sum_{i \geq 0} k_i = 21, \text{ and } \sum_{i \geq 1} i k_i \leq 19 . \tag{1}$$

Case analysis shows that every subgraph of K_7 with at least ten edges contains C_3 or H_1. By Obs. 1(b) and Obs. 2, the graphs C_3 and H_1 do not have 0-bend drawings. Hence $k_0 \leq 9$. Suppose $k_0 = 8$ or $k_0 = 9$. By (1)

$$12 \leq \sum_{i \geq 1} k_i \leq 19 - \sum_{i \geq 1} (i - 1) k_i$$

$$\sum_{i \geq 2} (i - 1) k_i \leq 7 . \tag{2}$$

Case analysis shows that every subgraph of K_7 with at least eight edges contains a cycle C_k $(k \neq 4)$, two chord-disjoint cycles, or an H_1 subgraph. Therefore the 0-bend subgraph contains a cycle C_k $(k \geq 5)$ or two chord-disjoint subgraphs (since C_3 and H_1 do not have 0-bend drawings by Obs. 1(b) and Obs. 2, respectively). If the 0-bend subgraph contains a cycle C_k $(k \geq 5)$ then by Lemma 1, Lemma 2 and Lemma 3 there are at least four chords of the cycle each with at least three bends. On the other hand if the 0-bend subgraph contains two chord-disjoint cycles then these cycles have length at least four; thus by Lemma 1, Lemma 2 and Lemma 3, two chords from each of these cycles each have at least three bends. In either case the drawing of K_7 has at least four edges each with at least three bends; that is,

$$4 \le \sum_{i \ge 3} k_i$$

$$k_2 + 8 \le k_2 + \sum_{i \ge 3} 2k_i \le \sum_{i \ge 2} (i-1)k_i$$

$$k_2 + 8 \le 7 \qquad \qquad \text{(by (2))}$$

Hence $k_2 \le -1$, which is a contradiction. Therefore $k_0 \le 7$. By (1) with $k_0 \le 7$

$$14 \le \sum_{i \ge 1} k_i \le 19 - \sum_{i \ge 1} (i-1)k_i$$

$$\sum_{i \ge 2} (i-1)k_i \le 5$$

$$k_2 + 2\sum_{i \ge 3} k_i \le 5 \qquad \qquad (3)$$

Let A be the set of edges of K_7 routed using an extremal port at exactly one end-vertex. Let B be the set of edges routed using extremal ports in the same direction at its end-vertices. Let C be the set of edges routed using extremal ports in differing directions at its end-vertices. K_7 is 6-regular, thus all ports are used. Since there is at least one extremal port in each direction, we have $|A| + |B| + 2|C| \ge 6$. It is easily seen that an edge in A or B has at least two bends, and that an edge in C has at least three bends. Hence

$$k_2 + 2\sum_{i \ge 3} k_i \ge 6 \ .$$

This contradicts (3). The result follows.   □

4   Constructing Large Graphs

In this section we use the lower bounds for the total number of bends in drawings of the complete graphs established in Sect. 3 as building blocks to construct infinite families of graph with lower bounds for the number of bends.

Given graphs G and H with $|V(G)| \ge \Delta(H)$, we define $H\langle G \rangle$ to be the graph obtained by replacing each vertex of H by a copy of G, and connecting the vertices adjacent to a vertex $v \in V(H)$ to different vertices in the copy of G corresponding to v. In most cases, H is regular and G is a complete graph, thus $H\langle G \rangle$ is well-defined. In other cases we shall specify the mapping between edges incident to v and the vertices in the copy of G corresponding to v. We also employ the *cartesian product* $G \times H$ of graphs G and H to construct larger graphs. $G \times H$ has vertex set $V(H) \times V(G)$ with (v_1, w_1) and (v_2, w_2) adjacent in $G \times H$ if either $v_1 = v_2$ and $w_1 w_2 \in E(H)$, or $w_1 = w_2$ and $v_1 v_2 \in E(G)$.

By taking disjoint copies of K_7, K_6, K_5 and K_4 the next result follows immediately from Theorem 1 and Theorem 2.

Theorem 3. *There exists infinite families of simple (disconnected) m-edge graphs with maximum degree six (respectively, five, four and three) with at least $\frac{20}{21}m$ ($\frac{4}{5}m$, $\frac{7}{10}m$ and $\frac{1}{2}m$) bends in every drawing.* $\square$

To obtain our lower bounds for 2-, 3- and 4-connected graphs consider the graphs $C_r \langle K_p \rangle$ $(p \geq 3)$, $(C_r \times K_2)\langle K_p \rangle$ $(p \geq 3)$ and $(C_r \times C_3)\langle K_p \rangle$ $(p \geq 4)$ for some $r \geq 3$, as illustrated in Fig. 6.

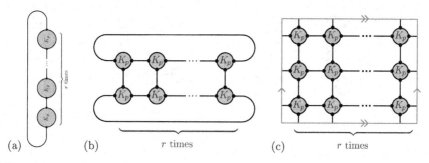

Fig. 6. (a) 2-connected $C_r \langle K_p \rangle$, (b) 3-connected $(C_r \times K_2)\langle K_p \rangle$, (c) 4-connected $(C_r \times C_3)\langle K_p \rangle$.

Theorem 4. *There exists infinite families of simple 2-connected m-edge graphs with maximum degree six (respectively, five, four and three) with at least $\frac{3}{4}m$ ($\frac{7}{11}m$, $\frac{3}{7}m$ and $\frac{1}{4}m$) bends in every drawing.*

Proof. The graphs $C_r \langle K_6 \rangle$ with $r \geq 2$ have maximum degree six and $m = 16r$ edges. By Theorem 1, K_6 has at least 12 bends in every drawing, thus $C_r \langle K_6 \rangle$ has at least $12r = \frac{3}{4}m$ bends in every drawing. The graphs $C_r \langle K_5 \rangle$ with $r \geq 2$ have maximum degree five and $m = 11r$ edges. By Theorem 1, K_5 has at least 7 bends in every drawing, thus $C_r \langle K_5 \rangle$ has at least $7r = \frac{7}{11}m$ bends in every drawing. The graphs $C_r \langle K_4 \rangle$ with $r \geq 2$ have maximum degree four and $m = 7r$ edges. By Theorem 1, K_4 has at least 3 bends in every drawing, thus $C_r \langle K_4 \rangle$ has at least $3r = \frac{3}{7}m$ bends in every drawing. The graphs $C_r \langle K_3 \rangle$ with $r \geq 2$ have maximum degree three and $m = 4r$ edges. By Obs. 1(b), K_3 has at least 1 bend in every drawing, thus $C_r \langle K_3 \rangle$ has at least $r = \frac{1}{4}m$ bends in every drawing. Clearly $C_r \langle K_p \rangle$ is 2-connected. $\square$

The proofs of the following results for 3- and 4-connected graphs are similar to the proof of Theorem 4.

Theorem 5. *There exists infinite families of simple 3-connected m-edge graphs with maximum degree six (respectively, five, four and three) with at least $\frac{8}{11}m$ ($\frac{14}{23}m$, $\frac{2}{5}m$ and $\frac{2}{9}m$) bends in every drawing.* □

Theorem 6. *There exists an infinite family of simple 4-connected m-edge graphs with maximum degree six (respectively, five and four) with at least $\frac{12}{17}m$ ($\frac{7}{12}m$ and $\frac{3}{8}m$) bends in every drawing.* □

To obtain our lower bounds for 5- and 6-connected graphs consider the graphs $(C_r \times C_3 \times K_2)\langle K_p \rangle$ $(p \geq 5)$ and $(C_r \times C_3 \times C_3)\langle K_6 \rangle$ for some $r \geq 3$, as illustrated in Fig. 7. Again the proofs are very similar to that of Theorem 4.

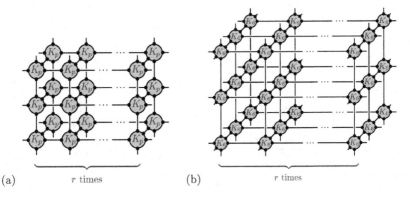

(a) r times   (b) r times

Fig. 7. (a) 5-connected $(C_r \times C_3 \times K_2)\langle K_p \rangle$, (b) 6-connected $(C_r \times C_3 \times C_3)\langle K_6 \rangle$.

Theorem 7. *There exists infinite families of simple 5-connected m-edge graphs with maximum degree six (respectively, five) with at least $\frac{24}{35}m$ ($\frac{14}{25}m$) bends in every drawing.* □

Theorem 8. *There exists an infinite family of simple 6-connected m-edge graphs with maximum degree six with at least $\frac{2}{3}m$ bends in every drawing.* □

5   General Position Drawings and the 2-Bends Problem

Recall that a 3-D orthogonal graph drawing is in general position if no two vertices lie in a common grid-plane. In this section we establish lower bounds for the number of bends in general position drawings.

Lemma 4. *If the graph G has at least k bends in every general position drawing then for any edge e of G the graph $G \setminus e$ has at least $k - 4$ bends in every general position drawing.*

Proof. If $G \setminus e$ has a drawing with b bends then, the edge e can be inserted into the drawing with at most four bends and the edges rerouted so that there are no edge crossings [12,14]. Thus, there is a general position drawing of G with $b + 4$ bends. Every general position drawing of G has at least k bends, hence $b + 4 \geq k$ and $b \geq k - 4$. □

Clearly every edge in a general position drawing has at least two bends. The following lower bounds for general position drawings are based on the observation that an edge routed using an extremal port in a general position drawing has at least three bends. For 6-regular m-edge graphs all ports must be used, thus such a graph has at least $2m + 6$ bends in every general position drawing. Hence the graphs consisting of disjoint copies of K_7 provide the following lower bound.

Lemma 5. *There exists an infinite family of n-vertex m-edge simple graphs, each with at least $2m + 6n/7$ bends in every general position drawing.*   □

Note that for 6-regular graphs the above lower bound is within $m/21$ of the upper bound of $7m/3$ for the total number of bends in general position drawings established by the DIAGONAL LAYOUT AND MOVEMENT algorithm [12].

To establish our lower bound for general position drawings of 2-connected graphs consider the graph $C_r\langle K_7 \setminus e\rangle$ for $r \geq 2$, where the non-adjacent vertices of each $K_7 \setminus e$ are incident to the edges of C_r, as illustrated in Fig. 8(a).

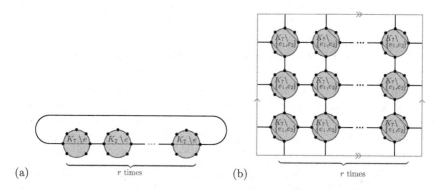

(a)   (b)

Fig. 8. (a) 2-connected $C_r\langle K_7 \setminus e\rangle$, (b) 4-connected $(C_r \times C_3)\langle K_7 \setminus \{e_1, e_2\}\rangle$.

Lemma 6. *There is an infinite family of n-vertex m-edge simple 2-connected graphs, each with at least $2m + 4n/7$ bends in every general position drawing.*

Proof. Clearly $C_r\langle K_7 \setminus e\rangle$ is 2-connected. K_7 has at least $2|E(K_7)| + 6$ bends in every general position drawing. Thus by Lemma 4, a general position drawing of $K_7 \setminus e$ has at least $2|E(K_7)| + 6 - 4 = 2|E(K_7 \setminus e)| + 4$ bends. The edges of C_r each have at least two bends, thus $C_r\langle K_7 \setminus e\rangle$ has at least $2m + 4n/7$ bends in every general position drawing.   □

To establish our lower bound for general position drawings of 4-connected graphs consider the graph $(C_r \times C_3)\langle K_7 \setminus \{e_1, e_2\}\rangle$ for $r \geq 2$, where for each copy of $K_7 \setminus \{e_1, e_2\}$, e_1 and e_2 are edges of K_7 with no common end-vertices, and these end-vertices are incident to different edges in $C_r \times C_3$, as illustrated in Fig. 8(b).

The proof of the next result is very similar to that of Lemma 6.

Lemma 7. *There is an infinite family of n-vertex m-edge simple 4-connected graphs, each with at least $2m + 2n/7$ bends in every general position drawing.*   □

We now look at the ramifications of the above general position lower bounds for the 2-bends problem. Edges with at most two bends can be classified as 0-bend, 1-bend, 2-bend planar or 2-bend non-planar. As illustrated in Fig. 9, a given 2-bend drawing can be transformed into a general position drawing whose number of bends depends on the number of 0-bend and 2-bend planar edge routes in the 2-bend drawing. We omit the proof.

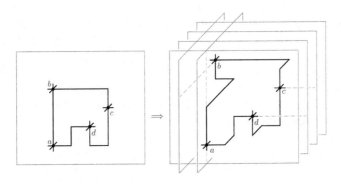

Fig. 9. Removing a plane containing many vertices.

Lemma 8. *If in a 2-bend drawing of an m-edge graph G the number of 0-bend edges is k_0 and the number of 2-bend planar edges is k_2', then there exists a general position 3-D orthogonal drawing of G with $2m + k_0 + k_2'$ bends.* $\square$

Corollary 1. *There exists an infinite family of 6-regular n-vertex graphs, such that in a 2-bend drawing of any of the graphs, $k_0 + k_2' \geq 6n/7$.*

Proof. By Lemma 5, there exists an infinite family of graphs, each with at least $2m + 6n/7$ bends in any general position drawing. If there is a 2-bend drawing of such a graph, then by Lemma 8 there is a general position drawing with $2m + k_0 + k_2'$ bends. Hence $2m + k_0 + k_2' \geq 2m + 6n/7$ and $k_0 + k_2' \geq 6n/7$. $\square$

The following two results are obtained using the same argument used in the proof of Corollary 1 applied with Lemma 6 and Lemma 7, respectively.

Corollary 2. *There exists an infinite family of 6-regular 2-connected n-vertex graphs, such that in a 2-bend drawing of any of the graphs, $k_0 + k_2' \geq 4n/7$.* $\square$

Corollary 3. *There exists an infinite family of 6-regular 4-connected n-vertex graphs, such that in a 2-bend drawing of any of the graphs, $k_0 + k_2' \geq 2n/7$.* $\square$

A natural variation of the general position model allows at most two vertices in any one grid-plane with each vertex being coplanar with at most one other vertex. In this model there is at most $n/2$ pairs of coplanar vertices and hence at most $n/2$ planar edge routes. Since $n/2 \leq 4n/7$, it follows from Corollary 2 that there exists graphs which do not have 2-bend drawings in this model.

Theorem 9. *There exists an infinite family of 2-connected graphs each of which does not have a 2-bend drawing with at most two vertices in any one grid-plane and with each vertex being coplanar with at most one other vertex.* $\square$

Acknowledgements: This research was partially completed while a PhD student in the School of Computer Science and Software Engineering at Monash University, Melbourne, Australia. The author gratefully acknowledges the suggestions of his supervisor Graham Farr, and of Therese Biedl and Antonios Symvonis. In particular, Lemma 6 and Lemma 8 were developed in conjunction with Therese Biedl and Antonios Symvonis, respectively.

References

1. T. C. Biedl. New lower bounds for orthogonal drawings. *J. Graph Algorithms Appl.*, 2(7):1–31, 1998. 260

2. M. Closson, S. Gartshore, J. Johansen, and S. K. Wismath. Fully dynamic 3-dimensional orthogonal graph drawing. In J. Kratochvil, editor, *Proc. Graph Drawing: 7th International Symp.* (GD'99), volume 1731 of *Lecture Notes in Comput. Sci.*, pages 49–58, Springer, 1999. 259, 261

3. G. Di Battista, M. Patrignani, and F. Vargiu. A split&push approach to 3D orthogonal drawing. In Whitesides [10], pages 87–101. 259, 261

4. P. Eades, C. Stirk, and S. Whitesides. The techniques of Komolgorov and Bardzin for three dimensional orthogonal graph drawings. *Inform. Proc. Lett.*, 60(2):97–103, 1996. 259, 260

5. P. Eades, A. Symvonis, and S. Whitesides. Three dimensional orthogonal graph drawing algorithms. *Discrete Applied Math.*, 103:55–87, 2000. 259, 260, 261, 261, 261, 261, 261

6. A. N. Kolmogorov and Ya. M. Barzdin. On the realization of nets in 3-dimensional space. *Problems in Cybernetics*, 8:261–268, March 1967. 259, 260, 261

7. A. Papakostas and I. G. Tollis. Algorithms for incremental orthogonal graph drawing in three dimensions. *J. Graph Algorithms Appl.*, 3(4):81–115, 1999. 259, 261, 261, 261

8. M. Patrignani and F. Vargiu. 3DCube: a tool for three dimensional graph drawing. In G. Di Battista, editor, *Proc. Graph Drawing: 5th International Symp.* (GD'97), volume 1353 of *Lecture Notes in Comput. Sci.*, pages 284–290, Springer, 1998. 259, 261

9. R. Tamassia, I. G. Tollis, and J. S. Vitter. Lower bounds for planar orthogonal drawings of graphs. *Inform. Process. Lett.*, 39(1):35–40, 1991. 260

10. S. Whitesides, editor. *Proc. Graph Drawing: 6th International Symp.* (GD'98), volume 1547 of *Lecture Notes in Comput. Sci.*, Springer, 1998. 271, 271

11. D. R. Wood. On higher-dimensional orthogonal graph drawing. In J. Harland, editor, *Proc. Computing: the Australasian Theory Symp.* (CATS'97), volume 19(2) of *Austral. Comput. Sci. Comm.*, pages 3–8, 1997. 260, 260

12. D. R. Wood. An algorithm for three-dimensional orthogonal graph drawing. In Whitesides [10], pages 332–346. 259, 261, 262, 268, 269

13. D. R. Wood. Lower bounds for the number of bends in three-dimensional orthogonal graph drawings. Technical Report CS-AAG-2000-01, Basser Department of Computer Science, The University of Sydney, 2000. 259, 262

14. D. R. Wood. *Three-Dimensional Orthogonal Graph Drawing*. PhD thesis, School of Computer Science and Software Engineering, Monash University, Australia, 2000. 259, 260, 261, 261, 261, 268

Orthogonal Drawings of Cycles in 3D Space [*]
(Extended Abstract)

Giuseppe Di Battista[1], Giuseppe Liotta[2], Anna Lubiw[3], and Sue Whitesides[4]

[1] Dipartimento di Informatica ed Automazione, Università di Roma Tre, Roma, Italy.
gdb@dia.uniroma3.it
[2] Dipartimento di Ingegneria Elettronica e dell'Informazione,
Università di Perugia, Perugia, Italy.
liotta@diei.unipg.it
[3] Department of Computer Science, University of Waterloo,Waterloo, Canada.
alubiw@daisy.uwaterloo.ca
[4] School of Computer Science, McGill University,Montreal, Canada.
sue@cs.mcgill.ca

Abstract. Let C be a directed cycle, whose edges have each been assigned a desired direction in $3D$ (East, West, North, South, Up, or Down) but no length. We say that C is a shape cycle. We consider the following problem. Does there exist an orthogonal drawing Γ of C in $3D$ such that each edge of Γ respects the direction assigned to it and such that Γ does not intersect itself? If the answer is positive, we say that C is simple. This problem arises in the context of extending orthogonal graph drawing techniques and VLSI rectilinear layout techniques from $2D$ to $3D$. We give a combinatorial characterization of simple shape cycles that yields linear time recognition and drawing algorithms.

1   Introduction

The *topology-shape-metrics approach* [4] for constructing an orthogonal drawing of a planar graph in $2D$ consists of three main steps, called planarization, orthogonalization, and compaction. The planarization step determines an embedding, i.e., the face cycles, for the graph in the plane. The orthogonalization step determines an orthogonal representation of the input graph, i.e. a labeling for each edge (u, v) of the graph that defines the shape of (u, v) in the final drawing. For example, (u, v) could be labeled $NESNE$, which would say "starting from u first go North, then go East, etc." Finally, the compaction step computes the drawing, giving coordinates to vertices and bends while preserving the shape of the edges determined in the orthogonalization step.

[*] Research partially supported by operating grants from the Natural Sciences and Engineering Research Council (NSERC) of Canada, by the project "Algorithms for Large Data Sets: Science and Engineering" of the Italian Ministry of University and Scientific and Technological Research (MURST 40%), and by the project "Geometria Computazionale Robusta con Applicazioni alla Grafica ed a CAD" of the Italian National Research Council (CNR).

J. Marks (Ed.): GD 2000, LNCS 1984, pp. 272–283, 2001.
© Springer-Verlag Berlin Heidelberg 2001

The topology-shape-metrics approach for $2D$ orthogonal drawings has been the subject of much literature. For each step of the approach, different optimization problems (for example minimizing the number of bends, minimizing the area, minimizing the maximum edge length) have been studied, and papers providing optimal algorithms and effective heuristics have been presented. An essential prerequisite of the topology-shape-metrics approach is a characterization of those graphs with edges labeled by orthogonal directions that can be drawn without crossings, while respecting the desired shapes for the edges. This problem has been studied in several papers, including [11,12]. The problem has been generalized to non-orthogonal polygons and to graphs in [6,9,13].

While the literature on $3D$ orthogonal drawings is quite rich (see, e.g. [1,7,8, 14,16]), the extension of the topology-shape-metrics approach to $3D$ has, as far as we know, not been previously explored. A major difficulty is that in $3D$, there is no counterpart to the $2D$ characterization of orthogonal representations. By studying orthogonal representations of cycles in $3D$, this paper represents a first step toward the goal of extending the topology-shape-metrics approach to $3D$.

A $3D$ *shape path* σ is an ordered sequence of labels for the edges of an oriented (graph theoretical) path P, where each label specifies a direction *East*, *West*, *North*, *South*, *Up*, or *Down* for the corresponding edge, and σ contains at least one of each oppositely directed pair of direction labels. Similarly, a *shape cycle* σ is a circularly ordered sequence of direction labels for the edges of an oriented cycle C, where each label specifies a direction for the corresponding edge, and each of the six directions occurs at least once as a label. The *simplicity testing problem for* σ is to decide whether there exists an orthogonal drawing Γ of C so that Γ is *simple* (i.e., no two edges of Γ share any points except common endpoints) and satisfies the direction constraints on its edges as specified by σ. If so, then the shape cycle σ is said to be *simple*.

Not all shape cycles are simple. For example, consider the shape cycle given by the circular sequence of labels $ESUNDWUN$, where E stands for *East*, U stands for *Up*, and so on. This shape cycle has no simple orthogonal drawing, even though each direction label appears at least once (see Figure 1). By contrast, its subcycle $ESDWUN$ *is* simple.

Our main result is a combinatorial characterization of simple shape cycles that yields linear time testing and drawing algorithms.

2 Overview

In $2D$, simple shape cycles were characterized by Vijayan and Widgerson [12] in terms of editing operations on the sequence of labels in the shape cycle. If their editing operations are carried out until no further application is possible, the result is a unique, reduced form for the shape cycle. A $2D$ shape cycle is simple if and only if it can be edited to give the shape cycle for a rectangle, that is, a sequence of the four distinct labels $\{E, W, N, S\}$, with consecutive labels orthogonal. The editing operations arise in the context of repeatedly taking shortcuts at U turns in a rectilinear polygon, where the shortcuts do not intersect

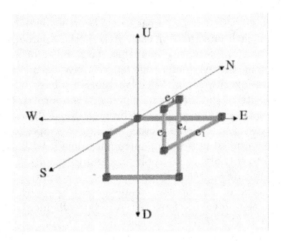

Fig. 1. In the shape cycle $\sigma = ESUNDWUN$, the labels assigned to the edges e_1, e_2, e_3, and e_4 define a "flat" . This shape cycle is *not* simple.

the boundary of the polygon. Their characterization can also be stated as follows: a $2D$ shape cycle is simple if and only if then number of right turns differs from the number of left turns by four. See also the paper by Tamassia [11].

Our recognition algorithm for simple shape cycles does *not* work by repeatedly applying editing operations to the shape cycle. Instead, it looks directly for a subsequence satisfying certain combinatorial properties, which were obtained by considering cycles of length six drawn on the edges of a cube. To state our characterization results precisely, we next introduce the concepts of *flat* and of *canonical sequence* for a shape cycle or path.

Let σ be a shape cycle or path. A *flat* of σ is a consecutive subsequence $F \subset \sigma$ that is maximal with respect to the property that any orthogonal drawing of F must consist of edges that lie on the same axis-aligned plane. For example, consider the shape cycle $\sigma = \sigma_1\sigma_2 \ldots \sigma_8 = ESUNDWUN$, for which Figure 1 gives a (non-simple) drawing. Labels $\sigma_1 = E$ and $\sigma_2 = S$ must lie in an $EWNS$ flat, i.e., a flat whose labels belong to $\{E, W, N, S\}$. Since σ is a circular sequence and since, by definition, flats are maximal, label $\sigma_8 = N$ also belongs to the flat F_1 containing $\sigma_1 = E$ and $\sigma_2 = S$. Since an $EWNS$ flat cannot contain a U label, $F_1 = \sigma_8\sigma_1\sigma_2$. Shape σ contains three additional flats, namely $F_2 = \sigma_2\sigma_3\sigma_4\sigma_5 = SUND$; $F_3 = \sigma_5\sigma_6\sigma_7 = DWU$; and $F_4 = \sigma_7\sigma_8 = UN$. Note that each pair of consecutive flats F_iF_{i+1} share a *transition label* ($\sigma_2, \sigma_5, \sigma_7, \sigma_8$ in the example).

A not necessarily consecutive subsequence $\tau \subseteq \sigma$, where τ consists of k labels, is a *canonical sequence* if: (1) $1 \leq k \leq 6$; (2) the labels of τ are distinct; (3) no flat of σ contains more than three labels of τ; and (4) if a flat F of σ contains one or more labels of τ, then $\tau \cap F$ forms a consecutive subsequence of σ.

For example, the shape cycle $\sigma = ESUNDWUN$ of Figure 1 does not contain a canonical sequence of any length containing an S and a D: each of these labels occurs only once, in flat F_2, where S and D are not consecutive as elements of σ. Hence according to our characterization of simple shape cycles below, cycle σ is not simple.

Theorem 1. *A 3D shape cycle with at least two flats is simple if and only if it contains a canonical sequence of length six.*

In a companion paper [5], we introduced a simpler notion of canonical sequence in solving the *shape path reachability problem,* which is to determine, given a shape path σ and a point p in an octant, whether σ admits a simple orthogonal drawing that starts at the origin and ends at p. The main result of [5] can be summarized in concrete terms for a particular octant as follows.

Theorem 2. *[5] Let σ be a shape path, and let p be any point of the UNE octant. Then σ admits a simple orthogonal drawing that starts at the origin and ends at p if and only if σ contains a canonical sequence of length 3 containing the labels U, N, E in some order.*

At a very first glance, the necessity of the condition of Theorem 1 may appear to be an immediate consequence of Theorem 2: if σ admits a simple orthogonal drawing, then σ can be shown to split into the concatenation of two shape paths that reach opposite octants. However, the flats where two such paths join require special study, as they contain labels from each of the two paths. Hence the union of canonical sequences for each path need not yield a canonical sequence of length six for the cycle.

3   Preliminaries

We assume from now on that two adjacent labels of a shape path or cycle are neither identical, in which case they could be replaced by a single label, nor oppositely directed, in which case the shape could not be simple. Also, we omit some straightforward special case handling by considering here only shape cycles that contain at least four flats.

We regard 3D space as partitioned into eight open octants, eight open quadrants, six open (semi)axes directed away from the origin, and the origin itself. A triple XYZ of distinct unordered labels no two of which are opposite defines the XYZ octant. Similarly, a pair XY of distinct orthogonal labels defines the XY quadrant in 2D or 3D, and a direction label X defines the X (semi)axis.

We sometimes use the term "shape" to refer either to a shape path or to a shape cycle. We sometimes say a shape is *drawable* if it is a simple shape.

To traverse a shape σ or a drawing $\Gamma(\sigma)$ of σ in the *positive sense* means to visit its labels or edges in the order specified by σ. If σ is a shape path, the starting point for a drawing $\Gamma(\sigma)$ of σ is regarded as the origin for that drawing. Visiting $\Gamma(\sigma)$ in the positive sense then orients each edge of $\Gamma(\sigma)$. An edge oriented in this way from u to v is denoted uv. It points in the direction specified by its associated label in σ.

Suppose that σ is a $3D$ shape path or cycle and that F is a flat of σ. Then *first* and *last* labels for F (also called its entry and exit labels) are determined by traversing the cyclic sequence σ in the positive sense. In a drawing of F, the *starting* or *entering point* for F is the vertex at the tail of the edge corresponding to the entry label of F.

A label Y of a shape σ is said to *occur between* two other labels X and Z if Y is met when traversing σ in the positive sense from X to Z.

Remark. Let $\phi()$ be a permutation of the six direction labels that maps opposite pairs of labels to possibly different opposite pairs (for example, ϕ might map N, S, E, W, U, D to E, W, N, S, D, U, respectively). Note that $\phi()$ defines a linear transformation of $3D$ space that determines a bijection between drawings of σ and drawings of $\phi(\sigma)$.

For concreteness, as in Theorem 2, we often state our results and proofs referring to some given octant, quadrant, or axis. However, the results can also be stated with respect to any other octant, quadrant, or axis since, by the Remark, they are preserved under the $\phi()$ transformation.

We sometimes specify the labels of a canonical sequence τ by using set notation. For example, $\{U, N, E, S\}$ might describe a canonical sequence whose directions labels are U, N, E, and S. In this notation, the order of the labels is not specified and is inherited from the shape σ once a particular subsequence τ has been chosen.

We sometimes distinguish the labels in a canonical sequence τ from the other labels of σ with special notation. To say that $\{\bar{U}, \bar{N}, \bar{E}, \bar{S}\}$ is a canonical sequence means not only that the canonical sequence contains a U, an S, an N, and an E direction label, but also that the U label in τ is a specific element $\bar{U}$ of σ, the S label is $\bar{S}$, and so on.

We say that the elements of σ that occur in a canonical sequence τ are *canonical labels*. It is useful to recall that a shape path is a sequence and that a shape cycle is a circular sequence. Thus a canonical sequence for a shape path is a sequence, and a canonical sequence for a shape cycle is a circular sequence.

4   Sufficiency

We now sketch a constructive proof that any shape cycle σ that contains a canonical sequence of length six admits a simple orthogonal drawing.

4.1   The Proof Technique

The intuition behind our construction of a drawing for a shape cycle is to imagine that it will be an elaboration of a cycle of six edges to be drawn along the edges of a box. A shape cycle of length six has one of two essentially different shapes, namely, a *chair* shape such as $UNDESW$, which has four flats, or a *skew* shape such as $UENDWS$, which has six flats. From a canonical sequence of length six we obtain (possibly after some modification) the six labels of a chair shape or a skew shape to follow the edges of a big box, with the remaining labels to be drawn as paths of short edges located near corners of the big box, serving to connect together the six long edges. The underlying chair or skew shape conveniently allows us to place the connecting paths of short segments in distinct octants by assigning long lengths to the canonical labels. In practice, the drawings we produce do not necessarily assign long lengths to the canonical labels, but this mental model gives the basic idea for the construction.

One difficulty is that, whereas σ does not contain any pairs of oppositely directed labels that are adjacent, a canonical sequence τ of σ may contain pairs of oppositely directed labels that are adjacent as elements of τ. Such a canonical sequence τ would then not provide the convenient underlying chair or skew shape for the construction. This motivates the following definition.

Definition 1. *A subsequence τ of a shape cycle σ is a* strong canonical cycle *if τ is a canonical cycle such that no two labels that are adjacent in τ are oppositely directed.*

The next lemma resolves the difficulty. Its proof is a technical case analysis in which new choices of canonical labels are substituted for old ones.

Lemma 1. *If a shape cycle σ contains a canonical sequence τ of length six, then σ contains a strong canonical sequence τ' of length six.*

Given a strong canonical sequence of length six (which can be found in linear time if one exists), we compute simple drawings for the connecting paths between the canonical labels, then assign lengths to the canonical labels so that these drawings remain in separate octants (this is made possible by the underlying chair or skew form of the canonical cycle). To ensure that this is the case, and to ensure that the cycle closes, we formulate and solve a system of linear inequalities expressing these constraints.

4.2   Constructing a Drawing

This subsection describes how to construct a drawing from a strong canonical sequence.

A drawing $\Gamma(\sigma)$ of a shape path σ is an *expanding drawing* if each segment travels one unit farther in its assigned direction than the extreme points, with respect to that direction, of the previous segments of $\Gamma(\sigma)$. A drawing $\Gamma(\sigma)$ of a shape path σ is a *doubly extensible drawing* if its first and last edges can be replaced by arbitrarily long edges without creating any intersections within the drawing of that shape path.

Lemma 2. *[5] Let σ be a shape path with n labels. Then σ admits an expanding drawing that can be computed in linear time on a real RAM. Also, if σ is such that either it consists of exactly two labels or it contains at least two flats, then σ has a doubly extensible drawing that can be computed in $O(n)$ time on a real RAM.*

We briefly review the proof, whose details are needed below for our cycle construction and its proof of correctness. The first part of the lemma follows from the algorithmic nature of the definition of expanding drawing. For the second part, note that if σ consists of exactly two labels, then it is clearly doubly extensible.

Now suppose σ has at least two flats. The subsequence strictly between the first and last elements thus contains a transition label. Place the tip of the first transition label at the origin. Working *backwards* through σ from this first transition label, create an expanding drawing for the initial subsequence of σ. Thus the transition label is the first label to be drawn, and has length 1. When eventually the first label of σ is reached, it can be drawn arbitrarily long.

To draw the remainder of σ, consider the label that immediately follows the first transition label. It must be drawn with its tail at the origin, and perpendicular to the plane of the previous flat. Working *forward* from this label, create an expanding drawing using the rule that when a new label is drawn, it extends farther by 1 in its direction than any previously drawn segment except the one that could be made arbitrarily long. Thus the first segment past the transition label is also assigned length 1, and when eventually the last segment of σ is drawn, it may be made arbitrarily long. This concludes the review of the proof.

Note that in the above proof sketch, the tip of the next-to-last segment of $\Gamma(\sigma)$, and hence the tail of its last segment, lies on the bounding box of the drawing of the remaining internal labels of σ.

Now we describe how to obtain a drawing for a shape cycle with a canonical subsequence τ. We assume, in accordance with Lemma 1, that τ has no adjacent, oppositely directed labels. Removal of τ from σ determines six *connecting* shape paths (some may be empty).

To each of these connecting paths, add back on the two elements of τ that bound it. Unless this path consists of just the two elements of τ, it must contain at least two flats. Otherwise, the two elements of τ, which are not adjacent in σ, would lie in the same flat of σ, contradicting the fact that τ is canonical.

The connecting parts of the doubly extensible drawings will be placed in separate octants. The segments of τ are precisely the end segments of these six doubly extensible drawings and can be drawn arbitrarily long. Their lengths will be chosen so long that they can connect the internal parts of the doubly extensible drawings isolated in distinct octants.

Make the drawings (and hence their bounding boxes) for the six connecting paths above. Some of these may be just points. Relative to a local origin of each drawing, we know the coordinates of all endpoints of the segments in that part of the drawing.

Now we determine lengths for the canonical segments and position the origins of the bounding boxes.

Look at the shape of τ. Since it is a strong canonical sequence, there are just two possibilities, the chair shape (e.g., $UNEDSW$) or the skew shape (e.g., $UNDESW$). Use this to determine octants for the placement of the bounding boxes containing the connecting drawings. Note that no two boxes are assigned to the same octant.

Let $l_E, l_N, \ldots$ denote the unknown lengths to be assigned to the canonical segments. A simple system of equations and inequalities must be satisfied by the unknowns for each oppositely directed pair of canonical segments: the total length of all segments directed E, say, must equal the total length of all segments directed W, and similarly for the other pairs. This will guarantee that the cycle closes.

Note that we have already determined the lengths of the segments that are not canonical, as well as the location of the endpoints of the canonical segments, in terms of the local coordinates of the boxes. Hence it easy to determine, for each local origin of a box, a system of three inequalities, one for each of the three orthogonal directions, that guarantees that the box stays strictly inside its assigned octant.

Satisfying these systems of inequalities implies that a corresponding system of inequalities on the lengths of the canonical segments must also be satisfied. This gives a lower bound on the length of each canonical segment of the form $l_E \geq c_E$ for some constant c_E, and so on.

To ensure that the cycle will close, we add to the system of inequalities on lengths three equations, one for each pair of opposite directions, as follows. The total length of segments directed E must equal the total length of segments directed W, and similarly for the other two pairs. The form for the E, W equation is either $l_E = l_W + c_{EW}$ or $l_W = l_E + c_{EW}$ for some positive constant c_{EW}, and similarly for the other two pairs.

Consider the constraints on the lengths of a particular oppositely directed pair, say on l_E and l_W. These constraints are

- for non-negative constant c_{EW} and for $l_E \geq l_W$, we have $l_E = l_W + c_{EW}$ (or if $l_E < l_W$, then $l_W = l_E + c_{EW}$);
- $l_E \geq c_E$;
- $l_W \geq c_W$.

These may be satisfied by assigning the value

$$l_W = max(c_W, c_E - c_{EW}) \text{ (or } l_E = max(c_E, c_W - c_{EW}) \text{ in case } l_E < l_W).$$

This determines the value of the length of the canonical segments directed E and W. The remaining lengths for the other directions may be determined similarly.

The lengths have now been chosen so that the path forms a closed cycle. To see that the cycle is simple, note that clearly, segments that are not canonical do not intersect each other. Hence it suffices to check that no canonical segment intersects another canonical segment or a non-canonical one (including ones in boxes not located at the endpoints of the canonical segment). This follows easily from the fact that the bounding boxes for the connecting paths are located in distinct octants.

4.3   Algorithmic Issues

To obtain a linear time algorithm for testing for the condition one must search for and produce, if one exists, a canonical sequence of length six in linear time. To do this, find, in linear time, a pair of parallel flats in σ. The proof of the necessity of the condition (see Section 5 for a sketch), reveals that if σ satisfies the condition, then it must contain one of a constant number of canonical sequences of special types defined by the relation of the labels in the canonical sequence to each other and to the two given parallel flats. Even though σ is a circular sequence, the fact that the pair of parallel flats can be chosen arbitrarily gives a starting label for σ, namely, the first label of one of these flats. Hence, it is not necessary to try each label of the entire sequence σ as a starting place when searching for a canonical sequence of one of the special types. Consequently, a linear time algorithm can be designed to check for the presence of one of these special canonical sequences. Given a strong canonical cycle, a simple orthogonal drawing for it can be constructed as described in the previous subsection. The computation of the coordinates of the endpoints of the segments of a drawing requires $O(n)$ time for the real RAM model of computation. Since the lengths of some segments might require $\Theta(lgn)$ bits to record, the running time becomes $O(nlgn)$ for a Turing machine model.

5   Necessity

Given a simple orthogonal drawing $\Gamma(\sigma)$ of a shape cycle σ, our goal is to show that σ contains a canonical sequence of length six. By slightly perturbing $\Gamma(\sigma)$ if necessary, we may assume without loss of generality that $\Gamma(\sigma)$ satisfies a *general position assumption*, namely, that no two vertices belonging to distinct flats of σ are drawn on the same axis-aligned plane. The lemmas and theorems that follow are based on this assumption.

5.1   The Proof Technique

The proof is based on the idea of cutting $\Gamma(\sigma)$ into two paths such that one reaches an octant and the other one goes back to the origin. We follow the two paths and look for canonical sequences on each path.

As mentioned in Section 2, a proof based on this approach does *not* follow easily from Theorem 2. It requires more elaborate machinery:

- We suitably choose the points a and b where we cut $\Gamma(\sigma)$ in order to define the two shape paths σ_{ab} and σ_{ba}.
- We find canonical sequences for $\sigma_{ab}\ell_{bb'}$ and $\sigma_{ba}\ell_{aa'}$ that may consist of three or four labels; $\ell_{bb'}$ is the label of σ after the last label of σ_{ab}, and $\ell_{aa'}$ is the label of σ after the last label of σ_{ba}.
- We use a certain necessary condition, together with Theorem 2 and the properties and lemmas of Subsection 5.2 to construct a canonical sequence of length six for σ.

5.2 Some Useful Properties

We now observe some basic properties of a canonical sequence τ of a shape cycle or path σ. These properties are useful and easy to prove. Unless specified otherwise, in this subsection σ is understood to denote either a shape path or a shape cycle. By the union $\tau_1 \cup \tau_2$ of two subsequences of σ, we mean the sequence whose elements are the elements of τ_1 and τ_2, ordered as in σ.

Property 1. If τ contains three labels XYZ such that they are consecutive on the same flat of σ, then X and Z define opposite directions.

Property 2. Let σ be a shape path. If we remove from τ its first or last label, the resulting sequence is still canonical for σ.

Property 3. If τ consists of three labels that define mutually orthogonal directions, then any subsequence of τ is a canonical sequence of σ .

The next property allows us to remove a label from a canonical sequence of length four; its proof is an immediate consequence of the definition of τ and of Property 1.

Property 4. If τ consists of four labels exactly two of which are oppositely directed, then a subsequence obtained by deleting from τ one of these two opposite labels is a canonical sequence of σ.

For example, if $\tau = \{\bar{D}, \bar{S}, \bar{W}, \bar{U}\}$ is a canonical sequence, then by Property 4, $\tau' = \{\bar{D}, \bar{S}, \bar{W}\}$ is also a canonical sequence.

The following lemmas allow us to merge two canonical sequences to obtain a new canonical sequence.

Lemma 3. *Let $\tau_1 \subset \sigma$ and $\tau_2 \subset \sigma$ be two canonical sequences such that (1) $\tau_1 \cap \tau_2 = \emptyset$, and (2) for all pairs of canonical labels X, Y such that $X \in \tau_1$ and $Y \in \tau_2$ there is no flat containing both X and Y. Then the sequence $\tau = \tau_1 \cup \tau_2$ is canonical for σ.*

Lemma 4. *Let σ have the form $\sigma = \sigma_2 X \sigma_1$ where X is a transition label for σ. Let $\tau_1 \subseteq X\sigma_1$ and $\tau_2 \subseteq \sigma_2 X$ be canonical sequences for σ such that (1) $\tau_1 \cap \tau_2 = X$ and (2) for all pairs of canonical labels $Y, Z \neq X$ such that $Y \in \tau_1$ and $Z \in \tau_2$, there is no flat containing both Y and Z. Then the sequence $\tau = \tau_1 \cup \tau_2$ is canonical for σ.*

Lemma 5. *Let σ be a shape cycle of the form $X\sigma_1 Y \sigma_2$, where X and Y are transition labels, and let τ_1 and τ_2 be canonical sequences for σ such that (1) $\tau_1 \subseteq X\sigma_1 Y$ and $\tau_2 \subseteq Y\sigma_2 X$, and (2) $\tau_1 \cap \tau_2 = X, Y$. Then $\tau_1 \cup \tau_2$ is a canonical sequence for σ.*

Next is a necessity result for $3D$ shape paths.

Lemma 6. *Let $\Gamma(\sigma)$ be a simple drawing of a shape path σ starting at the origin, and let uv be an edge of $\Gamma(\sigma)$. If u is in the DSW octant and v is in the DSE octant, then σ contains a canonical sequence $\tau = \{D, S, W, E\}$.*

5.3   Proof of Necessity

We sketch the proof in the case that shape cycle σ has at least four flats. Straightforward case analysis handles shape cycles with fewer than four flats.

Under the general position assumption, if σ has at least four flats, then it always has two flats F_a and F_b such that $\Gamma(F_a)$ and $\Gamma(F_b)$ lie on parallel planes. Let a be the starting point of F_a and let b be the starting point of F_b. Let aa' be the first edge of $\Gamma(F_a)$, and let bb' be the first edge of $\Gamma(F_b)$. Let $\ell_{aa'}$ be the direction label for aa' and let $\ell_{bb'}$ be the direction label for bb'. Observe that $\ell_{aa'}$ is a transition label shared by two flats of σ, the flat preceding F_a and flat F_a. Similarly, $\ell_{bb'}$ is a transition label shared by the flat preceding F_b and flat F_b. We denote with F_{a-} and F_{b-} the flats preceding F_a and F_b, respectively.

Observe that if the origin is chosen at a, then b is a point of an octant. We define two disjoint directed paths: $\Gamma(\sigma_{a'b})$ is the path from a' to b and $\Gamma(\sigma_{b'a})$ is the path from b' to a. We therefore have that $\sigma = \ell_{aa'}\sigma_{a'b}\ell_{bb'}\sigma_{b'a}$. We also have $\sigma_{ab} = \ell_{aa'}\sigma_{a'b}$ and $\sigma_{ba} = \ell_{bb'}\sigma_{b'a}$.

Suppose we locate the origin at a and let XYZ be the octant containing b. We say that a and a' are *equivalent with respect to* b if moving the origin from a to a' leaves b in the XYZ octant. A similar definition can be given for the relationship of b and b' with respect to a. Observe that if a and a' are not equivalent with respect to b, then when we locate the origin at b, we have that a' does not lie in the octant that contains a.

We consider four main cases, determined by whether or not a and a' are equivalent with respect to b, and by whether or not b and b' are equivalent with respect to a. For each case, we show how to choose a canonical sequence of length six for σ. This is done by using Properties 1, 3, 4, and Lemmas 3, 4, and 5 to perform merging operations on two canonical sequences, one defined in $\sigma_{ab}\ell_{bb'}$ and the other defined in $\sigma_{ba}\ell_{aa'}$. The canonical sequence of $\sigma_{ab}\ell_{bb'}$ ($\sigma_{ba}\ell_{aa'}$) can either consist of three labels if b and b' are equivalent with respect to a (a and a' are equivalent with respect to b), in which case Theorem 2 is used to define the canonical sequence; or it can consist of four labels if b and b' are not equivalent with respect to a (a and a' are not equivalent with respect to b), in which case Lemma 6 is used to define the canonical sequence. Since paths $\sigma_{ab}\ell_{bb'}$ and $\sigma_{ba}\ell_{aa'}$ are not disjoint, their canonical sequences need not be disjoint. However, the merging operations performed on these canonical sequences for paths produce a cyclic canonical sequence of length six for cycle σ.

We summarize the results of this section with the following theorem.

Theorem 3. *Let $\Gamma(\sigma)$ be a simple orthogonal drawing of a shape cycle σ. Then σ contains a canonical sequence of length six.*

6   Conclusion

This paper has characterized those shape cycles that admit a simple orthogonal drawing in $3D$. The characterization yields a linear time recognition algorithm, and a drawing algorithm that is linear in the real RAM model and $O(nlgn)$ in the Turing machine model. Interesting related problems that remain include: (1) characterizing simple shapes for graphs that are not just cycles, (2) minimizing the volume of bounding boxes of shape cycles that must be drawn with vertices at grid points (the coordinates of our drawing will be rational and can be scaled up to be integers; however, we have not attempted to minimize the volume of the drawing), and (3) extending the characterization of this paper to shape cycles with more than six directions and/or to dimension higher than three.

References

1. T. C. Biedl. Heuristics for *3d*-orthogonal graph drawings. In *Proc. 4th Twente Workshop on Graphs and Combinatorial Optimization*, pp. 41–44, 1995.
2. T. Biedl, T. Shermer, S. Wismath, and S. Whitesides. Orthogonal 3-D graph drawing. *J. Graph Algorithms and Applications*, 3(4):63–79, 1999.
3. R. F. Cohen, P. Eades, T. Lin and F. Ruskey. Three-dimensional graph drawing. *Algorithmica* , 17(2):199–208, 1997.
4. G. Di Battista, P. Eades, R. Tamassia, and I. Tollis. Graph Drawing. Prentice Hall, 1999.
5. G. Di Battista, G. Liotta, A. Lubiw, and S. Whitesides. Embedding problems for paths with direction constrained edges. In D.-Z. Du, P. Eades, V. Estivill-Castro, X. Lin, and A. Sharma, eds., *Computing and Combinatorics, 6th Ann. Int. Conf., COCOON 2000*, Springer-Verlag LNCS vol. 1858, pp. 64-73, 2000.
6. G. Di Battista and L. Vismara. Angles of planar triangular graphs. *SIAM J. Discrete Math.*, 9(3):349–359, 1996.
7. P. Eades, C. Stirk, and S. Whitesides. The techniques of Komolgorov and Bardzin for three dimensional orthogonal graph drawings. *Inform. Process. Lett.*, 60:97–103, 1996.
8. P. Eades, A. Symvonis, and S. Whitesides. Three-dimensional orthogonal graph drawing algorithms. *Discrete Applied Math.*, vol. 103, pp. 55-87, 2000.
9. A. Garg. New results on drawing angle graphs. *Comput. Geom. Theory Appl.*, 9(1–2):43–82, 1998. Special Issue on Geometric Representations of Graphs, G. Di Battista and R. Tamassia, *eds.*.
10. A. Papakostas and I. Tollis. Algorithms for incremental orthogonal graph drawing in three dimensions. *J. Graph Algorithms and Appl.*, 3(4):81-115, 1999.
11. R. Tamassia. On embedding a graph in the grid with the minimum number of bends. *SIAM J. Comput.*, 16(3):421–444, 1987.
12. G. Vijayan and A. Wigderson. Rectilinear graphs and their embeddings. *SIAM J. Comput.*, 14:355–372, 1985.
13. V. Vijayan. Geometry of planar graphs with angles. In *Proc. 2nd Annu. ACM Sympos. Comput. Geom.*, pp. 116–124, 1986.
14. D. R. Wood. Two-bend three-dimensional orthogonal grid drawing of maximum degree five graphs. TR 98/03, School of Computer Science and Software Engineering, Monash University, 1998.
15. D. R. Wood. An algorithm for three-dimensional orthogonal graph drawing. In S. Whitesides, ed., *Graph Drawing (6th Int. Symp., GD '98)*, Springer-Verlag, LNCS vol. 1547, pp. 332-346, 1998.
16. D. R. Wood. Three-Dimensional Orthogonal Graph Drawing. Ph.D. thesis, School of Computer Science and Software Engineering, Monash University, 2000.

Three-Dimensional Orthogonal Graph Drawing with Optimal Volume

Therese Biedl [*1], Torsten Thiele[2], and David R. Wood [**3]

[1] Department of Computer Science
University of Waterloo
Waterloo, ON N2L 3G1, Canada
biedl@uwaterloo.ca
[2] Institut für Informatik
Freie Universität Berlin
Takustraße 19, 14195 Berlin, Germany
thiele@inf.fu-berlin.de
[3] Basser Department of Computer Science
The University of Sydney
Sydney NSW 2006, Australia
davidw@cs.usyd.edu.au

Abstract. In this paper, we study three-dimensional orthogonal box-drawings of graphs without loops. We provide lower bounds for three scenarios: (1) drawings where vertices have bounded aspect ratio, (2) drawings where the surface of vertices is proportional to their degree, and (3) drawings without any such restrictions. Then we give constructions that match the lower bounds in all scenarios within an order of magnitude.

1 Introduction

In this paper we consider three-dimensional orthogonal drawings of an n-vertex m-edge graph $G = (V, E)$ with maximum degree Δ (allowed to have parallel edges but no self loops). An *orthogonal (box-)drawing* of G represents vertices by pairwise non-intersecting boxes in the three-dimensional grid. An edge $vw \in E$ is represented by a sequence of contiguous segments of grid lines possibly bent at grid points, between *ports* (points extremal in a particular direction) on the boxes of v and w. Any two edge routes are disjoint except possibly at endpoints. The graph-theoretic terms 'vertex' and 'edge' will also refer to their representation in an orthogonal drawing. The number of ports on a box will be called its *surface*, and the number of grid points in a box is its *volume*. The *aspect ratio* of a box is its largest side length divided by its smallest side length.

---

[*] Research supported by NSERC.
[**] Supported by the Australian Research Council Large Grant A49906214, and partially completed while a PhD student in the School of Computer Science and Software Engineering at Monash University under the supervision of Dr Graham Farr.

J. Marks (Ed.): GD 2000, LNCS 1984, pp. 284–295, 2001.

An orthogonal drawing with a particular shape of box representing every vertex, e.g. point, line, or cube, is called an orthogonal *shape*-drawing for each particular shape. Orthogonal point-drawings have been studied in [7,8,9,10,16, 18]. However, orthogonal point-drawings can only exist for graphs with maximum degree at most six. Overcoming this restriction has motivated recent interest in orthogonal box-drawings [3,5,16,19,20].

The smallest box enclosing an orthogonal drawing is called the *bounding box*. The bounding box volume and the maximum number of bends per edge are the most commonly proposed measures for determining the aesthetic quality of an orthogonal drawing. For box-drawings the size and shape of a vertex with respect to its degree are also considered an important measure of aesthetic quality.

An orthogonal drawing is said to be *strictly α-degree-restricted* if for some constant $\alpha \geq 1$, the surface of v is at most $\alpha \cdot \deg(v)$ for all vertices v; it is *α-degree-restricted* if the surface of v is at most $\alpha \cdot \deg(v) + o(\deg(v))$ for all v.

Lower Bounds: Let $vol(G, r, \alpha)$ denote the minimum, taken over all orthogonal drawings of a graph G that have aspect ratios at most r and are strictly α-degree restricted. Let $vol(n, m, r, \alpha)$ be the maximum, taken over all graphs G with n vertices and m edges, of $vol(G, r, \alpha)$. Thus, $vol(n, m, r, \alpha)$ describes a volume bound within which we can draw all graphs with n vertices and m edges such that each vertex v has aspect ratio at most r and surface at most $\alpha \cdot \deg(v)$.

The first lower bounds for 3-D orthogonal box-drawings were due to Hagihara *et al.* [12]. They show that, in the above notation, $vol(n, m, 1, 1) = \Omega(\max \{mn, (m/\log n)^{3/2}\})$. In this paper, we show that

- $vol(n, m, \infty, \infty) = \Omega(m\sqrt{n})$,
- $vol(n, m, r, \infty) = \Omega(m^{3/2}/\sqrt{r})$,
- $vol(n, m, \infty, \alpha) = \Omega(m^{3/2}/\alpha)$.

We thus improve the results of [12] in three ways: Firstly, we remove the log-factor, to establish $vol(n, m, 1, 1) = \Omega(m^{3/2})$ as the lower bound. Secondly, we prove this result even if only one of having bounded aspect ratios and being strictly degree-restricted holds. Finally, we also study the case when neither of these two conditions hold, and establish a weaker, but optimal, lower bound. Our first result includes the lower bound of $\Omega(n^{5/2})$ for orthogonal drawings of K_n established by Biedl *et al.* [5]. In fact, the proof of our lower bounds are based on techniques developed in this paper, though we generalize them to graphs with an arbitrary number of edges and include considerations of bounded aspect ratios and degree restriction.

Algorithms: A trade-off between the maximum number of bends per edge route and the bounding box volume is apparent in algorithms for orthogonal graph drawing; see Table 1 for an overview of the known results.

In this paper, we present two optimal algorithms for orthogonal graph drawing. The first algorithm produces degree-restricted orthogonal cube-drawings with $\mathcal{O}(m^{3/2})$ bounding box volume and at most six bends per edge. The technique used is a generalization of the COMPACT algorithm of Eades *et al.* [10]

Table 1. The tradeoff between bounding box volume and the maximum number of bends in orthogonal graph drawings. All lower bounds are proved in Theorem 2.

lower bound	volume	bends	model	graphs	reference
bounded aspect ratio / degree-restricted					
$\Omega(m^{3/2})$	$\mathcal{O}((nm)^{3/2})$	2	general position	simple	[3,19]
$\Omega(m^{3/2})$	$\mathcal{O}(nm\sqrt{\Delta})$	2	lifting $\frac{1}{2}$-edges	simple	[3]
$\Omega(m^{3/2})$	$\mathcal{O}(m^2)$	5	plane layout	multigraphs	Thm. 4
$\Omega(m^{3/2})$	$\mathcal{O}((n\Delta)^{3/2})$	10 [1]	plane layout	simple	[12]
$\Omega(m^{3/2})$	$\mathcal{O}(m^{3/2})$	6	plane layout	multigraphs	Thm. 3
no bounds on aspect ratio / degree-restricted					
$\Omega(m^{3/2})$	$\mathcal{O}(n^2m)$	2	general position	simple	[3,19]
$\Omega(m^{3/2})$	$\mathcal{O}(n^2\Delta)$	2	lifting $\frac{1}{2}$-edges	simple	[3]
$\Omega(m^{3/2})$	$\mathcal{O}(m^2)$	5	plane layout	multigraphs	Thm. 4
$\Omega(m^{3/2})$	$\mathcal{O}(m^{3/2})$	6	plane layout	multigraphs	Thm. 3
no bounds on aspect ratio / not necessarily degree-restricted					
$\Omega(m\sqrt{n})$	$\mathcal{O}(n^3)$	1	lifting edges	simple	[5]
$\Omega(m\sqrt{n})$	$\mathcal{O}(n^{5/2})$	3	lifting edges	simple	[5]
$\Omega(m\sqrt{n})$	$\mathcal{O}(mn)$	3	plane layout	multigraphs	Thm. 6
$\Omega(m\sqrt{n})$	$\mathcal{O}(m\sqrt{n})$	4	plane layout	simple	Thm. 5

for orthogonal point-drawing, and is an improvement on the algorithms of Hagihara *et al.* [12] and Wood [20], who obtained upper bounds of $\mathcal{O}((n\Delta)^{3/2})$ and $\mathcal{O}(m^2/\sqrt{n})$, respectively. Our second algorithm produces orthogonal rectangle-drawings with $\mathcal{O}(m\sqrt{n})$ bounding box volume and at most four bends per edge; the drawings are not necessarily degree-restricted nor do the vertices have bounded aspect ratios. Both upper bounds are therefore within an order of magnitude of the lower bound. We also present refinements of both our algorithms with one less bend per edge, at the cost of an increase in the volume.

2   Lower Bounds

In this section we prove lower bounds using graphs that have $\Omega(m)$ edges in any cut with at least $n/6$ vertices on each side of the cut.

Lemma 1. *If $p \neq q$ are primes, $p \equiv 1 \bmod 4$, $q \equiv 1 \bmod 4$, $144 \leq p < q(q-1)/2$, then there exists a simple graph $G_{p,q}$ with the following properties:*

- *$G_{p,q}$ is d-regular for $d = p + 1$,*
- *the number n of vertices of $G_{p,q}$ is at least $q(q-1)/2$ and at most $q(q-1)$.*
- *for any disjoint sets $S, T \subset V(G_{p,q})$ with $|S||T| \geq n^2/36$ there are at least $C \cdot dn$ edges between S and T, where $C > 0.00009$ is a constant.*

---

[1] Hagihara *et al.* [12] did not count the number of bends per edge; we deduce the bound of 10 from their construction.

Proof. (Sketch) Let $G_{p,q}$ be the Ramanujan graph $X^{p,q}$ defined in [15]; the first two properties of the graph were shown in this paper. It was also shown that $\lambda \le 2\sqrt{d-1}$, where λ denotes the second-largest eigenvalue of $G_{p,q}$. Assume S and T are disjoint vertex sets with $|S||T| \ge n^2/36$. We know from [2] that the number of edges between S and T is at least $\frac{d|S||T|}{n} - \lambda\sqrt{|S||T|}$. Using $6\sqrt{|S||T|}/n \ge 1$ and $\sqrt{d}/\sqrt{145} \ge 1$, one can show that $\lambda\sqrt{|S||T|} \le 12/\sqrt{145} \cdot d|S||T|/n$, and hence the number of edges between S and T is at least $(1 - 12/\sqrt{145}) \cdot d|S||T|/n \ge \frac{1}{36}(1 - 12/\sqrt{145})dn$. $\square$

Our lower bound proof is based on the technique developed in [5], which distinguishes three cases: either many vertices are intersected by one grid line, or many vertices are intersected by one plane, or neither of these is the case.

Theorem 1. *Let $G = (V, E)$ be an n-vertex d-regular simple graph ($n \ge 8$) such that for any disjoint sets $V_1, V_2 \subset V$ with $|V_1||V_2| \ge n^2/36$ there are at least $C \cdot dn$ edges, $0 < C \le 1$ between V_1 and V_2. Then*

- $vol(G, \infty, \infty) \ge \frac{1}{3}C^{3/2} \cdot dn^{3/2}$.
- $vol(G, r, \infty) \ge \frac{1}{3}C^{3/2} \cdot (dn)^{3/2}/\sqrt{r}$, $r \ge 1$.
- $vol(G, \infty, \alpha) \ge \frac{1}{3}C^2 \cdot (dn)^{3/2}/\alpha$, $\alpha \ge 1$.

Proof. Consider a drawing of G in a grid of dimensions $X \times Y \times Z$. For ease of proof we assume that n is divisible by 6; see [6] for a proof without this assumption.

Case 1: A line intersects many vertices. Assume that a Z-*line* (i.e., a grid-line parallel to the Z-axis) intersects at least $\frac{1}{3}n$ vertices. Let Z_0 be such that the $(Z = Z_0)$-plane intersects none of these vertices and separates them into two groups of at least $\frac{1}{6}n$ vertices each. By assumption at least $C \cdot dn$ edges connect these two groups. These edges cross the $(Z = Z_0)$-plane, which thus must contain at least $C \cdot dn$ points with integer X- and Y-coordinates. Hence $XY \ge C \cdot dn$. Since the Z-line intersects at least $\frac{1}{3}n$ vertices, we have $Z \ge \frac{1}{3}n$, so $XYZ \ge \frac{1}{3}C \cdot dn^2$, which proves all claims by $C \le 1$ and $d \le n$.

Case 2: No plane intersects many vertices. Assume that any X-*plane*, Y-*plane* or Z-*plane* (i.e., a plane perpendicular to the X-axis, Y-axis or Z-axis, respectively) intersects at most $\frac{2}{3}n$ vertices. A vertex is *left* of an $(X = X_0)$-plane if all the points in its grid box have X-coordinates less than X_0. The notion of *right* of an $(X = X_0)$-plane is analogous. Let X' be the largest integral value such that fewer than $\frac{1}{6}n$ vertices are left of the $(X = X')$-plane. Since the $(X = X')$-plane intersects at most $\frac{2}{3}n$ vertices, there are at least $\frac{1}{6}n$ vertices right of the $(X = X')$-plane. All these vertices also lie to the right of $(X = X'+\frac{1}{2})$-plane. By definition of X', at least $\frac{1}{6}n$ vertices lie left of the $(X = X'+1)$-plane. All these vertices also lie to the left of $(X = X'+\frac{1}{2})$-plane.

By assumption there are at least $C \cdot dn$ edges between the vertices on the left and the vertices on the right of the $(X = X'+\frac{1}{2})$-plane, so $YZ \ge C \cdot dn$. Since the same argument holds for the other two directions, $XYZ = \sqrt{XY \cdot YZ \cdot XZ} \ge (C \cdot dn)^{3/2}$, which proves all claims.

Case 3: Neither of the above. Assume now that any X-line, Y-line or Z-line intersects at most $\frac{1}{3}n$ vertices, but there exists, say, a $(Z = Z_0)$-plane that intersects at least $\frac{2}{3}n$ vertices. As an $(X = X_0)$-plane is swept from smaller to larger values of X_0, the Y-line determined by the intersection of this $(X = X_0)$-plane with the $(Z = Z_0)$-plane sweeps the $(Z = Z_0)$-plane. At any time, this Y-line intersects at most $\frac{1}{3}n$ vertices by assumption. Similarly as above, one can therefore find a $(X = X_0)$-plane that splits the vertices intersected by the $(Z = Z_0)$-plane into two sets V_- and V_+ to the left and right of it which each contain at least $\frac{1}{6}n$ vertices. By assumption there are at least $C \cdot dn$ edges between V_- and V_+, so $YZ \geq C \cdot dn$. Apply exactly the same argument in the Y-direction to obtain $XZ \geq C \cdot dn$. We get the three lower bounds as follows:

- The $(Z = Z_0)$-plane intersects at least $\frac{2}{3}n$ vertices, so $XY \geq \frac{2}{3}n$. This implies
  $$XYZ = \sqrt{XY \cdot YZ \cdot XZ} \geq \sqrt{\tfrac{2}{3}n \cdot (C \cdot dn)^2} = \sqrt{\tfrac{2}{3}}C \cdot dn^{3/2},$$ which proves
  the first lower bound.

- Assume that every vertex has aspect ratio at most r ($r \geq 1$). In particular therefore, $Z(v) \leq rX(v)$ and $Z(v) \leq rY(v)$ for every vertex v represented by an $X(v) \times Y(v) \times Z(v)$-box. Since the surface of v is at least $\deg(v)$, this implies $X(v)Y(v) \geq \deg(v)/6r = d/6r$. Since the $(Z = Z_0)$-plane intersects at least $\frac{2}{3}n$ vertices, and these intersections are disjoint, there must be at least $\frac{2}{3}n \cdot d/6r$ integral points in the $(Z = Z_0)$-plane. So $XY \geq \frac{1}{9} \cdot dn/r$ and
  $$XYZ = \sqrt{XY \cdot YZ \cdot XZ} \geq \sqrt{\tfrac{1}{9} \cdot dn/r \cdot (C \cdot dn)^2} = \tfrac{1}{3}C \cdot (dn)^{3/2}/\sqrt{r},$$ which
  proves the second lower bound.

- Assume that the surface of every vertex v is at most $\alpha \deg(v) = \alpha d$ ($\alpha \geq 1$), which implies $Z(v) < \alpha d/4$. Define $Z_- = Z_0 - \alpha d/4$ and $Z_+ = Z_0 + \alpha d/4$. We say that a point is *inside* if its Z-coordinate z satisfies $Z_- < z < Z_+$, and *outside* otherwise. Note that all vertices in V_- and V_+ cross the $(Z = Z_0)$-plane, hence they can cross neither the $(Z = Z_-)$-plane nor the $(Z = Z_+)$-plane, and all grid points of all vertices in V_- and V_+ are inside.
  Recall that the $(X = X_0)$-plane separates the vertices in V_+ and V_- and hence is crossed by at least $C \cdot dn$ edges. There are only $Y \cdot \alpha d/2$ integral points that are inside and on the $(X = X_0)$-plane, so at least $C \cdot dn - Y \cdot \alpha d/2$ edges cross the $(X = X_0)$-plane at an outside point.
  Each of these $C \cdot dn - Y \cdot \alpha d/2$ edges starts at a vertex in V_- at an inside point, crosses the $(X = X_0)$-plane at an outside point, and ends at a vertex in V_+ at an inside point. Hence each such edge crosses the $(Z = Z_-)$-plane or the $(Z = Z_+)$-plane at least twice. These two planes together therefore must have at least $2(C \cdot dn - Y \cdot \alpha d/2)$ grid points, therefore $XY \geq C \cdot dn - Y \cdot \alpha d/2$. Applying the exact same argument in the Y-direction, we obtain $XY \geq C \cdot dn - X \cdot \alpha d/2$.
  Now, if $X \leq Cn/\alpha$ or $Y \leq Cn/\alpha$ then $XY \geq C \cdot dn/2$, and therefore $XYZ = \sqrt{XY \cdot YZ \cdot XZ} \geq \sqrt{\tfrac{1}{2}C \cdot dn \cdot (C \cdot dn)^2} = \tfrac{1}{\sqrt{2}}C^{3/2} \cdot (dn)^{3/2}$. If $X > Cn/\alpha$ and $Y > Cn/\alpha$, then $XY > (Cn)^2/\alpha^2$, and $XYZ = \sqrt{XY \cdot YZ \cdot XZ} \geq \sqrt{(C^2 \cdot n^2/\alpha^2) \cdot (C \cdot dn)^2} = C^2 \cdot dn^2/\alpha \geq C^2 \cdot (dn)^{3/2}/\alpha$ by $d \leq n$. Either way, the third claim is proved.   □

Now we extend the result to arbitrary values of m and n, as long as both are big enough.

Lemma 2. *There exists constants x_1 and $k \geq 2$ such that for all $x \geq x_1$, the interval $[\frac{1}{k}x, x]$ contains a prime number p with $p \equiv 1 \bmod 4$.*

Proof. (Sketch) This follows from a famous theorem by de la Vallée Poussin that establishes that the number of primes $p \leq x$ with $p \equiv 1 \bmod 4$ is proportional to $x/\log x$ (see e.g. [11] for a proof). □

Lemma 3. *For any sufficiently large n, and any sufficiently large $m \leq \binom{n}{2}$, there exists a simple graph G with n vertices and m edges that has a d'-regular subgraph G' with n' vertices and m' edges such that*

- *G' satisfies the conditions of Theorem 1 with constant $C' \geq C$,*
- *$n \geq n' \geq n/8k^2$,*
- *$m \geq m' = d'n'/2 \geq m/64k^4$,*
- *any vertex of G' has degree $\leq 8k^2 d'$ in G,*

where $k \geq 2$ is the constant of Lemma 2.

Proof. (Sketch) Assume $n \geq \max\{32k^4, 2x_1^2\}$ and $m \geq \max\{\frac{1}{4}(x_1+1)n, \frac{145}{4}kn\}$, where x_1 is the lower bound on x in Lemma 2. The proof splits into two cases, depending on whether m is big or not.

If $m \geq n^2/32k^4$, then let G' be the complete graph on $n' = \lceil n/8k^2 \rceil + 1$ vertices. To obtain G, add $n - n'$ additional vertices and $m - \binom{n'}{2}$ arbitrary edges such that the resulting graph is simple. Using the assumptions on n and m one can verify all conditions.

If $m < n^2/32k^4$, then let $q' = \frac{1}{2} + \sqrt{\frac{n}{2} + \frac{1}{4}}$, and find a prime q with $q \equiv 1 \bmod 4$ such that $\frac{1}{k}q' \leq q \leq q'$. Let $p' = 2m/q(q-1) - 1$ and find a prime p with $p \equiv 1 \bmod 4$ such that $\frac{1}{k}p' \leq p \leq p'$. Using the conditions on n and m one verifies $q', p' \geq x_1$, so this is possible. Also, $p \neq q$, $p \geq 144$ and $p < q(q-1)/2$. Let G' be the graph $G_{p,q}$ as in Lemma 1 and suppose it has n' vertices and m' edges. Create G by adding $n - n'$ vertices and $m - m'$ edges between them; by $m < n^2/32k^4$ one can show $m - m' \leq \binom{n-n'}{2}$ and this can be done such that G is simple and the degrees in G' are unchanged. Using the assumptions on n and m one can verify all conditions. □

Theorem 2. $vol(n, m, \infty, \infty) = \Omega(m\sqrt{n})$, $vol(n, m, r, \infty) = \Omega(m^{3/2}/\sqrt{r})$, and $vol(n, m, \infty, \alpha) = \Omega(m^{3/2}/\alpha)$.

Proof. (Sketch) Let G be the graph of Lemma 3. Any drawing of G contains a drawing of G', and thus has volume $\geq \frac{1}{3}C'^{3/2}d'(n')^{3/2}$ by Theorem 1. If the drawing of G has aspect ratios bounded by r, then so does the drawing of G', which thus has volume $\geq \frac{1}{3}C'^{3/2}(d'n')^{3/2}/\sqrt{r}$. If the drawing of G is strictly α-degree restricted, then every vertex in G' has surface $\leq \alpha 8k^2 d'$. Thus the drawing of G' is strictly $(\alpha 8k^2)$-degree restricted, and has volume at $\geq \frac{1}{3}C'^{2}(d'n')^{3/2}/\alpha 8k^2$. Using the bounds on C', n' and m' known from Lemma 3, one obtains the results since k is a constant. □

3   Constructions

In the following, we give two constructions. The first one creates cube-drawings with asymptotically optimal volume, the second one creates drawings without restrictions on vertex boxes that have asymptotically optimal volume.

3.1   Cube-Drawings

In the following algorithm for producing orthogonal drawings, vertices are initially represented by squares in the $(Z = 0)$-plane, and edges are routed above the plane. Vertices are then extended in the Z-dimension to form cubes. For space reasons, we give here only a simplified construction with larger constants.

1. Represent each vertex $v \in V$ by a square S_v of sidelength $2\lceil\sqrt{\deg(v)}\rceil + 2$. [2]
2. Position the squares $\{S_v : v \in V\}$ in the $(Z = 0)$-plane with the square-packing algorithm of Kleitman and Krieger [13].
3. For each vertex $v \in V$, remove the top two rows from S_v and the two rightmost columns from S_v. Vertices are now disjoint; see Fig. 1).
4. Direct edges arbitrarily, and assign each edge $vw \in E$ unique Z^+ ports (ports extreme in the Z^+ direction) at v and w both with even X-coordinate and even Y-coordinate.
5. Construct a graph H with $V(H) = E$. For oriented edges $vw, xy \in E$, add the edge $\{vw, xy\}$ to $E(H)$ if the ports of vw at v and of xy at x are in the same column, or if the ports of vw at w and of xy at y are in the same row.
6. Vertex colour the graph H with $\Delta(H) + 1$ colours, and for each vertex $v \in V(H)$ coloured i corresponding to an edge vw, set $h(vw) \leftarrow i$.
7. For each oriented edge $vw \in E$ construct an edge route for vw as follows. Suppose the ports on v and w assigned to vw have coordinates $(v_X, v_Y, 0)$ and $(w_X, w_Y, 0)$, respectively. Route the edge vw with one of the following four or six bend routes, as illustrated in Fig. 1.

    - $v_X = w_X$:
      $(v_X, v_Y, 0) \rightarrow (v_X, v_Y, 2h(vw)) \rightarrow (v_X + 1, v_Y, 2h(vw)) \rightarrow$
      $(v_X + 1, w_Y, 2h(vw)) \rightarrow (v_X, w_Y, 2h(vw)) \rightarrow (v_X, w_Y, 0)$
    - $v_Y = w_Y$:
      $(v_X, v_Y, 0) \rightarrow (v_X, v_Y, 2h(vw) + 1) \rightarrow (v_X, v_Y + 1, 2h(vw) + 1) \rightarrow$
      $(w_X, v_Y + 1, 2h(vw) + 1) \rightarrow (w_X, v_Y, 2h(vw) + 1) \rightarrow (w_X, v_Y, 0)$
    - $v_X \neq w_X$ and $v_Y \neq w_Y$:
      $(v_X, v_Y, 0) \rightarrow (v_X, v_Y, 2h(vw)) \rightarrow (v_X + 1, v_Y, 2h(vw)) \rightarrow$
      $(v_X + 1, w_Y + 1, 2h(vw)) \rightarrow (v_X + 1, w_Y + 1, 2h(vw) + 1) \rightarrow$
      $(w_X, w_Y + 1, 2h(vw) + 1) \rightarrow (w_X, w_Y, 2h(vw) + 1) \rightarrow (w_X, w_Y, 0)$

8. Enlarge squares representing vertices into cubes by extending their side parallel to the Z-axis.

---

[2] The algorithm by Hagihara *et al.* [12] is similar in spirit, but uses a square of size $\mathcal{O}(\sqrt{\Delta})$ for each vertex, hence resulting in an $\mathcal{O}((n\Delta)^{3/2})$ volume drawing.

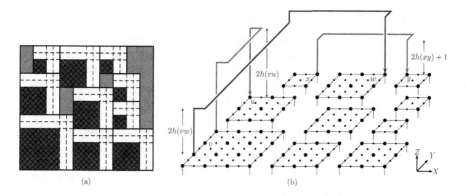

Fig. 1. (a) Square packing, (b) Routing edges.

Theorem 3. *The above algorithm determines a 24-degree-restricted orthogonal cube-drawing of any loopless graph G in $\mathcal{O}(m^{3/2})$ time, with $\mathcal{O}(m^{3/2})$ bounding box volume (assuming $m = \Omega(n)$) and at most six bends per edge route.*

Proof. (Sketch) The squares $\{S_v : v \in V\}$ have area $\sum_v (2\lceil\sqrt{\deg(v)}\rceil + 2)^2 = 8m + o(m)$. The algorithm of [13] packs squares with a total area of 1 in a $2/\sqrt{3} \times \sqrt{2}$ rectangle. So the squares $\{S_v : v \in V(G)\}$ can be packed in a rectangle with size $(2\sqrt{8m}/\sqrt{3} + o(\sqrt{m})) \times (\sqrt{8m} \cdot \sqrt{2} + o(\sqrt{m})) = \mathcal{O}(\sqrt{m}) \times \mathcal{O}(\sqrt{m})$. Hence the maximum degree of H is $\mathcal{O}(\sqrt{m})$, the height of the drawing above the $(Z = 0)$-plane is also $\mathcal{O}(\sqrt{m})$. Expanding vertices adds height $\leq 2\lceil\sqrt{\Delta}\rceil \leq \sqrt{m}$. Hence the total height is $\mathcal{O}(\sqrt{m})$, which proves the bound on the volume.

The surface of a vertex v is $6(2\lceil\sqrt{\deg(v)}\rceil)^2 = 24\deg(v) + o(\deg(v))$. By construction, there are at most six bends per edge route.

The time-consuming stage of the algorithm is the vertex colouring of H, which takes $\mathcal{O}(|E(H)|) = \mathcal{O}(|V(H)|\Delta(H)) = \mathcal{O}(m\sqrt{m}) = \mathcal{O}(m^{3/2})$ time. $\square$

By routing the edges above and below the vertices, 12-degree-restricted vertices are possible; see [6] for details. Very recently, Biedl and Chan [4] developed a technique to implement steps 5 and 6 of our algorithm more efficiently. With this technique, the time complexity of our algorithm reduces to $\mathcal{O}(m \log m)$, and the volume decreases to $\approx 83m^{3/2} + o(m^{3/2})$.

If we remove the middle segment from each 6-bend edge route and route each edge with unique height, then the overall height is $\mathcal{O}(m)$ and we obtain the following result.

Theorem 4. *Every loopless graph has a 12-degree-restricted orthogonal cube-drawing, which can be computed in $\mathcal{O}(m)$ time, with $\mathcal{O}(m^2)$ bounding box volume and at most five bends per edge route.*

3.2   Drawings with Unbounded Aspect Ratio

We now show how to create orthogonal drawings of simple graphs that have volume $\mathcal{O}(m\sqrt{n})$; the vertices have unbounded aspect ratios and are not necessarily degree-restricted; Vertices are initially represented by points or line segments in the $(Z = 0)$-plane. The vertices are then extended in the Z-dimension to form lines or rectangles. Let $N = \lceil \sqrt{n} \rceil$.

1. Define $V_{big} = \{v \in V : \deg(v) \geq 4m/N\}$, $V_{small} = V - V_{big}$. Define E_{big} to be the set of edges with both endpoints in V_{big}, E_{cut} to be the set of edges with exactly one endpoint in V_{big}, and E_{small} to be the remaining edges.
2. For each vertex $v \in V_{small}$, represent v by a 2×2-rectangle S_v.
3. For each vertex $v \in V_{big}$, define $k(v) = \lceil \deg(v)N/4m \rceil$, and represent v by a $2k(v) \times 2$-rectangle S_v.
4. Let $w_1, \ldots, w_l$ be the vertices in V_{big}. For $j = 1, \ldots, l$, position S_{w_j} in the $(Z = 0)$-plane such that its top left corner is at $(\sum_{i=1}^{j-1} 2k(w_i), 2)$. So all these rectangles attach to the X-axis and abut each other.
5. Add extra vertices of degree 0 such that V_{small} has exactly N^2 vertices. (This step is only done to ease the description.)
6. Sort the vertices in V_{small} by non-increasing degree as $v_0, v_1, v_2, \ldots, v_{N^2-1}$. For $k = 0, \ldots, N^2 - 1$, if $k = iN + j$, then place the bottom left corner of S_{v_k} at $(2(i+j) \bmod 2N, 2j)$.
7. For each vertex $v \in V$, remove the top two rows from S_v and the two rightmost columns from S_v. Vertices are now disjoint; see Fig. 2.
8. Orient each edge in E_{cut} from the endpoint in V_{big} to the other endpoint. Orient all edges in E_{small} arbitrarily.
9. For each oriented edge $e = vw \in E_{cut}$, assign a Z^+-port at S_v with even coordinates to e in such a way that at most $4m/N$ edges are assigned to any Z^+-port of any vertex in V_{big}.
   For any edge $e \in E_{cut} \cup E_{small}$ with an endpoint $v \in V_{small}$, edge e uses the unique Z^+-port with even coordinates at S_v.
10. Construct a graph H with $V(H) = E_{cut} \cup E_{small}$. For oriented edges $vw, xy \in V(H)$, add the edge $\{vw, xy\}$ to $E(H)$ if the ports of vw at v and of xy at x are in the same column, or if the ports of vw and w and of xy at y are in the same row.
11. Vertex colour the graph H with $\Delta(H) + 1$ colours, and for each vertex $v \in V(H)$ coloured i corresponding to an edge vw, set $h(vw) \leftarrow i$.
12. For each oriented edge $vw \in E_{cut} \cup E_{small}$ construct an edge route for vw exactly as described in the algorithm in Sect. 3.1.
13. Construct edge routes for edges in E_{big} by copying the 2-bend layout of the complete graph developed in [5]. More precisely, split E_{big} into at most $|V_{big}| - 1$ matchings such that no two edges within one matching are forced to cross, assign one Z-plane for each matching, and route each edge in the matching with at most 2 bends in this Z-plane.
14. Enlarge points/lines representing vertices into lines/rectangles by extending their sides parallel to the Z-axis from the minimum to the maximum Z-coordinate used in the drawing. Clip all segments of edges that overlap the box of an endpoint.

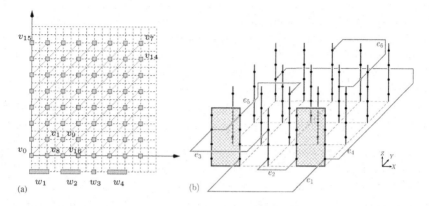

Fig. 2. (a) Vertex layout: w_1, w_2, w_3, w_4 belong to V_{big}. (b) Edge routes: e_1, e_2 and e_3 belong to E_{big}, e_4 and e_5 belong to E_{cut}, and e_6 belongs to E_{small}.

Lemma 4. *The maximum degree of H is at most $16m/N - 2$.*

Proof. (Sketch) By construction there are at most $4m/N$ edges in E_{cut} assigned to the same Z^+-port in the row with Y-coordinate -2. Let a *proper column* be a column with even X-coordinate after taking away the points with negative Y-coordinates. Let a *proper row* be a row that has non-negative even Y-coordinate. The result holds if any proper row/column has at most $6m/N$ edges that use a Z^+-port in that row/column. This can be shown by analyzing the way vertices in V_{small} have been assigned to points in proper rows/columns, and using the fact that the maximum degree of any vertex in V_{small} is at most $4m/N$.   □

Theorem 5. *The above algorithm determines an orthogonal drawing of any simple graph G in $\mathcal{O}(m^2/\sqrt{n})$ time, with $\mathcal{O}(m\sqrt{n})$ bounding box volume and at most four bends per edge route.*

Proof. (Sketch) We start with $2N$ X-planes and $2N+2$ Y-planes. The edges in E_{big} may add up to $|V_{big}|/2$ Y-planes, but one can show that $|V_{big}| \leq N/2$.

A greedy vertex colouring of H requires at most $\Delta(H)+1$ colours, so to route the edges in $E_{cut} \cup E_{small}$, we need at most $2 \cdot 16m/N$ Z-planes. To route the edges in E_{big}, we need at most $|V_{big}| \leq N/2$ Z-planes, which is smaller by $m \geq n/2$, so the height of the drawing above the $Z = 0$ plane is at most $32m/N$. The bounding box is therefore $2N \times (\frac{9}{4}N + 2) \times 32m/N = 144mN + \mathcal{O}(m) = \mathcal{O}(m\sqrt{n})$.

The time-consuming stage of the algorithm is the vertex colouring of H, which takes $\mathcal{O}(|E(H)|) = \mathcal{O}(|V(H)|\Delta(H)) = \mathcal{O}(m \cdot m/N) = \mathcal{O}(m^2/\sqrt{n})$ time.

Originally there are at most six bends per edge route, and at most four bends per edge route for edges in V_{big}. During the clipping step, the first and last segment of each edge is clipped, thus every edge has at most four bends.   □

The technique of Biedl and Chan [4] works for steps 10 and 11 of this algorithm as well, reducing the time complexity of our algorithm to $\mathcal{O}(m \log n)$, and decreasing the volume to $90m\sqrt{n} + o(m\sqrt{n})$.

We can decrease the number of bends to three bends if we allow an increase in volume. In fact, the construction greatly simplifies: position the vertices as Z-lines in an $N \times N$-grid, assign to each edge a unique height, and route each edge as before, but omitting the middle segment. The height is then m, and the width and depth are both $2N$.

Theorem 6. *Every loopless graph has an orthogonal drawing, which can be computed in $\mathcal{O}(m)$ time, with $\mathcal{O}(mn)$ bounding box volume and three bends per edge route.*

This algorithm is particularly appropriate for multilayer VLSI as there are no vertical edge segments ('cross-cuts'); see [1].

4   Conclusions and Open Problems

In this paper we have provided matching upper and lower bounds for three-dimensional orthogonal box-drawings.

In particular, we showed that any algorithm to create three-dimensional orthogonal drawings that have bounded aspect ratios or are degree-restricted cannot do better than a bounding box volume of $\Omega(m^{3/2})$. We then gave an algorithm that matches this bound, i.e., constructs three-dimensional degree-restricted orthogonal cube-drawings with bounding box volume $\mathcal{O}(m^{3/2})$.

If there are no restrictions on the drawing, then we showed that no algorithm can do better than a bounding box volume of $\Omega(m\sqrt{n})$. We gave a second algorithm that matches this bound, i.e., constructs three-dimensional orthogonal drawings with bounding box volume $\mathcal{O}(m\sqrt{n})$.

The following open problems remain to be studied:

- Table 1 suggests a trade-off between the number of bends per edge and the bounding box volume. Can such a trade-off be proved? What are lower bounds for drawings where edges are allowed to have only one or two bends?
- In [5], Biedl *et al.* constructed drawings of K_n with at most one bend per edge such that each vertex box has surface $2n + 4$. For any simple graph, this yields a drawing with at most one bend per edge, but this drawing is degree-restricted only if the minimum degree is $\theta(n)$. Does every graph have a three-dimensional orthogonal drawing that is both degree-restricted and has at most one bend per edge?
- Can the upper bounds on the maximum number of bends per edge in orthogonal drawings with optimal volume be improved? In particular, (a) does every graph have a 5-bend degree-restricted cube-drawing with $\mathcal{O}(m^{3/2})$ volume, and (b) does every graph have a 3-bend drawing with $\mathcal{O}(m\sqrt{n})$ volume? Note that K_n *does* have a $\mathcal{O}(n^{5/2}) = \mathcal{O}(m\sqrt{n})$ volume 3-bend drawing [5].

References

1. A. Aggarwal, M. Klawe, and P. Shor. Multilayer grid embeddings for VLSI. *Algorithmica*, 6(1):129–151, 1991.
2. N. Alon and J. Spencer. *The Probabilistic Method*. John Wiley & Sons, 1992.
3. T. Biedl. Three approaches to 3D-orthogonal box-drawings. In Whitesides [17], pages 30–43.
4. T. Biedl and T. Chan. Cross-coloring: improving the technique by Kolmogorov and Barzdin. Technical Report CS-2000-13. Department of Computer Science, University of Waterloo, Canada, 2000.
5. T. Biedl, T. Shermer, S. Whitesides, and S. Wismath. Bounds for orthogonal 3-D graph drawing. *J. Graph Alg. Appl.*, 3(4):63–79, 1999.
6. T. Biedl, T. Thiele and D. R. Wood. Three-Dimensional Orthogonal Graph Drawing with Optimal Volume. Technical Report CS-2000-12. Department of Computer Science, University of Waterloo, Canada, 2000.
7. M. Closson, S. Gartshore, J. Johansen, and S. K. Wismath. Fully dynamic 3-dimensional orthogonal graph drawing. In Kratochvíl [14], pages 49–58.
8. G. Di Battista, M. Patrignani, and F. Vargiu. A split&push approach to 3D orthogonal drawing. In Whitesides [17], pages 87–101.
9. P. Eades, C. Stirk, and S. Whitesides. The techniques of Kolmogorov and Bardzin for three-dimensional orthogonal graph drawings. *Information Processing Letters*, 60(2):97–103, 1996.
10. P. Eades, A. Symvonis, and S. Whitesides. Three dimensional orthogonal graph drawing algorithms. *Discrete Applied Math.*, 103(1-3):55–87, 2000.
11. É. K. Fogel. An elementary proof of formulae of de la Vallée Poussin (In Russian). *Latvijas PSR Zinātņu Akad. Vestis*, 11(40):123–130, 1950.
12. K. Hagihara, N. Tokura, and N. Suzuki. Graph embedding on a three-dimensional model. *Systems-Comput.-Controls*, 14(6):58–66, 1983.
13. D. Kleitman and M. Krieger. An optimal bound for two dimensional bin packing. In *16th Annual Symposium on Foundations of Computer Science (FOCS'75)*, pages 163–168. IEEE, 1975.
14. J. Kratochvíl, editor. *Symposium on Graph Drawing 99*, volume 1731 of *Lecture Notes in Computer Science*. Springer-Verlag, 1999.
15. A. Lubotzky, R. Phillips, and P. Sarnak. Ramanujan graphs. *Combinatorica*, 8:261–277, 1988.
16. A. Papakostas and I. Tollis. Incremental orthogonal graph drawing in three dimensions. *J. Graph Alg. Appl.*, 3(4):81–115, 1999.
17. S. Whitesides, editor. *Symposium on Graph Drawing 98*, volume 1547 of *Lecture Notes in Computer Science*. Springer-Verlag, 1998.
18. D. R. Wood. An algorithm for three-dimensional orthogonal graph drawing. In Whitesides [17], pages 332–346.
19. D. R. Wood. Multi-dimensional orthogonal graph drawing with small boxes. In Kratochvíl [14], pages 311–322.
20. D. R. Wood. A new algorithm and open problems in three-dimensional orthogonal graph drawing. In R. Raman and J. Simpson, editors, *Proc. Australasian Workshop on Combinatorial Algorithms (AWOCA'99)*, pages 157–167. Curtin University of Technology, Perth, 1999.

A Linear-Time Algorithm
for Bend-Optimal Orthogonal Drawings
of Biconnected Cubic Plane Graphs
(Extended Abstract)

Shin-ichi Nakano and Makiko Yoshikawa

Gunma University, Kiryu 376-8515, Japan
nakano@cs.gunma-u.ac.jp

Abstract. An orthogonal drawing of a plane graph G is a drawing of G with the given planar embedding in which each vertex is mapped to a point, each edge is drawn as a sequence of alternate horizontal and vertical line segments, and any two edges do not cross except at their common end. Observe that only a planar graph with the maximum degree four or less has an orthogonal drawing. The best known algorithm to find an orthogonal drawing runs in time $O(n^{7/4}\sqrt{\log n})$ for any plane graph with n vertices. In this paper we give a linear-time algorithm to find an orthogonal drawing of a given biconnected cubic plane graph with the minimum number of bends.

1 Introduction

An *orthogonal drawing* of a plane graph G is a drawing of G with the given planar embedding in which each vertex is mapped to a point, each edge is drawn as a sequence of alternate horizontal and vertical line segments, and any two edges do not cross except at their common end. Orthogonal drawings have attracted much attention due to its numerous practical applications in circuit schematics, etc. [BLV93,K96,T87]. In particular, we wish to find an orthogonal drawing with the minimum number of bends.

For a given planar graph G, if it is allowed to choose its planar embedding, then finding an orthogonal drawing of G with the minimum number of bends is NP-complete[GT94]. However, Tamassia[T87] and Garg and Tamassia [GT96] presented algorithms which find an orthogonal drawing of a given plane graph G with the minimum number of bends in $O(n^2 \log n)$ and $O(n^{7/4}\sqrt{\log n})$ time respectively unless it is allowed to choose its planar embedding, where n is the number of vertices in G. They reduce the minimum-bend orthogonal drawing problem to a minimum cost flow problem. On the other hand, several linear-time algorithms are known for finding an orthogonal drawing of a plane graph with a presumably small number of bends[K96], and for 3-connected cubic plane graphs a linear-time algorithm is known for finding an orthogonal drawing with the minimum number of bends[RNN99]. Observe that only a planar graph with the maximum degree four or less has an orthogonal drawing.

J. Marks (Ed.): GD 2000, LNCS 1984, pp. 296–307, 2001.

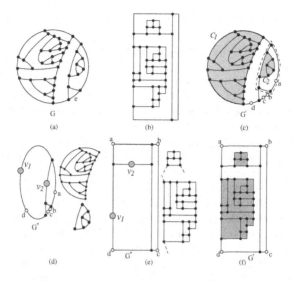

Fig. 1. A plane graph and its orthogonal drawing.

In this paper, generalizing the result in [RNN99], we give a linear-time algorithm to find an orthogonal drawing of a biconnected cubic plane graph with the minimum number of bends.

An orthogonal drawing in which there is no bend and each face is drawn as a rectangle is called a *rectangular drawing*. Given a plane graph G such that every vertex has degree either two or three, in linear-time we can find a rectangular drawing of G whenever such a graph has a rectangular drawing [KH94,RNN96, RNN00]. The key idea of our algorithm is to reduce the orthogonal drawing problem to the rectangular drawing problem.

An outline of our algorithm is illustrated in Fig. 1. Given a plane graph G as shown in Fig. 1(a), we first find a tree structure among some cycles in G, then by analyzing the tree structure we put four dummy vertices a, b, c and d of degree two on the outer boundary of G, and let G' be the resulting graph. The four dummy vertices are drawn by white circles in Fig. 1(c). We then contract each of some cycles $C_1, C_2, \cdots$ and their insides (shaded in Fig. 1(c)) into a single vertex as shown in Fig. 1(d) so that the resulting graph G'' has a rectangular drawing as shown in Fig. 1(e). We also find orthogonal drawings of those cycles $C_1, C_2, \cdots$ and their insides recursively (See Figs. 1(d) and (e)). Patching the obtained drawings, we get an orthogonal drawing of G' as shown in Fig. 1(f). Replacing the dummy vertices a, b, c and d in the drawing of G' with bends, we finally obtain an orthogonal drawing of G as shown in Fig. 1(b).

The rest of the paper is organized as follows. Section 2 gives some definitions and presents a known result. Section 3 shows a tree structure among some cycles in G. Section 4 presents an algorithm to find an orthogonal drawing with the minimum number of bends.

2   Preliminaries

Let G be a connected graph with n vertices. An edge connecting vertices x and y is denoted by (x, y). The *degree* of a vertex v is the number of neighbors of v in G. If every vertex of G has degree three, then G is called a *cubic graph*. The *connectivity* $\kappa(G)$ of a graph G is the minimum number of vertices whose removal results in a disconnected graph or a single-vertex graph K_1. We say that G is k-connected if $\kappa(G) \geq k$.

A graph is *planar* if it can be embedded in the plane so that no two edges intersect geometrically except at a vertex to which they are both incident. A *plane* graph is a planar graph with a fixed planar embedding. A plane graph divides the plane into connected regions called *faces*. We regard the *contour* of a face as a clockwise cycle formed by the edges on the boundary of the face. We denote the contour of the outer face of graph G by $C_o(G)$.

For a simple cycle C in a plane graph G, we denote by $G(C)$ the plane subgraph of G inside C (including C). We say that cycles C_1 and C_2 in a plane graph G are *independent* if $G(C_1)$ and $G(C_2)$ have no common vertex. Cycles C_1 and C_2 are *vertex-disjoint* if C_1 and C_2 have no common vertex. An edge which is incident to exactly one vertex of a simple cycle C and located outside of C is called a *leg* of the cycle C, and the vertex on C to which the leg is incident is called a *leg-vertex* of C. A simple cycle with exactly k legs is called a *k-legged cycle*. For k-legged cycle C the k subpaths of C dividing C at the k leg-vertices are called *the contour paths* of C.

An *orthogonal drawing* of a plane graph G is a drawing of G with the given planar embedding in which each vertex is mapped to a point, each edge is drawn as a sequence of alternate horizontal and vertical line segments, and any two edges do not cross except at their common end. A point where an edge changes its direction in a drawing is called a *bend*. We denote by $b(G)$ the minimum number of bends for orthogonal drawings of G. An orthogonal drawing of G with exactly $b(G)$ bends is *bend-optimal*.

A *rectangular drawing* of a plane graph G is a drawing of G such that each edge is drawn as a horizontal or vertical line segment, and each face is drawn as a rectangle. Thus a rectangular drawing is an orthogonal drawing in which there is no bend and each face is drawn as a rectangle. The drawing of G'' in Fig. 1(e) is a rectangular drawing. The drawing of G' in Fig. 1(f) is not a rectangular drawing, but is an orthogonal drawing. In any rectangular drawing D of G, the four corners of the rectangle corresponding to $C_o(G)$ are vertices of degree two on $C_o(G)$. We call these four vertices *the corner vertices* of D. The following result on rectangular drawings is known.

Lemma 1. *Let G be a connected plane graph such that every vertex has degree either two or three, and let a, b, c, d be four designated vertices of degree two on $C_o(G)$. Then G has a rectangular drawing with the corner vertices a, b, c, d if and only if G has none of the following three types of simple cycles [T84]:*

(r1) 1-legged cycles,
(r2) 2-legged cycles which contain at most one designated vertex of degree two, and

(r3) 3-legged cycles which contain no designated vertex of degree two.

Furthermore one can check in linear time whether G satisfies the condition above, and if G does then one can find a rectangular drawing of G in linear time [RNN96,RNN00].

3   Genealogical Tree

Let G be a biconnected cubic plane graph. For a pair of distinct cycles C_a and C_d in G, C_d is called a *descendant-cycle* of C_a if (i) C_d is either 2- or 3-legged cycle, and (ii) $G(C_d)$ is a proper subgraph of $G(C_a)$. Note that since G is biconnected there is neither 0- nor 1-legged cycle except the only 0-legged cycle $C_o(G)$. Now we choose an edge $e = (x, y)$ on $C_o(G)$, and replace e with two edges (x, z) and (z, y). Let G' be the resulting plane graph. (Note that, for $G - e$, that is a plane subgraph of G obtained from G by deleting e, $C_o(G - e)$ is a 2-legged cycle of G', however, $C_o(G - e)$ is not a 2-legged cycle of G.) Let $D_e(C_o) = \{C | C$ is a descendant cycle of $C_o(G')$ not containing $z\}$. A cycle C_c in $D_e(C_o)$ is called a *child-cycle* of $C_o(G')$ (with respect to edge e) if C_c is not located inside of any other cycle in $D_e(C_o)$. Since G is a biconnected cubic plane graph, $C_o(G')$ has exactly one child-cycle $C_o(G - e)$ (with respect to edge e). (See Fig 2.) Then, recursively, for each child-cycle C_c we define its child-cycle as follows. We have the following two cases.

Case 1: C_c is a 2-legged cycle.

Choose a leg-vertex of C_c as z. Let $D_z(C_c) = \{C | C$ is a descendant cycle of C_c not containing $z \}$. A cycle C_{cc} in $D_z(C_c)$ is called a *child-cycle* of C_c (with respect to z) if C_{cc} is not located inside of any other cycle in $D_z(C_c)$. Since G is a biconnected cubic plane graph, C_c has at most one 3-legged child-cycle. (C_c has no 3-legged child-cycle if $G(C)$ has an inner face F containing the two leg-vertices, and C_c has exactly one 3-legged child-cycle otherwise.)

Case 2: Otherwise, C_c is a 3-legged cycle.

Let $D(C_c)$ be the set of all descendant cycles of C_c. A cycle C_{cc} in $D(C_c)$ is called a *child-cycle* of C_c if C_{cc} is not located inside of any other cycle in $D(C_c)$.

In both cases above all child-cycles of C_c are independent each other.

By the definition above we can find child-cycles of each child-cycle recursively, and eventually we get a (hierarchical) tree structure of cycles in G represented by a "genealogical tree" T_g, as shown in Fig 2. Because of the choices for e and z, T_g may have some variations. We choose an arbitrary (but fixed) one as T_g.

Using a method similar to one in [RNN96,RNN99,RNN00], in linear time one can find such a tree structure T_g among cycles by traversing the contour of each face a constant number of times.

Now we observe the following. In any orthogonal drawing of G, every cycle C in G has at least four convex corners, i.e., polygonal vertices of inner angle $90°$. Since G is cubic, such a corner must be a bend if it is not a leg-vertex of C. Thus we have the following facts for any orthogonal drawing of G.

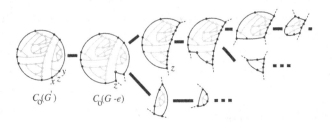

Fig. 2. cycles in G' and a genealogical tree T_g.

Fact 1 At least four bends must appear on $C_o(G)$.
Fact 2 At least two bend must appear on each 2-legged cycle in G.
Fact 3 At least one bend must appear on each 3-legged cycle in G.

4   Orthogonal Drawing

In this section we give a linear-time algorithm to find a bend-optimal orthogonal drawing of a biconnected cubic plane graph. Assume that we have a genealogical tree T_g of a biconnected cubic plane graph G. We need some definitions.

Let C be a 2-legged cycle with the two leg-vertices x and y, and P_1 and P_2 be the clockwise contour paths from x to y and from y to x, respectively. A bend-optimal orthogonal drawing D of $G(C)$ is *feasible* for (P_1, P_1) if none of the following four open halflines intersects D. (See Fig. 3(a). Intuitively D needs two convex bends on P_1.)

the vertical open halfline with the upper end at x.
the horizontal open halfline with the left end at x.
the vertical open halfline with the lower end at y.
the horizontal open halfline with the left end at y.

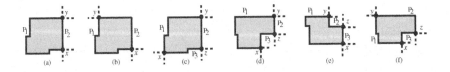

Fig. 3. Illustration for feasible drawings.

Also, a bend-optimal orthogonal drawing D of $G(C)$ is *feasible* for (P_1, P_2) if none of the four open halflines depicted in dashed lines in Fig. 3(b) intersects D.

Let C be a 3-legged cycle with the three leg-vertices x, y and z appearing clockwise in this order, and P_1, P_2 and P_3 be the clockwise contour path from x to y, from y to z, and from z to x, respectively. A bend-optimal orthogonal drawing D of $G(C)$ is *feasible* for (P_1) if none of the six open halflines depicted in dashed lines in Fig. 3(c) intersects D. Similarly, we define *feasible* orthogonal drawings for $(P_1, P_1, -P_3)$, $(P_1, P_1, -P_2)$ and $(P_1, P_2, -P_3)$.(See Fig. 3(d)–(f).)

Now, for each cycle $C \neq C_o(G)$ corresponding to a vertex in T_g, we determine whether $G(C)$ has each type of feasible drawings by a bottom-up computation on T_g. For the bottom-up computation we also compute a set S_C of vertex-disjoint cycles in $G(C)$ consisting of ℓ_2 2-legged cycles and ℓ_3 3-legged cycles for some ℓ_2 and ℓ_3. Thus $b(G(C)) \geq 2 \cdot \ell_2 + \ell_3$ by Facts 3.2 and 3.3. We then show that $G(C)$ always has at least one feasible drawing using $2 \cdot \ell_2 + \ell_3$ bends. Thus $b(G(C)) = 2 \cdot \ell_2 + \ell_3$ holds.

In the bottom-up computation we classify each contour path of each cycle as either *0-, 1-,* or *2-corner path*. Intuitively k-corner path has a chance to have k convex bends. And we define $P_1 P_2$-strain by those corner paths as follows. Let

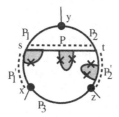

Fig. 4. Illustration for $P_1 P_2$-strain.

x, y, z be the three leg-vertices of a 3-legged cycle C, P_1 and P_2 be the clockwise contour paths from x to y and y to z, respectively. Assume that s and t are vertices on P_1 and P_2, respectively, and let P_1' be the subpath of P_1 from x to s, and P_2' be the subpath of P_2 from t to z. If (i) there is a path P from s to t such that the left side of P is an inner face of $G(C)$, and (ii) $G(C)$ has no child cycle having 1- or 2-corner path on P, P_1' or P_2', then the path consisting of P_1', P, P_2' are called $P_1 P_2$-*strain*. An example is illustrated in Fig. 4. Intutively, we have only two chance to turn right at s and t on $P_1 P_2$-strain from x to z.

In the bottom-up computation we show that the following conditions (c1) – (c9) hold.

(c1) Any cycle C has at least one 1- or 2-corner path.
(c2) No cycle in S_C contains any edge on any 0-corner path of C.
(c3) For any 2-legged cycle C if C has a 1-corner path P_1, then $G(C)$ has a set S_C' of vertex-disjoint cycles containing no edge on P_1 and consisting of ℓ_2' 2-legged cycles and ℓ_3' 3-legged cycles such that $2 \cdot \ell_2' + \ell_3' = b(G(C)) - 1$.
(c4) For any 2-legged cycle C if C has a 0-corner path P_1, then the other contour path P_2 is a 2-corner path, and $G(C)$ has an orthogonal drawing feasible for (P_2, P_2).
(c5) For any 3-legged cycle C if C has a 1-corner path P_1, then $G(C)$ has a set S_C' of vertex-disjoint cycles containing no edge on P_1, and consisting of ℓ_2' 2-legged cycles and ℓ_3' 3-legged cycles such that $2 \cdot \ell_2' + \ell_3' = b(G(C)) - 1$.
(c6) For any 3-legged cycle C if C has a 1- or 2-corner path P_1, then $G(C)$ has an orthogonal drawing feasible for (P_1).
(c7) For any 3-legged cycle C if C has a 2-corner path P_1 and no $P_1 P_2$-strain , then $G(C)$ has an orthogonal drawing feasible for $(P_1, P_1, -P_3)$,

(c8) For any 3-legged cycle C if C has a 2-corner path P_1 and no P_3P_1-strain, then $G(C)$ has an orthogonal drawing feasible for $(P_1, P_1, -P_2)$,

(c9) For any 3-legged cycle C if C has 1-corner paths P_1 and P_2, and no P_1P_2-strain, then $G(C)$ has an orthogonal drawing feasible for $(P_1, P_2, -P_3)$.

Now we explain the bottom-up computation in the following four cases.

Case 1: C is a 2-legged cycle having no child-cycle.

Let x, y be the two leg-vertices of C, let P_1 and P_2 be the clockwise contour paths from x to y and from y to x, respectively. Now $G(C) = C$, since for any 2-legged cycle C if $G(C)$ has an edge in proper inside of C then C always has a child-cycle.

Computation for S_C: Set $S_C = \{C\}$. By Fact 3.2 any orthogonal drawing of $G(C)$ has at least two bends.

Feasible drawings: By introducing two bends on P_1, we can easily construct an orthogonal drawing of $G(C)$ feasible for (P_1, P_1). Similarly we can construct orthogonal drawings of $G(C)$ feasible for (P_2, P_2) and (P_1, P_2), respectively. Thus $G(C)$ has each type of feasible orthogonal drawings.

Classification and proof for (c1)–(c9): In this case every contour path of C is classified as a 2-corner paths. Conditions (c1)–(c4) hold since every contour path of C is 2-corner, and (c5)–(c9) hold since C is not a 3-legged cycle.

Case 2: C is a 3-legged cycle having no child-cycle.

Let x, y, z be the three leg-vertices of C, let P_1, P_2, P_3 be the clockwise contour path from x to y, from y to z, and from z to x, respectively. Now if we remove all edges on C from $G(C)$, then either $G(C) = C$ or the remaining edges induce a connected graph containing at least one vertex on each P_1, P_2, P_3, since otherwise C has a child-cycle, a contradiction.

Computation for S_C: Set $S_C = \{C\}$. By Fact 3.3 any orthogonal drawing of $G(C)$ has at least one bend.

Feasible drawings: Construct a new graph G' from $G(C)$ by adding one dummy vertices v on P_1. Now the resulting graph G' has no bad cycle (since G has no child-cycle) with respect to corner vertices x, v, y, z, and then G' has a rectangular drawing with the corner vertices x, v, y, z. The rectangular drawing is also an orthogonal drawing of $G(C)$ feasible for (P_1) using exactly one bend (corresponding to v). Similarly we can easily construct orthogonal drawings of $G(C)$ feasible for (P_2) and (P_3).

Now $G(C)$ has no orthogonal drawing feasible for $(P_1, P_1, -P_2)$, since it needs at least two bends only on P_1. Similarly $G(C)$ has no orthogonal drawing feasible for $(P_i, P_j, -P_k)$ for any $i, j, k \in \{1, 2, 3\}$.

Classification and proof for (c1)–(c9): In this case every contour path of C is classified as a 1-corner path. Conditions (c1),(c2) hold since every contour path of C is 1-corner, (c3),(c4) hold since C is not a 2-legged cycle, (c5) holds by choosing $S'_C = \phi$, (c6) holds since $G(C)$ has orthogonal drawings feasible for $(P_1), (P_2), (P_3)$, respectively, as mentioned above, and (c7)–(c9) hold since $G(C)$ has no 2-corner path.

Case 3: C is a 2-legged cycle having one or more child-cycles.

Let x, y be the two leg-vertices of C, and let P_1 and P_2 be the clockwise contour paths from x to y and from y to x, respectively. If $G(C)$ has an inner face containing x and y, then C has no 3-legged child-cycle, otherwise, C has exactly one 3-legged child-cycle, which contains exactly one leg-vertices of C. Thus C has at most one 3-legged child-cycle.

Let $C_1, C_2, \cdots, C_\ell$ be the child-cycle of C. Assume that for C_i, $1 \leq i \leq l$, we already have S_{C_i}, we know whether $G(C_i)$ has each type of feasible drawings, and conditions (c1)–(c9) holds. We have the following four subcases. Proofs for (c1)–(c9) are omitted.

Case 3(a): C has no child-cycle having a 1- or 2- corner path on C.

Computation for S_C: Condition (c2) means that no cycle in $S_{C_1}, S_{C_2}, \cdots, S_{C_\ell}$ contains any edge on C. Also since G is cubic, C is vertex-disjoint to any cycle in $S_{C_1}, S_{C_2}, \cdots, S_{C_\ell}$. Set $S_C = \{C\} \cup S_{C_1} \cup S_{C_2} \cup \cdots \cup S_{C_\ell}$. Thus we need to introduce two new bends.

Feasible drawings: We first consider whether $G(C)$ has an orthogonal drawing feasible for (P_1, P_1). Construct a new graph from $G(C)$ by adding two dummy vertices v, w on P_1 but not on any child cycle of C. Then contract each $G(C_1), G(C_2), \cdots, G(C_\ell)$ to vertices $v_1, v_2, \cdots, v_\ell$, respectively. See Figs. 5(a) and (b). Now the resulting graph is a cycle and has a rectangular drawing D with the corner vertices x, v, w, y. See Fig. 5(c). Next, if C has a 3-legged child-cycle, say C', then find an orthogonal drawing of $G(C')$ feasible for (P') where P' is the contour path of C' not on C, in a recursive manner. By conditions (c1) and (c6) $G(C')$ always has such a drawing. Next, find an orthogonal drawing of each 2-legged child-cycle $G(C_i)$ feasible for (P_i'', P_i'') where P_i'' is the contour path of C_i not on C, in a recursive manner. By condition (c4) $G(C)$ always has such a drawing. Finally patch the drawings of $G(C_1), G(C_2), \cdots, G(C_\ell)$ into D. See Fig. 5(d). The patching for 2- and 3-legged child-cycles always works correctly as shown in Fig. 6 and Fig. 7. Thus we can construct an orthogonal drawing of $G(C)$ feasible for (P_1, P_1). Similarly we can construct orthogonal drawings feasible for (P_2, P_2) and (P_1, P_2), respectively.

Classification: In this case every contour path of C is classified as a 2-corner path.

Case 3(b): C has exactly one child-cycles having a 1- or 2- corner path on C, and the child-cycle is a 2-legged cycle.

Computation for S_C: Let C_1 be the 2-legged child-cycle having a corner path on C. We consider two cases. If C_1 has a 2-corner path on C, then set $S_C = S_{C_1} \cup S_{C_2} \cup \cdots \cup S_{C_\ell}$. In this case we do not need to introduce any new bends. If C_1 has a 1-corner path on C, then, by (c3), $G(C_1)$ has a set S'_{C_1} of vertex-disjoint cycles containing no edge on C, and consisting of ℓ_2' 2-legged cycles and ℓ_3' 3-legged cycles such that $2 \cdot \ell_2' + \ell_3' = b(G(C_1)) - 1$. Condition (c2) means that no cycle in $S_{C_2}, S_{C_3}, \cdots, S_{C_\ell}$ contains any edge on C. Set $S_C = \{C\} \cup S'_{C_1} \cup S_{C_2} \cup \cdots \cup S_{C_\ell}$. In this case we need to introduce one new bend.

Feasible drawings: Omitted. Similar to the previous case.

Classification: If C_1 has a 2-corner path on P_1, then P_1 is a 2-corner path and P_2 is a 0-corner path. If C_1 has a 2-corner path on P_2, then P_1 is a 0-corner path and P_2 is a 2-corner path. If C_1 has a 1-corner path on P_1, then P_1 is a 2-corner path and P_2 is a 1-corner path. (In this case we can add one new bend either on

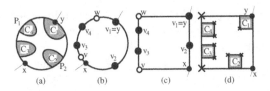

Fig. 5. Illustration for Case 3(a).

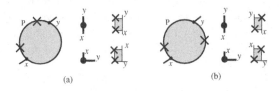

Fig. 6. Illustration for patchings.

P_1 or P_2.) If C_2 has a 1-corner path on P_2, then P_1 is a 1-corner path and P_2 is a 2-corner path.

Case 3(c): C has exactly one child-cycles having a 1- or 2-corner path on C, and the child-cycle is a 3-legged cycle.

Let C_1 be the 3-legged child-cycle having a 1- or 2-corner path on C. Assume that C_1 shares y with C as a leg-vertex. Let P_{11} be the contour path of C_1 on P_1 and P_{12} be the contour path of C_1 on P_2.

Computation for S_C: We consider three cases.

If C_1 has a $P_{11}P_{12}$-strain, then set $S_C = \{C_S\} \cup S_{C_1} \cup S_{C_2} \cup \cdots \cup S_{C_\ell}$, where C_S is the 3-legged cycle consisting of the $P_{11}P_{12}$-strain and the edges on P_1 and P_2 not contained in C_1. By the definition of strain and (c2), C_S is vertex-disjoint to any cycle in S_{C_1}. In this case we need to introduce one new bend for C_S. (See Figs. 8(a)–(d).)

Otherwise, if C_1 has no $P_{11}P_{12}$-strain and either (i)P_{11} is a 2-corner path, (ii)P_{12} is a 2-corner path or (iii) P_{11} is a 1-corner path and P_{12} is a 1-corner path, then set $S_C = S_{C_1} \cup S_{C_2} \cup \cdots \cup S_{C_\ell}$. In this case we do not need to introduce any new bends. (See Figs. 8(e)–(g).)

Otherwise, C_1 has no $P_{11}P_{12}$-strain, and either (i) P_{11} is a 1-corner path and P_{12} is a 0-corner path, or (ii) P_{11} is a 0-corner path and P_{12} is a 1-corner path. By (c5) $G(C_1)$ has a set S'_{C_1} of vertex-disjoint cycles containing no edge on C,

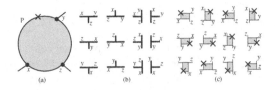

Fig. 7. Illustration for patchings. (Rotated cases are omitted.)

and consisting of ℓ'_2 2-legged cycles and ℓ'_3 3-legged cycles such that $2 \cdot \ell'_2 + \ell'_3 = b(G(C_1)) - 1$. Set $S_C = \{C\} \cup S'_{C_1} \cup S_{C_2} \cup \cdots \cup S_{C_\ell}$. Thus in this case we need to introduce one new bend. (See Figs. 8(a)–(d).)

Feasible drawings: Omitted. Similar to the previous case.

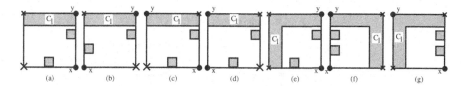

Fig. 8. Illustration for Case 3(c).

Classification: If either (i) P_{11} is a 2-corner path and C_1 has no $P_{11}P_{12}$-strain, (ii) P_{11} is a 1- or 2-corner path and C_1 has a $P_{11}P_{12}$-strain, or (iii) P_{11} is a 1-corner path, P_{12} is a 0-corner path and C_1 has no $P_{11}P_{12}$-strain, then P_1 is a 2-corner path. (See Figs. 8(e),(a),(a), respectively.) Otherwise if (i) P_{11} is a 1-corner path, P_{12} is a 1-corner path and C_1 has no $P_{11}P_{12}$-strain, (ii) P_{11} is a 0-corner path, P_{12} is a 1- or 2-corner path and C_1 has a $P_{11}P_{12}$-strain, or (iii) P_{11} is a 0-corner path, P_{12} is a 1-corner path and C_1 has no $P_{11}P_{12}$-strain, then P_1 is a 1-corner path. (See Figs. 8(g),(c),(c), respectively.) Otherwise, P_{11} is a 0-corner path, P_{12} is a 2-corner path and C_1 has no $P_{11}P_{12}$-strain, then P_1 is a 0-corner path. (See Fig. 8(f).) Classify P_2 similarly.

Case 3(d): C has two or more child-cycles having a 1- or 2- corner path on C. Omitted

Case 4: C is a 3-legged cycle having one or more child-cycles.

Let x, y, z be the three leg-vertices of C, and let P_1, P_2, P_3 be the clockwise contour path from x to y, from y to z, and from z to x, respectively.

Computation for S_C: If C has no child-cycle having a 1- or 2-corner path on C then set $S_C = \{C\} \cup S_{C_1} \cup S_{C_2} \cup \cdots \cup S_{C_\ell}$. In this case we need to introduce one new bend. Otherwise set $S_C = S_{C_1} \cup S_{C_2} \cup \cdots \cup S_{C_\ell}$. In this case we do not need to introduce any new bend.

Feasible drawings: If $G(C)$ has no child-cycle having a 1- or 2-corner path on C then $G(C)$ has orthogonal drawings feasible for (P_1), (P_2), (P_3), respectively. (In this case we need to introduce one new bend.)

Otherwise, $G(C)$ has an orthogonal drawing feasible for (P_1) if and only if $G(C)$ has a child-cycle having a 1- or 2-corner path on P_1. Similarly we can determine whether $G(C)$ has orthogonal drawings feasible for (P_2) and (P_3).

If C has no child-cycle having a 1- or 2-corner path on C then $G(C)$ has no orthogonal drawing feasible for $(P_1, P_1, -P_3)$, since we have no chance to have two bend on P_1 even if we introduce one new bend on P_1.

$G(C)$ has an orthogonal drawing feasible for $(P_1, P_1, -P_3)$ if and only if (i) C has two child-cycle having a 1- or 2-corner path on P_1, or C has a child-cycle having a 2-corner path on P_1, and (ii) C has no P_1P_2-strain. (Construction is omitted. See Figs. 9 and 10.)

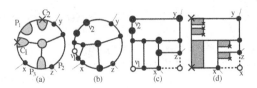

Fig. 9. Illustration for Case 4.

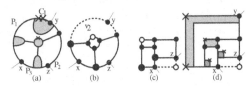

Fig. 10. Illustration for Case 4.

Classification: If C has no child-cycle having a 1- or 2-corner path on C, then P_1, P_2 and P_3 are 1-corner paths. Otherwise, if either (i) C has two or more child-cycles having a 1- or 2-corner path on P_1, or C has a child-cycle having a 2-corner path on P_1, then P_1 is classified as a 2-corner path. Otherwise if C has exactly one child-cycle having 1-corner path on P_1, then P_1 is classified as a 1-corner path. Otherwise P_1 is classified as a 0-corner path. We classify P_2 similarly.

Now we give our algorithm to find a bend-optimal orthogonal drawing. Using a method similar to one in [RNN96,RNN99,RNN00] the algorithm above runs in linear time.

> **Algorithm** Orthogonal-Draw(G)
> **begin**
> 1 Choose an edge e on $C_o(G)$; Find a genealogical tree T_g;
> 2 Do the bottom-up computation;
> 3 Find minimal cycles having 1- or 2-corner path on $C_o(G')$ as many as possible;
> 4 Do the following until G_0 has exactly four vertices of degree two.
> > For each minimal 2-legged cycle C having 2-corner path on G_0 replace $G(C)$ with a quadrangle containing two vertices of degree two on G_0.
> > For each minimal 2-legged cycle C having 1-corner path on G_0 replace $G(C)$ with a vertex of degree two.
> > For each minimal 3-legged cycle C having 1-corner path on G_0 replace $G(C)$ with a quadrangle containing one vertex of degree two on G_0.
> > Put vertices of degree two on the edge e.
> 5 Find maximal bad cycles $C_1, C_2, \cdots, C_\ell$;
> 6 Let G'' be the graph derived from G' by contracting each $G(C_i)$, $i = 1, 2, \cdots, \ell$ into a vertex v_i;
> 7 Find a rectangular drawing $D(G'')$ of G'';
> 8 For each $i = 1, 2, \cdots, \ell$, find a feasible orthogonal drawing $D(G(C_i))$ of $G(C_i)$;
> 9 Patch the drawings $D(G(C_i))$, $i = 1, 2, \cdots, \ell$, into $D(G'')$ to get an orthogonal drawing of G; (See Figs. 1(e) and (f).)
> **end.**

Theorem 1. *The algorithm above find a bend-optimal orthogonal drawing of a biconnected cubic plane graph in linear time.*

References

[BLV93]   G. Di Battista, G. Liotta and F. Vargiu, *Spirality of orthogonal represen-tations and optimal drawings of series-parallel graphs and 3-planar graphs*, Proc. of Workshop on Algorithms and Data structures, LNCS 709, Springer (1993) 151-162.

[GT94]   A. Garg and R. Tamassia, *On the computational complexity of upward and rectilinear planarity testing*, Proc. of Graph Drawing'94, LNCS 894, Springer (1995) 286-297.

[GT96]   A. Garg and R. Tamassia, *A new minimum cost flow algorithm with appli-cations to graph drawing*, Proc. of Graph Drawing'96, LNCS 1190, Springer (1997) 201-226.

[K96]   G. Kant, *Drawing planar graphs using the canonical ordering*, Algorith-mica, 16 (1996) 4-32.

[KH94]   G. Kant and X. He, *Two algorithms for finding rectangular duals of planar graphs*, Proc. of WG'93, LNCS 790, Springer (1994) 396-410.

[RNN96]   M. S. Rahman, S. Nakano and T. Nishizeki, *Rectangular grid drawings of plane graphs*, Proc. of COCOON'96, LNCS 1090, Springer (1996) 92-105. Also, Computational Geometry: Theory and Applications, 10 (1998) 203-220.

[RNN99]   M. S. Rahman, S. Nakano and T. Nishizeki, *A linear algorithm for bend-optimal orthogonal drawings of triconnected cubic plane graphs*, Journal of Graph Algorithms and Applications, 3 (1999) 31-62.

[RNN00]   M. S. Rahman, S. Nakano and T. Nishizeki, *Rectangular Drawings of Plane Graphs without Designated Corners*, Proc. of COCOON'00, LNCS 1858, Springer (2000) 85-94.

[T87]   R. Tamassia, *On embedding a graph in the grid with the minimum number of bends*, SIAM J. Comput., 16 (1987) 421-444.

[T84]   C. Thomassen, *Plane representations of graphs*, (Eds.) J.A. Bondy and U.S.R. Murty, Progress in Graph Theory, Academic Press Canada (1984) 43-69.

Refinement of Three-Dimensional Orthogonal Graph Drawings

Benjamin Y. S. Lynn, Antonios Symvonis, and David R. Wood

Basser Department of Computer Science
The University of Sydney
Sydney NSW 2006, Australia
{ben,symvonis,davidw}@cs.usyd.edu.au

Abstract. In this paper we introduce a number of techniques for the refinement of three-dimensional orthogonal drawings of maximum degree six graphs. We have implemented several existing algorithms for three-dimensional orthogonal graph drawing including a number of heuristics to improve their performance. The performance of the refinements on the produced drawings is then evaluated in an extensive experimental study. We measure the aesthetic criteria of the bounding box volume, the average and maximum number of bends per edge, and the average and maximum edge length. On the same set of graphs used in Di Battista *et al.* [3], our main refinement algorithm improves the above aesthetic criteria by 80%, 38%, 10%, 54% and 49%, respectively.

1 Introduction

The *3-D orthogonal grid* consists of *grid-points* in 3-space with integer coordinates, together with the axis-parallel *grid-lines* determined by these points. A *3-D orthogonal drawing* of a graph places the vertices at grid-points and routes the edges along sequences of contiguous segments of grid-lines. Edges are allowed to contain bends and can only intersect at a common vertex.

For brevity we say a 3-D orthogonal drawing of a graph G, denoted by $D(G)$, is a *drawing*. A drawing with no more than b bends per edge is called a *b-bend drawing*. The graph-theoretic terms 'vertex' and 'edge' also refer to their representation in a drawing. At a vertex v, the six directions the edges incident with v can use are called *ports*. Clearly, orthogonal drawings can only exist for graphs with maximum degree six. 3-D orthogonal graph drawings have been studied in [1,2,3,4,5,7,9,10,13,14]. By representing a vertex by a grid-box, 3-D orthogonal drawings of arbitrary degree graphs have also been considered (see [14]).

The bounding box of a given drawing is the minimum axis-parallel box which encloses the drawing. The following aesthetic criteria are the most commonly proposed measures for the quality of a given drawing.

- minimise the bounding box volume.
- minimise the maximum or average number of bends per edge.
- minimise the maximum or average length of edges.

Using straightforward extensions of the corresponding 2-D NP-hardness results, optimising any of these criteria is NP-hard [4]. In the existing algorithms

J. Marks (Ed.): GD 2000, LNCS 1984, pp. 308–320, 2001.

for 3-D orthogonal graph drawing there is an apparent tradeoff between these aesthetic criteria, in particular, between the bounding box volume and the maximum number of bends per edge (see [5]).

Despite the fact that the drawings produced by the existing algorithms possess several desirable theoretical properties, they largely fail to communicate to the user the semantic properties of the graph being visualised. The poor visual quality of drawings produced by current algorithms can be attributed to the graph-theoretic methods which they employ. In their effort to guarantee intersection-free drawings for worst-case input graphs, they produce worst-case drawings even when the graph can be drawn in a much better way.

Post-processing refinement techniques can help rectify this situation. Here we simplify the drawings, while maintaining desired theoretic properties such as the maximum number of bends per edge route and the bounding box volume. In this paper we introduce a number of techniques for the refinement of 3-D orthogonal graph drawings. The performance of these refinements on drawings produced by several existing algorithms is then evaluated in an extensive experimental study. Refinement techniques for 2-D orthogonal drawings have been developed by Fößmeier $et\ al.$ [6] and Six $et\ al.$ [11].

We use the following definitions. A $direction$ is an element of $\{\pm X, \pm Y, \pm Z\}$. We speak of $positive$ and $negative$ directions in the obvious sense. For each dimension $I \in \{X, Y, Z\}$ and direction $d = \pm I$, we say $a <_d b$, for two grid-points a and b if $I(a) < I(b)$ and d is positive, or $I(b) < I(a)$ and d is negative. A k-bend edge route vw is represented by the list $(v = b_0, b_1, b_2, \ldots, b_{k+1} = w)$, where $b_1, b_2, \ldots, b_k$ are the bends of vw. So that consecutive bends differ by at most one coordinate and there are no redundant bends, it is necessary that (1) $b_{i+1} - b_i$, $0 \le i \le k$, is an axis-parallel vector and (2) $b_{i+1} - b_i$ is in a different direction to $b_i - b_{i-1}$, $1 \le i \le k$. The length of a vector x is denoted by $|x|$.

2 Implementation of Algorithms

This section describes the algorithms which we use to construct 3-D orthogonal drawings and particular aspects of the implementation of these algorithms which are pertinent to our experiment. For many of these algorithms, the authors were only interested in establishing asymptotic worst-case bounds for their performance, and numerous obvious improvements can be made to the algorithms, which in practice give a constant-factor improvement in some aesthetic criteria. Wherever possible, our implementations have included such improvements. For example, we remove grid-planes not containing a vertex or a bend from a given drawing, thus reducing its volume and the length of edges.

The Compact Algorithms: We now describe the COMPACT family of algorithms due to Eades $et\ al.$ [5], and discuss issues relevant to their implementation. The COMPACT-7 algorithm positions the vertices in a $O(\sqrt{n}) \times O(\sqrt{n})$ grid in the $(Z = 0)$-plane, and produces drawings with $O(\sqrt{n}) \times O(\sqrt{n}) \times O(\sqrt{n})$ volume and at most seven bends per edge.

A critical component of the implementation of the algorithm is the construction of a graph H whose vertices correspond to the edges to be routed above the $(Z = 0)$-plane, and similarly for edges routed below the $(Z = 0)$-plane. In [5] vertices are adjacent in H if the corresponding edges start in the same row or end in the same column. A vertex-colouring of H determines the height at which edges are routed. This ensures that edges routed at the same height do not intersect. In our implementation, vertices are adjacent in H if the corresponding edge routes will intersect if routed at the same height. In general, there are less edges in H using this approach; hence in practice less colours and therefore less volume is used. We use the sequential greedy algorithm to vertex colour the graph H. Note that this method will in practice use less colours than the method of Biedl and Chan [1] which necessarily assigns a different colour to edges which start in the same row or end in the same column, even if they will not intersect if routed at the same height.

We have also implemented the COMPACT-6 and COMPACT-5 variations of the COMPACT algorithm, which produce drawings with at most six and five bends per edge, respectively, and with $O(\sqrt{n}) \times O(\sqrt{n}) \times O(n)$ and $O(\sqrt{n}) \times O(n) \times O(n)$ volume, respectively. As described above for the COMPACT-7 algorithm we again employ an enhanced colouring method to determine the heights of edges.

Algorithms in the General Position Model: A 3-D orthogonal graph drawing is said to be in *general position* if no two vertices are in the same grid-plane. We have implemented the 3-BENDS algorithm of Eades *et al.* [5] and the DLM (Diagonal Layout plus Movement) algorithm of Wood [13], both of which produce general position drawings with $O(n^3)$ volume. Drawings produced by the 3-BENDS algorithm have at most three bends per edge. Drawings produced by the DLM algorithm have at most four bends per edge and an average of at most $2\frac{1}{3}$ bends per edge. For graphs with maximum degree at most five the DLM algorithm produces drawings with two bends per edge.

Ad-Hoc Algorithms: We have implemented the INCREMENTAL algorithm of Papakostas and Tollis [9] and the REDUCE-FORKS algorithm of Di Battista *et al.* [3]. The INCREMENTAL algorithm, which supports the on-line insertion of vertices, produces drawings with $O(n^3)$ volume and at most three bends per edge. No bounds on the volume or the number of bends have been established for the REDUCE-FORKS algorithm. Both of these algorithms involve a number of arbitrary decisions, thus the drawings produced may differ from one implementation to another.

Note that, due to time constraints, we have not implemented a number of algorithms in Wood [14] nor the DYNAMIC algorithm of Closson *et al.* [2] which produces 6-bend drawings with $O(n^2)$ volume, and supports the on-line insertion and deletion of vertices and edges.

3   Refinements

This section describes a number of techniques for the local refinement of 3-D orthogonal graph drawings. Each refinement, which can be applied to an arbitrary drawing, is aimed at improving at least one of the aesthetic criteria; namely the bounding box volume, number of bends, or the length of edges.

The MoveVertex refinement attempts to remove a bend from a given drawing by moving a vertex v to the first bend of an edge route incident to v. Applied to a drawing $D(G)$, MoveVertex(v,d) is applied for each vertex $v \in V(G)$ and direction $d \in \{\pm X, \pm Y, \pm Z\}$.

MoveVertex(vertex v, direction d)

Let $(v = a_0, a_1, a_2, \ldots)$ be the edge route, if any, using the d-port v. Let d' be the direction of the edge segment (a_1, a_2). If there is no route at v using the d' port, or if there is such an edge route $(v = b_0, b_1, b_2, \ldots)$, and $a_1 <_d b_2$, then as long as doing so does not create any new edge route intersections, move v to a_1 and reroute the edges incident to v as illustrated in Fig. 1.

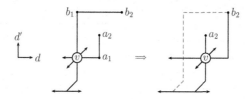

Fig. 1. The MoveVertex refinement.

The PermutePorts refinement attempts to remove bends from a given drawing by reassigning the ports at a given vertex to its incident edges. Applied to a drawing $D(G)$, PermutePorts is applied to each vertex $v \in V(G)$.

PermutePorts (vertex v)

Let $vw_1, vw_2, \ldots, vw_d$ be the edge routes incident to v, where $d = \deg(v)$. Split each edge route vw_i $1 \leq i \leq d$, into components vx_i and x_iw_i, where vx_i is the maximal subroute entirely contained in a grid-plane also containing v, as in Fig. 2(c). Let $S = \{vx_i : 1 \leq i \leq d\}$. Add to S any 0- or 1-bend edge route from v to x_i, $1 \leq i \leq d$, which does not intersect the remainder of the graph, as in Fig. 2(b). Find d pairwise non-intersecting edge routes in S, one for each vx_i, with the minimum total number of bends. (Such edge routes must exist since the original edge routes are in S.) Concatenate each of these edge routes with the appropriate x_iw, as in Fig. 2(c).

The RemoveSegment refinement aims to remove a bend by removing a given segment in an edge route. Applied to a drawing $D(G)$, RemoveSegment is applied for each edge $vw \in E(G)$ and pair of parallel segments (b_i, b_{i+1}) and (b_j, b_{j+1}) in vw.

RemoveSegment(edge vw, segment (b_i, b_{i+1}), segment (b_j, b_{j+1}))
Input Conditions: (b_i, b_{i+1}) and (b_j, b_{j+1}) are parallel segments of $vw = (b_0, b_1, \ldots, b_{k+1})$ such that $j > i$.

Let $\boldsymbol{x}$ be the vector $b_{i+1} - b_i$. Consider the edge route $(b_0, b_1, \ldots, b_i, b_{i+2} -$

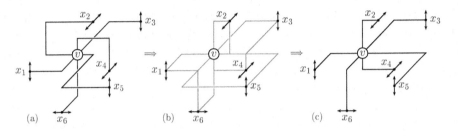

Fig. 2. Example of the PERMUTEPORTS refinement.

$\boldsymbol{x}, b_{i+3} - \boldsymbol{x}, \ldots, b_j - \boldsymbol{x}, b_{j+1}, \ldots, b_{k+1}$); that is, b_{i+1} is removed, and all grid points from b_{i+2} to b_j are translated by $-\boldsymbol{x}$. If this edge route does not intersect the remainder of the drawing, then replace vw by this route, and remove any self-intersections and redundant bends from vw.

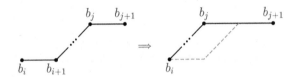

Fig. 3. The REMOVESEGMENT refinement.

The COMBINEPLANES refinement aims to reduce the volume of a given drawing by combining adjacent planes. Applied to a drawing $D(G)$, COMBINEPLANES is applied to each grid-plane in the drawing.

COMBINEPLANES(dimension I, integer x)

If the ($I = x$)-plane and the ($I = x + 1$)-plane can be combined without invalid edge or vertex intersections then do so.

The SHORTENU-TURN refinement, which is somewhat similar to the REMOVESEGMENT refinement, aims to reduce the length of a given edge by shortening parallel segments in the edge route. Applied to a drawing $D(G)$, SHORTENU-TURN($vw, (b_i, b_{i+1}), (b_j, b_{j+1})$) is applied for each edge $vw \in E(G)$ and pair of parallel segments (b_i, b_{i+1}) and (b_j, b_{j+1}) in vw satisfying the input conditions.

SHORTENU-TURN(edge vw, segment (b_i, b_{i+1}), segment (b_j, b_{j+1}))
Input Conditions: (b_i, b_{i+1}) and (b_j, b_{j+1}) are parallel segments of $vw = (b_0, b_1, \ldots, b_{k+1})$ such that $j > i$ and the vectors $\boldsymbol{p} = b_{i+1} - b_i$ and $\boldsymbol{q} = b_j - b_{j+1}$ are in the same direction.

Consider the route $(b_0, b_1, \ldots, b_i, b_{i+2} - \boldsymbol{x}, \ldots, b_j - \boldsymbol{x}, b_{j+1}, \ldots, b_{k+1})$ (with redundant bends also removed), where $\boldsymbol{x}$ is a vector pointing in the same direction as $\boldsymbol{p}$ with length in the range $\{0, 1, \ldots, |\boldsymbol{p}| + |\boldsymbol{q}| - 1\}$, Replace the edge route

vw with the shortest of such routes (and then with the least bends) that do not intersect the remainder of the drawing. Remove any self-intersections in vw.

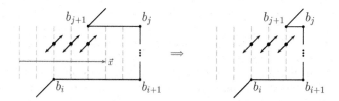

Fig. 4. An example of the SHORTENU-TURN refinement.

Our final refinement, called DRAWTREES&CHAINS, is different in nature to the previous refinements. It consists of two phases. In the first phase certain vertices are removed and certain paths are replaced by a single edge. In the second phase, which is designed to occur after other refinements have been applied, the removed vertices are reinserted, and the edge route for the single edges are replaced by paths (hopefully) with fewer edges. We now describe the first phase of the refinement.

REMOVETREES&CHAINS(drawing $D(G)$)

1. Repeatedly remove vertices with degree one (in the current drawing); that is, remove 'attached' trees from the graph.
2. We say a *chain* is a maximal path $(v_1v_2, v_2v_3, \ldots, v_{k-1}v_k)$ where every vertex v_i, except possibly for v_1 and v_k, has degree two. Replace each chain $(v_1v_2, v_2v_3, \ldots, v_{k-1}v_k)$ by the edge v_1v_k, as in Fig. 5(b).

The second phase of the refinement is as follows.

REDRAWTREES&CHAINS(drawing $D(G)$)

1. Insert each vertex v removed by REMOVETREES&CHAINS in the opposite order to their removal as follows. Let w be the adjacent vertex to v. Choose a free port at w, if any, such that vw can be routed as a unit-length segment. If such a port exists then route vw as a unit-length segment. Otherwise choose an arbitrary free port at w, insert a plane adjacent to w such that vw can be routed as a unit-length segment.
2. For each chain $(v_1v_2, v_2v_3, \ldots, v_{k-1}v_k)$ replaced by the edge route v_1v_k in the REMOVETREES&CHAINS phase, place the vertices $v_2, v_3, \ldots, v_{k-1}$ at the bends of the edge route v_1v_k as evenly spaced as possible. If there are more vertices than bends then position the remaining vertices arbitrarily on the edge route; see Fig. 5(d). If the edge route has less grid points than vertices then insert planes to accommodate the vertices.

We now describe how all the refinements are combined into one algorithm which we call 3D-REFINE. An important decision to be made is the order

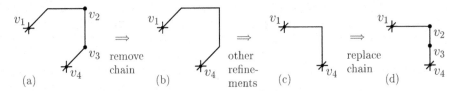

Fig. 5. An example of the REDRAWTREES&CHAINS refinement.

of application of the individual refinements. To determine an optimal order, we ran five different combinations of the refinements on 36 of the *Degree-4* drawings (see Sect. 4). These drawings were determined by the COMPACT-7, 3-BENDS, INCREMENTAL and REDUCE-FORKS algorithms applied to 9 graphs with $n = 15, 25, \ldots, 95$ vertices. The average percentage improvement in average bounding box side length differed by at most 4% across the five orderings of the refinements, and the average percentage improvement in average bends per edge differed by at most 5% across the five orderings of the refinements. We conclude that the ordering of the refinements is not 'significant'. The arbitrary ordering of the refinements which we chose is presented in the following algorithm.

> 3D-REFINE(drawing $D(G)$)
> **begin**
>   REMOVETREES&CHAINS($D(G)$);
>   **repeat**
>   MOVEVERTEX($D(G)$); PERMUTEPORTS($D(G)$);
>   REMOVESEGMENT($D(G)$); COMBINEPLANES($D(G)$);
>   SHORTENU-TURN($D(G)$);
>   **until** no changes;
>   REDRAWTREES&CHAINS($D(G)$);
> **end**

The only refinement which can possibly increase any of volume, total number of bends or total edge length is MOVEVERTEX, which can increase the total edge length. Therefore the algorithm 3D-REFINE will continue until the volume and the number of bends cannot be reduced any further, and will then only reduce the total edge length; hence the algorithm 3D-REFINE will terminate.

4   The Experiment

The first set of input drawings which we use are those generated by each of the seven algorithms applied to the same graphs used in the experiment of Di Battista *et al.* [3]. These randomly generated simple connected graphs have average degree four. The authors argue that in practical graph drawing applications it is unusual to have graphs with greater density than average degree four. We call these graphs and the set of drawings produced by the algorithms applied to these graphs the *Degree-4* graphs and drawings. There are 20 graphs with n vertices for each $n = 5, 6, \ldots, 100$. Hence there are 1920 graphs and 13440 drawings.

The second set of input drawings which we use consist of randomly generated simple connected graphs which are 'almost' 6-regular. Note that this is the first experiment measuring the performance of 3-D orthogonal graph drawing algorithms on high degree graphs. The generator used here continues to add random edges to the graph until there are no pairs of non-adjacent vertices both with degree less than six. Again we have 20 graphs for each $n = 5, 6, \ldots, 100$ vertices. We call these graphs and the set of drawings produces by the algorithms applied to these graphs the *6-Regular* graphs and drawings. One would expect that 6-regular graphs are the most difficult to draw, as all ports must be used.

Performance of Individual Refinements: We now report the effects of each refinement on the input drawings. For each of the *Degree-4* and *6-Regular* drawings, we repeatedly executed each individual refinement until it cannot be applied to any portion of the drawing. This process is guaranteed to terminate since each refinement, when applied, reduces the volume of the drawing. Table 1 reports the percentage improvement of the aesthetic criteria under each individual refinement and under the 3D-REFINE algorithm, averaged over the *Degree-4* drawings and over the *6-Regular* drawings.

Table 1. Average improvements for each refinement.

Refinement	Degree-4 Drawings					6-Regular Drawings				
	Vol.	Avg. Bends	Max. Bends	Avg. Edge Len.	Max. Edge Len.	Vol.	Avg. Bends	Max. Bends	Avg. Edge Len.	Max. Edge Len.
MOVEVERTEX	35	18	0.6	20	15	5.1	1.1	0.0	2.3	2.0
PERMUTEPORTS	4.1	4.6	1.0	1.3	1.1	1.5	3.6	0.6	0.5	0.5
REMOVESEGMENT	20	18	7.3	11	9.3	11	13	6.2	6.9	6.0
COMBINEPLANES	64	4.4	0.7	31	32	50	0.9	0.3	23	24
SHORTENU-TURN	36	11	2.0	20	20	30	6.7	1.8	18	18
3D-REFINE	**80**	**38**	**10**	**53**	**49**	**57**	**15**	**7.0**	**32**	**32**

The results in Table 1 show that the refinements are considerably more effective when applied to the *Degree-4* drawings than to the *6-Regular* drawings. This is expected since the *Degree-4* graphs have average degree four, whereas the *6-Regular* graphs have high degree, and therefore the free ports at each vertex allow the refinements to be applied more often. Perhaps a more surprising observation is that the COMBINEPLANES refinement is the most successful of the refinements in terms of reducing the volume and the length of edges.

Performance of Combined Refinements: We now report the improvement in the aesthetic criteria gained by applying the 3D-REFINE algorithm to the input drawings. We firstly consider the number of bends. For most of the algorithms the number of bends (both average and maximum) per edge is consistent across all values of n — for the REDUCE-FORKS algorithm we found that the number of bends gradually increases with n. We therefore describe our results in the following manner. Table 2 reports: (1) the average and maximum number of bends per edge in drawings produced by each algorithm (averaged over the *Degree-4* graphs and the *6-Regular* graphs); (2) the average and maximum number of bends per edge in drawings obtained after applying the 3D-REFINE algorithm to the drawings produced by each algorithm (averaged over the *Degree-4* graphs and the *6-Regular* graphs); (3) the percentage improvement in the average and maximum number of bends per edge gained by applying the 3D-REFINE algorithm to drawings produced by each algorithm (averaged over the *Degree-4* graphs and the *6-Regular* graphs). Note that (3) is not simply the percentage improvement in (2) from (1).

Table 2. Improvements in the number of bends before and after 3D-REFINE.

	Degree-4 Graphs		6-Regular Graphs	
	Avg. Bends	Max. Bends	Avg. Bends	Max. Bends
COMPACT-7	5.0→2.4 (52%)	7.0→5.7 (19%)	4.8→3.3 (32%)	7.0→6.5 (7.0%)
COMPACT-6	4.3→2.3 (46%)	6.0→5.4 (10%)	4.2→3.3 (22%)	6.0→5.9 (2.0%)
COMPACT-5	3.6→2.3 (36%)	5.0→4.9 (1.3%)	3.5→3.1 (9.8%)	5.0→5.0 (0.2%)
3-BENDS	2.7→1.8 (31%)	3.0→3.0 (0.2%)	2.6→2.6 (1.2%)	3.0→3.0 (0.1%)
DLM	2.1→1.5 (25%)	3.0→3.0 (1.0%)	2.2→2.1 (4.0%)	3.7→3.7 (0.0%)
INCREMENTAL	2.1→1.6 (27%)	3.0→3.0 (0.4%)	2.2→2.1 (3.1%)	3.0→3.0 (0.0%)
REDUCE-FORKS	4.2→2.0 (48%)	10→5.7 (41%)	4.8→3.1 (33%)	11→6.3 (37%)
Average	3.4→2.0 (**38%**)	5.3→4.4 (**10%**)	3.5→2.8 (**15%**)	5.5→4.8 (**7.0%**)

The results in Table 2 show that the 3D-REFINE algorithm considerably reduces the average number of bends in all of the *Degree-4* drawings and to a lesser extent in the *6-Regular* drawings. There is a significant improvement in the maximum number of bends per edge only in the drawings with relatively many bends prior to the application of the refinement algorithm.

Fig. 6 and Fig. 7 presents, for each algorithm, the average side length of the bounding box and the average edge length before and after the application of the 3D-REFINE algorithm (averaged over the *Degree-4* and the *6-Regular* graphs, for each value of n). We chart the data for the average side length of the bounding box rather than the bounding box volume for ease of presentation — of course the average side length is the cube root of the volume. The results for the maximum edge length have been omitted; these closely resemble the analogous results for the average edge length. For each algorithm, Fig. 6 and Fig. 7 also show (next to the algorithm's name) the average percentage improvement over all graphs. This provides a measure of the susceptibility of the drawings produced by that algorithm to improvement by the 3D-REFINE algorithm.

As was the case with the number of bends, the results in Fig. 6 and Fig. 7 show that the 3D-REFINE algorithm considerably reduces the bounding box volume and the length of edges in all of the *Degree-4* drawings and to a lesser

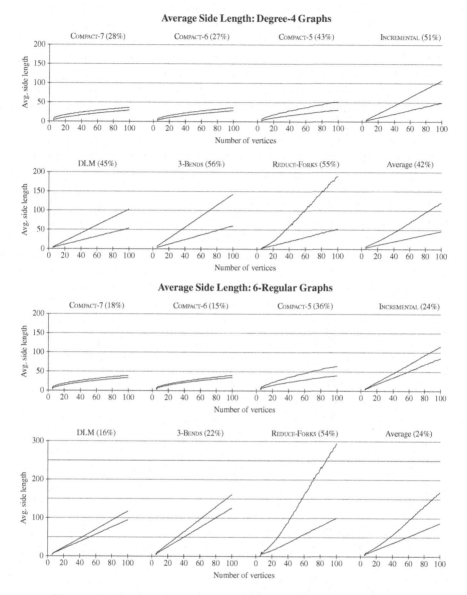

Fig. 6. Average bounding box side length before and after 3D-REFINE.

extent in the *6-Regular* drawings. Note that the average improvement of 42% for the average bounding box side length obtained for the *Degree-4* drawings corresponds to an improvement of 80% in the bounding box volume.

Comparison of Algorithms: We now compare the performance of the drawing algorithms, firstly without refinements and then following the application of the 3D-REFINE algorithm; see Table 2, Fig. 6 and Fig. 7. Our results generally

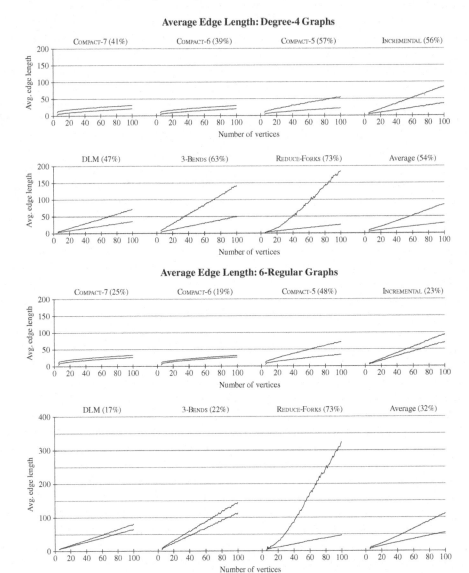

Fig. 7. Average edge length before and after 3D-REFINE.

confirm the worst case upper bounds established for each algorithm, and confirm the results in [3] for the *Degree-4* graphs. The performance of the algorithms on the *6-Regular* graphs continue the trends observed for the *Degree-4* graphs with one marked exception. The REDUCE-FORKS algorithm performs noticeably worse on the *6-Regular* graphs relative to the other algorithms compared with the *Degree-4* graphs.

In terms of volume and edge lengths the best algorithms (without refinements) are COMPACT7, COMPACT6 and COMPACT5. Note that the worst-case volume bounds for COMPACT7 and COMPACT6 are $O(n^{3/2})$ and $O(n^2)$, respectively. That in practice the difference in their performance is negligible is due to the enhanced colouring method discussed in Sect. 2. DLM and INCREMENTAL are the next best performing algorithms, followed by 3-BENDS and REDUCE-FORKS. A similar pattern emerges when comparing the algorithms after refinements, except that REDUCE-FORKS performs almost as well as the COMPACT algorithms; that is, the drawings produced by the REDUCE-FORKS algorithm are highly susceptible to improvements by 3D-REFINE.

5 Conclusion

In this paper we have described a number of post-processing techniques for the refinement of 3-D orthogonal graph drawings. Our main algorithm makes substantial improvements to all of the aesthetic criteria measured, especially when applied to relatively low degree graphs. This experiment has contributed to the ongoing research efforts to make 3-D orthogonal graph drawings more appropriate for visualisation purposes. An important future step toward this goal is the development of efficient data structures for the implementation of algorithms and refinements for 3-D orthogonal graph drawing.

Acknowledgements. Thanks to Chi Nguyen for technical support, and to Maurizio Patrignani for kindly supplying the code from the 3DCube implementations of the INCREMENTAL and REDUCE-FORKS algorithms.

References

1. T. Biedl and T. Chan. Cross-coloring: improving the technique by Kolmogorov and Barzdin. Technical Report CS-2000-13, University of Waterloo, Canada, 2000.
2. M. Closson, S. Gartshore, J. Johansen, and S. K. Wismath. Fully dynamic 3-dimensional orthogonal graph drawing. In Kratochvil [8], pages 49–58.
3. G. Di Battista, M. Patrignani, and F. Vargiu. A split&push approach to 3D orthogonal drawing. In Whitesides [12], pages 87–101.
4. P. Eades, C. Stirk, and S. Whitesides. The techniques of Komolgorov and Bardzin for three dimensional orthogonal graph drawings. *Inform. Proc. Lett.*, 60(2):97–103, 1996.
5. P. Eades, A. Symvonis, and S. Whitesides. Three dimensional orthogonal graph drawing algorithms. *Discrete Applied Math.*, 103:55–87, 2000.
6. U. Fößmeier, C. Heß, and M. Kaufmann. On improving orthogonal drawings: the 4M-algorithm. In Whitesides [12], pages 125–137.
7. A. N. Kolmogorov and Ya. M. Barzdin. On the realization of nets in 3-dimensional space. *Problems in Cybernetics*, 8:261–268, March 1967.
8. J. Kratochvil, editor. *Proc. Graph Drawing: 7th International Symp. (GD'99)*, volume 1731 of *Lecture Notes in Comput. Sci.*, Springer, 1999.
9. A. Papakostas and I. G. Tollis. Algorithms for incremental orthogonal graph drawing in three dimensions. *J. Graph Algorithms Appl.*, 3(4):81–115, 1999.

10. M. Patrignani and F. Vargiu. 3DCube: a tool for three dimensional graph drawing. In G. Di Battista, editor, *Proc. Graph Drawing: 5th International Symp. (GD'97)*, volume 1353 of *Lecture Notes in Comput. Sci.*, pages 284–290, Springer, 1998.

11. J. M. Six, K. G. Kakoulis, and I. G. Tollis. Refinement of orthogonal graph drawings. In Whitesides [12], pages 302–315.

12. S. Whitesides, editor. *Proc. Graph Drawing: 6th International Symp. (GD'98)*, volume 1547 of *Lecture Notes in Comput. Sci.*, Springer, 1998.

13. D. R. Wood. An algorithm for three-dimensional orthogonal graph drawing. In Whitesides [12], pages 332–346.

14. D. R. Wood. *Three-Dimensional Orthogonal Graph Drawing*. PhD thesis, School of Computer Science and Software Engineering, Monash University, Australia, 2000.

ω-Searchlight Obedient Graph Drawings

Gill Barequet

The Technion—Israel Institute of Technology, Haifa 32000, Israel,
`barequet@cs.technion.ac.il`,
WWW home page: `http://www.cs.technion.ac.il/~barequet`

Abstract. A drawing of a graph in the plane is *ω-searchlight obedient* if every vertex of the graph is located on the centerline of some strip of width ω, which does not contain any other vertex of the graph. We estimate the maximum possible value $\omega(n)$ of an ω-searchlight obedient drawing of a graph with n vertices, which is contained in the unit square. We show a lower bound and an upper bound on $\omega(n)$, namely, $\omega(n) = \Omega(\log n/n)$ and $\omega(n) = O(1/n^{4/7-\varepsilon})$, for an arbitrarily small $\varepsilon > 0$. Any improvement for either bound will also carry on to the famous Heilbronn's triangle problem.

Keywords: Geometric optimization, Heilbronn's triangle problem.

1 Introduction

In this paper we represent a graph-drawing problem as an optimization problem in combinatorial geometry, and establish a lower and an upper bound for the solution of this problem. Such "Erdős-type" problems attracted much attention throughout the last century. The interested reader is referred to the rich literature on combinatorial geometry; see, e.g., [2,3,6]. Specifically, we would like to draw a graph within the unit square, such that each vertex of the graph is covered by a strip associated with it, and no vertex is covered by a strip associated with another vertex. Our goal is to maximize the width of the strips.

A *searchlight* is a strip in $\mathbb{R}^2$, that is characterized by three parameters:

1. A point, the *source* of the searchlight, which lies on the centerline of the strip.
2. A direction which defines the orientation of the strip.
3. The width of the strip.

We focus on searchlights whose source points are located in the unit square. A valid collection of searchlights has the property that no searchlight source point is covered by any other searchlight. We are interested in the following problem:

Given a valid collection of n searchlights of width ω in the unit square, what is the maximum possible width $\omega(n)$ of the searchlights?

J. Marks (Ed.): GD 2000, LNCS 1984, pp. 321–327, 2001.

The quantity $\omega(n)$ is the maximum possible value of ω for an ω-*searchlight obedient* drawing within the unit square of a graph with n vertices. It is possible to superimpose such a graph with n strips of width ω, such that every strip contains one vertex on its centerline, but it does not contain any other vertex of the graph; see Figure 1. The edges of the graph in the figure are drawn entirely

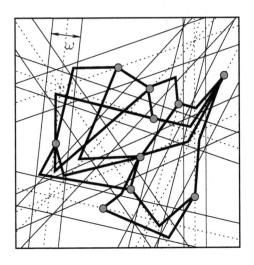

Fig. 1. An ω-searchlight obedient graph drawing

within the searchlights (bounded by the unit square), but requiring this does not make the problem harder. Any ω-searchlight obedient drawing can be modified so as to make all the clipped searchlights connected. This is done by rotating sufficiently the searchlights around their source points; before a strip hits another vertex, it must have a nonempty intersection (within the unit square) with the strip associated with the vertex. Such a drawing solves a routing problem in electronic chip design, where the vertices are ports and the edges are wires of the chip, and where all the wires pass through long skinny containers, where every container is associated with one port. (This is somewhat similar, but unrelated, to the notion of a *joint box* used in some graph-drawing algorithms; see, e.g., [1].)

We derive lower and upper bounds for the searchlights problem from its relation to the famous Heilbronn's triangle problem [7]:

> Given n points in the unit square, what is $\mathcal{H}(n)$, the maximum possible area of the *smallest* triangle defined by some three of these points?

Heilbronn posed the triangle problem about 50 years ago, and since then it intrigued the imagination of some of the best mathematicians. Yet there is still a large gap between the best currently-known lower and upper bounds for $\mathcal{H}(n)$,

$\Omega(\log n/n^2)$ [5] and $O(1/n^{8/7-\varepsilon})$ (for any $\varepsilon > 0$) [4].[1] A comprehensive survey of the history of this problem (excluding the results of Komlós et al.) is given by Roth in [9]. The relation between the triangle problem and the searchlights problem imposes a similar gap between the bounds shown in this paper for the latter problem. Improving either the lower or upper bound for one of the problems will also improve the respective bound for the other problem.

The problem discussed in this paper was also motivated by the following. Suppose that the actual value of $\mathcal{H}(n)$ is Δ. Consider the set S that realizes Δ. Every pair of points $p, q \in S$ defines a strip of width $4\Delta/d(p,q)$ which contains no other points of S, where $d(p,q)$ is the distance between p and q. Thus, the width of the "forbidden strip" defined by p, q is inversely proportional to $d(p,q)$. Roth showed in [8] that $\mathcal{H}(n) = O(\frac{1}{n^{1.117\ldots}})$. In this work he made the distinction between 'bad' and 'good' strips according to a relation between their width to the number of points of S they contain. We would like to demonstrate here that the strip-width effect (caused by the distances between points) on Heilbronn's triangle problem is minor. We do that by showing a tight relation between Heilbronn's problem to the searchlights problem, where all the searchlights are strips of the same width.

2   Easy Bounds on $\omega(n)$

We first establish easy lower and upper bounds on $\omega(n)$.

2.1   Lower Bound

Theorem 1. $\omega(n) = \Omega(1/n)$.

Proof. We establish the lower bound from a simple example. Put n vertical searchlights of width $1/n$ so that their interiors do not intersect. The source points may be located anywhere along the searchlight centerlines, and obviously no source is covered by any other searchlight.[2]

2.2   Upper Bound

Theorem 2. $\omega(n) = O(1/\sqrt{n})$.

Proof. Let $k = \lfloor \sqrt{n-1} \rfloor$. Partition the unit square into a full grid of $k \times k$ small squares each of sidelength $1/k$. Now locate n searchlights in the unit square. Since the grid contains at most $(n-1)$ small squares, there must exist one such square that contains two searchlight source points. The distance between these two sources is at most $\sqrt{2}/k$, therefore the searchlight width cannot exceed $2\sqrt{2}/k = 2\sqrt{2}/\lfloor \sqrt{n-1} \rfloor = O(1/\sqrt{n})$.

---

[1] Actually, Komlós et al. showed in this work that $\mathcal{H}(n) = O(e^{c\sqrt{\log n}}/n^{8/7})$ for some constant $c > 0$.

[2] In fact, it is rather easy, as an anonymous referee noted, to improve (increase) the constant of proportionality in this bound.

3   Improved Bounds on $\omega(n)$

In this section we improve the bounds on $\omega(n)$ obtained in the previous section. We begin with stating the relation between Heilbronn's triangle problem and the searchlights problem.

Theorem 3. *Assume that $c_1 f_1(n) \le \mathcal{H}(n) \le c_2 f_2(n)$, for some monotonically-growing functions $f_1(n), f_2(n)$ and constants $c_1, c_2 > 0$. Then $c_3 f_1(\sqrt{2n}) \le \omega(n) \le c_4 \sqrt{f_2(2n)}$, for some constants $c_3, c_4 > 0$.*

Proof. We first show the lower bound on $\omega(n)$. We put N points in the unit square such that the area of the smallest triangle defined by some three of these points assumes its maximum $\mathcal{H}(N)$. (See Figure 2(a); Heilbronn's points appear

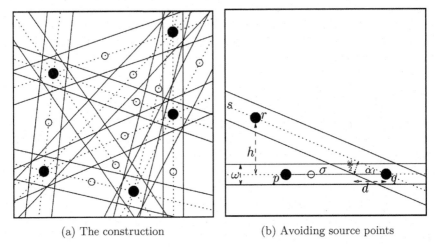

(a) The construction   (b) Avoiding source points

Fig. 2. Connecting Heilbronn's points ($\bullet$) with searchlights

as black circles.) We then draw the $\binom{N}{2}$ lines defined by the N points. These will be searchlight centerlines; the locations of the source points (shown as white circles in Figure 2(a)) will be defined later. Every such line is now split by the other lines into $O(N^2)$ segments. We select one of these segments, $\sigma = pq$ (p and q are the endpoints of the segment σ), and locate on it a source point of a searchlight of width ω (see Figure 2(b)). (We repeat this for all the $\Theta(N^2)$ lines.) For ease of exposition we assume without loss of generality that σ is horizontal.

Refer to the leftmost searchlight s emanating upwards from q, the right endpoint of σ. This is the strip that overlaps the most of the right side of σ among all searchlights whose centerline passes through q. (The rightmost searchlight emanating upwards from q can also overlap a significant portion of the right side of σ, in which case we apply the same analysis for searchlights that emanate downwards from σ.) Let r be the other point which, together with q, defined the strip s. Without loss of generality we assume that r is above σ. By the assumption, the area of the triangle defined by the points p, q, r is at least $c_1 f_1(N)$. Thus, the altitude h of r relative to σ satisfies $|\sigma| h/2 \geq c_1 f_1(N)$, that is,

$$h \geq 2c_1 f_1(N)/|\sigma|. \tag{1}$$

We now compute d, the amount of overlap between s and σ. Obviously $|qr| \leq \sqrt{2}$ and $\sin \alpha = h/|qr| = w/(2d)$. Thus

$$d = \frac{w|qr|}{2h} \leq \frac{\sqrt{2}w}{2h}. \tag{2}$$

By substituting Equation (1) in Equation (2) we obtain

$$d \leq \frac{\sqrt{2}w|\sigma|}{4c_1 f_1(N)}. \tag{3}$$

A similar analysis shows the same upper bound on the length of the portion of σ overlapped by the rightmost (or leftmost) searchlight emanating upwards from p. Therefore, the total length of the two overlapped portions is at most $\sqrt{2}w|\sigma|/(2c_1 f_1(N))$. Hence, in order to preclude a searchlight source point on σ we must have $\sqrt{2}w|\sigma|/(2c_1 f_1(N)) \geq |\sigma|$, that is,

$$w \geq \sqrt{2}c_1 f_1(N). \tag{4}$$

Finally, we set $N = \sqrt{2n}$ in order to have $n(1 + o(1))$ searchlights, and obtain $w \geq \sqrt{2}c_1 f_1(\sqrt{2n})$. Setting $c_3 = \sqrt{2}c_1$ completes the argument.

A reduction in the opposite direction shows the upper bound on $w(n)$. Refer to Figure 3. Here we put n searchlights of the maximum possible width w such that their source points are located in the unit square. The n centerlines of the searchlights intersect inside the unit square in at most $\binom{n}{2}$ points. On each centerline we position two "Heilbronn points" at the two intersection points which define the segment on which the source point of the respective searchlight lies. (For this purpose we also consider the intersections of the searchlight centerlines with the unit square.) In total we thus mark $N = 2n$ points.[3]

Refer to the smallest-area triangle defined by some triple p, q, r of these N points (see again Figure 2(b)). Here r is one of the points marked on the searchlight s whose intersection with the searchlight containing σ defines q. By the assumption, the area of this triangle is at most $c_2 f_2(N)$. Fix again the segment σ whose endpoints are p and q. The altitude h of r relative to σ must satisfy $|\sigma| h/2 \leq c_2 f_2(N)$, that is,

$$h \leq \frac{2c_2 f_2(N)}{|\sigma|}. \tag{5}$$

---

[3] Figure 3 is misleading in the sense that this example contains three or more collinear Heilbronn points, which obviously define a triangle of area 0. Note, however, that we use here an *upper* bound on the area of Heilbronn's triangles.

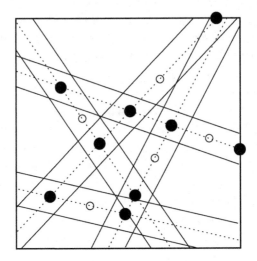

Fig. 3. Locating Heilbronn points on searchlights

We now compute the amount of overlap between s and σ. Again, $\sin \alpha = h/|qr| = \omega/(2d)$. This time we use the fact that $|qr| \geq \omega/2$ (otherwise the source point of the searchlight on σ would be covered by the searchlight centered either at pr or at qr).[4] Thus,

$$d = \frac{\omega|qr|}{2h} \geq \frac{\omega^2}{4h}. \tag{6}$$

By substituting Equation (5) in Equation (6) we obtain

$$d \geq \frac{\omega^2|\sigma|}{8c_2 f_2(N)}. \tag{7}$$

A similar analysis applies for the left side of σ. Therefore, the total length of the two overlapped portions of σ is at least $\omega^2|\sigma|/(4c_2 f_2(N))$. On the other hand, this total is at most $|\sigma|$ (since σ contains a searchlight source point), so we must have $\omega^2|\sigma|/(4c_2 f_2(N)) \leq |\sigma|$, that is,

$$\omega \leq 2\sqrt{c_2}\sqrt{f_2(N)} = 2\sqrt{c_2}\sqrt{f_2(2n)}. \tag{8}$$

Setting $c_4 = 2\sqrt{c_2}$ completes the argument.

Next we quote the best currently-known bounds for Heilbronn's triangle problem:

Theorem 4. [5] $\mathcal{H}(n) = \Omega(\log n/n^2)$.

---

[4] Note the difference between this lower bound on $|qr|$ and the upper bound of $\sqrt{2}$ used above.

Theorem 5. [4] $\mathcal{H}(n) = O(1/n^{8/7-\varepsilon})$ *for any* $\varepsilon > 0$.

Finally we combine Theorems 3, 4, and 5 to obtain our main result:

Theorem 6. $\omega(n) = \Omega(\log n/n)$ *and* $\omega(n) = O(1/n^{4/7-\varepsilon})$ *for any* $\varepsilon > 0$. $\square$

We also have the opposite dependence:

Theorem 7. *Assume that* $d_1 g_1(n) < \omega(n) < d_2 g_2(n)$, *for some monotonically-growing functions* $g_1(n), g_2(n)$ *and constants* $d_1, d_2 > 0$. *Then* $d_3 g_1^2(n/2) < \mathcal{H}(n) < d_4 g_2(n^2/2)$, *for some constants* $d_3, d_4 > 0$.

Proof. First we show that there exists a constant d_3 for which $\mathcal{H}(n) > d_3 g_1^2(n/2)$. Assume to the contrary that no such constant exists, that is, for any $d_3 > 0$ and for sufficiently-large values of n, we have $\mathcal{H}(n) \leq d_3 g_1^2(n/2)$. Then by Theorem 3 there exists a constant $d_3' > 0$ for which $\omega(n) \leq d_3' \sqrt{g_1^2((2n)/2)} = d_3' g_1(n)$, which is a contradiction.

Similarly we show that there exists a constant d_4 for which $\mathcal{H}(n) < d_4 g_2 (n^2/2)$. Assume to the contrary that no such constant exists, that if, for any $d_4 > 0$ and for sufficiently-large values of n, we have $\mathcal{H}(n) \geq d_4 g_2(n^2/2)$. Then by Theorem 3 there exists a constant $d_4' > 0$ for which $\omega(n) \geq d_4' g_2(\sqrt{2(n^2/2)}) = d_4' g_2(n)$, which is also a contradiction.

Theorem 7 implies that any improvement of the lower or the upper bound for the searchlights problem will also carry on to Heilbronn's triangle problem, which, as mentioned in the introduction, has puzzled many mathematicians throughout the last half of century.

References

1. Cheng, C.C., Duncan, C.A., Goodrich, M.T., Kobourov, S.G.: Drawing planar graphs with circular arcs. Proc. 7th Graph Darwing Conference, Prague, Czech, 1999, 117–126. Lecture Notes in Computer Science 1731, Springer-Verlag
2. Goodman, J.E., Lutwak, E., Malkevitch, J., Pollack, R. (eds.): Discrete Geometry and Convexity. Ann. New York Acad. Sci. (440), 1985
3. Hadwiger, H., Debrunner, H. (translation by V. Klee): Combinatorial Geometry in the Plane. Holt, Rinehart, and Winston, New York, 1964
4. Komlós, J., Pintz, J., Szemerédi, E.: On Heilbronn's triangle problem. J. London Mathematical Society (2) **24** (1981) 385–396
5. Komlós, J., Pintz, J., Szemerédi, E.: A lower bound for Heilbronn's problem. J. London Mathematical Society (2) **25** (1982) 13–24
6. Moser W., Pach, J.: Research problems in Discrete Geometry. Mineographed Notes, 1985
7. Roth, K.F.: On a problem of Heilbronn. Proc. London Mathematical Society **26** (1951) 198–204
8. Roth, K.F.: On a problem of Heilbronn, III. Proc. London Mathematical Society (3) **25** (1972) 543–549
9. Roth, K.F.: Developments in Heilbronn's triangle problem. Advances in Mathematics **22** (1976) 364–385

Unavoidable Configurations in Complete Topological Graphs

János Pach[1] and Géza Tóth[2]

[1] Courant Institute, NYU and Hungarian Academy of Sciences
[2] Massachusetts Institute of Technology and Hungarian Academy of Sciences

Abstract. A *topological graph* is a graph drawn in the plane so that its vertices are represented by points, and its edges are represented by Jordan curves connecting the corresponding points, with the property that any two curves have at most one point in common. We define two canonical classes of topological complete graphs, and prove that every topological complete graph with n vertices has a canonical subgraph of size at least $c \log \log n$, which belongs to one of these classes. We also show that every complete topological graph with n vertices has a non-crossing subgraph isomorphic to any fixed tree with at most $c \log^{1/6} n$ vertices.

1   Introduction, Results

A *topological graph* G is a graph drawn in the plane by Jordan curves, any two of which have at most one point in common. That is, it is defined as a pair $\{V(G), E(G)\}$, where $V(G)$ is a set of points in the plane and $E(G)$ is a set of simple continuous arcs connecting them so that they satisfy the following conditions:

1. no arc passes through any other element of $V(G)$ different from its endpoints;
2. any two arcs have at most one point in common, which is either a common endpoint or a proper crossing.

$V(G)$ and $E(G)$ are the *vertex-set* and *edge set* of G, respectively. We say that H is a *(topological) subgraph* of G if $V(H) \subseteq V(G)$ and $E(H) \subseteq E(G)$. Two topological graphs, G and H, are called *weakly isomorphic* if there is an incidence preserving one-to-one correspondence between $\{V(G), E(G)\}$ and $\{V(H), E(H)\}$ such that two edges of G intersect if and only if the corresponding edges of H do (see [C81]). If all edges of a topological graph are straight-line segments, then it is called a *geometric graph*. A geometric graph, whose vertices are in convex position, is called *convex*. Obviously, any two complete convex geometric graphs with m vertices are weakly isomorphic to each other, and to the convex geometric graph C_m, whose edge set consists of all sides and chords of a regular m-gon. (See Fig. 1.)

J. Marks (Ed.): GD 2000, LNCS 1984, pp. 328–337, 2001.

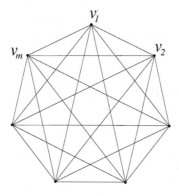

Figure 1: *The convex geometric graph C_m.*

The fairly extensive literature on topological graphs focuses on very few special questions, and there is no standard terminology. For topological graphs, Erdős and Guy [EG73] (see also [AR88]) use the term "good drawings," while Gronau, Harborth, Mengersen, and Thürmann [GH90], [HM74], [HM90], [HT94] simply call them "drawings." For a complete topological graph, Ringel [R64] and Mengersen [M78] use the term "immersion." The most popular problems in this field are Turán's Brick Factory Problem [T77] (Zarankiewicz's Conjecture [G69] and other problems about *crossing numbers*, i.e., about the *minimum* number of crossings in certain drawings of a graph [PT98]) and Conway's Thrackle Conjecture [W71], [LPS97], [CN00] (and other problems about the *maximum* number of crossings in certain drawings of a graph [HM92]).

The systematic study of geometric graphs was initiated by Erdős, Avital–Hanani [AH66], Kupitz [K79], and Perles. (See [P99] and [PA95], Chapter 14, for the most recent surveys on the subject.) It is not hard to see that every *complete geometric graph K_n* of n vertices has a non-crossing subgraph isomorphic to any triangulation of a cycle of length n (cf. [GMPP91]). Consequently, K_n has a non-crossing subtree isomorphic to any fixed tree of n vertices. In particular, K_n has a non-crossing path of n vertices and a non-crossing matching of size $\lfloor n/2 \rfloor$.

On the other hand, it is known that K_n has at least constant times $\sqrt{n}$ pairwise crossing edges.

Our aim is to establish analogous results for topological graphs.

Theorem 1. *Every topological complete graph of n vertices has a non-crossing subgraph isomorphic to any fixed tree T with at most $c\log^{1/6} n$ vertices. In particular, it contains a non-crossing path with at least $c\log^{1/6} n$ vertices.*

According to a wellknown theorem of Erdős and Szekeres [ES35],[ES60], any set of n points in general position in the plane contains a subset with at least $c\log n$ elements which form the vertex set of a convex polygon. (Throughout this note, the letter c appearing in different assertions denote unrelated positive constants. The best known bound in the last statement is due to Tóth and Valtr [TV98].) The Erdős-Szekeres Theorem can be reformulated, as follows.

Erdős-Szekeres Theorem. *Every complete geometric graph with n vertices has a complete geometric subgraph, weakly isomorphic to a convex complete graph C_m with $m \geq c \log n$ vertices.*

The situation is more complicated for *topological* graphs. In their study of topological complete graphs with m vertices and with the *maximum* possible number, $\binom{m}{4}$, of edge crossings, Harborth and Mengersen [HM92] found a drawing which contains no subgraph weakly isomorphic to C_5. We call this drawing, depicted in Figure 2, *twisted*, and denote it by T_m.

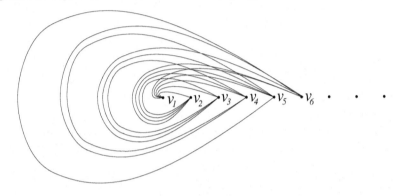

Figure 2: *The twisted drawing T_m.*

We show that one cannot avoid *both* C_m and T_m in a sufficiently large complete topological graph.

Theorem 2. *Every complete topological graph with n vertices has a complete topological subgraph with $m \geq c \log \log n$ vertices, which is weakly isomorphic either to a convex complete graph C_m or to a twisted complete graph T_m.*

2   An Erdős–Szekeres-Type Theorem

Before we turn to the proof of Theorem 2, we rephrase the definitions of convex and twisted complete topological graphs.

Definition 2.1. Let K_m be a complete topological graph on m vertices. If there is an enumeration of the vertices, $\{u_1, u_2, \ldots, u_m\}$, such that

(i) two edges, $u_i u_j$ $(i < j)$ and $u_k u_l$ $(k < l)$, cross each other if and only if $i < k < j < l$ or $k < i < l < j$, then K_m is called *convex*;

(ii) two edges, $u_i u_j$ $(i < j)$ and $u_k u_l$ $(k < l)$, cross each other if and only if $i < k < l < j$ or $k < i < j < l$, then K_m is called *twisted*.

Let K be a fixed complete topological graph with $n + 1$ vertices. The edges of K divide the plane into several cells, precisely one of which is unbounded. Without loss of generality, we can assume that there is a vertex $v_0 \in V(K)$ on the boundary of the unbounded cell. Otherwise, we can apply a stereographic projection to transform K into a drawing on a sphere, and then, by another projection, we can turn it into a topological graph weakly isomorphic to K, which satisfies the required property.

Consider all edges emanating from v_0, and denote their other endpoints by $v_1, v_2, \ldots, v_n$, in clockwise order.

Color the triples $v_i v_j v_k$, $1 \leq i < j < k \leq n$ with eight different colors, according to the following rules. Each color is represented by a zero-one sequence abc of length 3. For any $i < j < k$,

1. set $a = 0$ if the edges $v_i v_j, v_0 v_k \in E(K)$ do not cross, and let $a = 1$ otherwise;
2. set $b = 0$ if the edges $v_i v_k, v_0 v_j \in E(K)$ do not cross, and let $b = 1$ otherwise;
3. set $c = 0$ if the edges $v_j v_k, v_0 v_i \in E(K)$ do not cross, and let $c = 1$ otherwise.

It is easy to see that the complete topological subgraph of K induced by the vertices v_0, v_i, v_j, v_k (as any other complete topological graph with 4 vertices) has at most one pair of crossing edges. Therefore, we have

Claim 2.2. None of the colors 011, 101, 110, or 111 can occur.

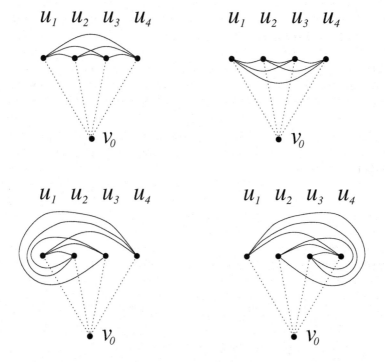

Figure 3: *All triples are of type* 000, 010, 001, *and* 100, *respectively.*

Proof of Theorem 2: By Ramsey's Theorem, there is an m-element subsequence, $(u_1, u_2, \ldots, u_m) \subseteq (v_1, v_2, \ldots, v_n)$, $m \geq c \log \log n$ such that all triples $u_i u_j u_k$ are of the same color (c is a positive constant).

Suppose first that all triples $u_i u_j u_k$ are of color 000 (see Fig. 3). Then, two edges, $u_i u_j$ $(i < j)$ and $u_k u_l$ $(k < l)$, cross each other if and only if $i < k < j < l$ or $k < i < l < j$. That is, $u_1, u_2, \ldots, u_m$ induce a *convex* complete topological subgraph in K.

We obtain exactly the same crossing pattern between the edges induced by $\{u_1, u_2, \ldots, u_m\}$ if all triples are of color 010 (see Fig. 3).

Suppose next that all triples in $\{u_1, u_2, \ldots, u_m\}$ are colored 001. Then two edges, $u_i u_j$ $(i < j)$ and $u_k u_l$ $(k < l)$, cross each other if and only if $i < k < l < j$ or $k < i < j < l$. In this case, $u_1, u_2, \ldots, u_m$ induce a *twisted* complete topological subgraph.

Finally, the case when all triples in $\{u_1, u_2, \ldots, u_m\}$ are of color 100, is isomorphic to the previous one, under a reflection reversing the orientation of the plane (and hence the numbering of the vertices v_i $(1 \le i \le n)$). □

It is very easy to check that both C_m and T_m contain non-crossing copies of every tree with m vertices. Thus, a weaker version of Theorem 1, with $c \log \log n$ instead of $c \log^{1/6} n$, readily follows from Theorem 2. In the next section, we apply a somewhat more delicate argument to improve this bound.

3   Proof of Theorem 1

Let G be a topological complete graph with an $(n+1)$-element vertex set V. Use the same numbering, $v_0, v_1, \ldots, v_n$, of the vertices as in the previous section. For any $0 < i < j$, we say that v_i *precedes* v_j (in notation, $v_i \prec v_j$). As before, color the triples $v_i v_j v_k$ $(1 \le i < j < k \le n)$ with *four* colors, 000, 100, 010, and 001.

Claim 3.1. There exists an m-element subset $U := \{u_1, u_2, \ldots, u_m\} \subset \{v_1, v_2, \ldots, v_n\}$, $m \ge \sqrt{\log_4 (n+1)}$ such that the triples $u_i u_j u_k$ and $u_i u_j u_l$ have the same color for any $i < j < k < l$.

Proof: The construction is recursive. Let $U_2 := \{v_1, v_2\}$ and $V_2 := V \setminus \{v_1, v_2\}$. Suppose that, for some $2 \le p < m$, we have already found two subsets $U_p = \{u_1, u_2, \ldots, u_p\}$ and $V_p \subset V$ with the properties

1. $u_1 \prec u_2 \prec \cdots \prec u_p$,
2. every element of U_p precedes all elements of V_p,
3. $|V_p| \ge \frac{|V_{p-1}| - 1}{4^p}$.

Let u_{p+1} be the smallest element of V_p with respect to the ordering '$\prec$.' Since we used *four* colors for coloring the triples, there is a subset $W \subset V_p \setminus \{u_{p+1}\}$ with $|W| \ge (|V_p| - 1)/4^p$ such that, for each $1 \le i \le p$, all triples $u_i u_{p+1} w$ $(w \in W)$ have the same color. Let $U_{p+1} := U_p \cup \{u_{p+1}\}$ and $V_{p+1} := W$. An easy computation shows that this procedure can be repeated at least $\lceil \sqrt{\log_4 (n+1)} \rceil$ times. □

Define the *type of an edge* $u_i u_j$ $(i < j < m)$ as the color of a triple $u_i u_j u_k$ for any $k > j$. The type of $u_i u_m$ can be defined arbitrarily.

Let $G(100)$ and $G(001)$ denote the topological subgraphs of G consisting of all edges of type 100 and 001, resp., whose both endpoints belong to $U = \{u_1, u_2, \ldots, u_m\}$. The topological subgraph consisting of all other edges of G induced by U (of types 000 and 010) is denoted by G'.

Claim 3.2. Let $i < j < k < m$.

    (i) If $u_i u_j$ and $u_j u_k$ belong to $G(100)$, then so does $u_i u_k$.

    (ii) If $u_i u_j$ and $u_i u_k$ belong to $G(001)$, then so does $u_j u_k$.

Proof: If $u_i u_j$ is of type 100, it must cross both $v_0 u_k$ and $v_0 u_m$. If the type of $u_j u_k$ is also 100, it must cross $v_0 u_m$, too. Using the assumption that two edges that share an endpoint cannot have any other point in common, we obtain that $u_i u_k$ must cross $v_0 u_m$, which implies that its type is also 100 (see Fig. 4). This proves part (i). Part (ii) can be established similarly. $\square$

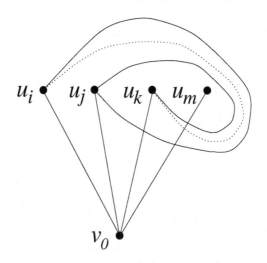

Figure 4: $u_i u_k$ *must cross* $v_0 u_m$.

Claim 3.3. If $G(100)$, $G(001)$, or G' contains a complete subgraph of size $r := \lceil m^{1/3} \rceil$, then G has a non-crossing subgraph isomorphic to any tree of r vertices.

Proof: Suppose that $w_1 \prec w_2 \prec \ldots \prec w_r$ induce a complete (topological) subgraph in $G(100)$. It is easy to see that this subgraph is *twisted*, i.e., it is weakly isomorphic to T_r. Take an arbitrary tree T with r vertices. Starting at any vertex $z_1 \in V(T)$, explore all other vertices of T using *breadth-first search*. Let $z_1, z_2, \ldots, z_r$ be a numbering of the elements of $V(T)$, in the order in which they are encountered by the algorithm. Then the embedding $f(z_i) = w_i$ $(1 \le i \le r)$ maps T into a non-crossing copy of T in $G(100)$, and we are done.

The case when $G(001)$ contains a complete subgraph of size r can be treated similarly.

Assume now that G' has a complete subgraph with r vertices, $w_1 \prec w_2 \prec \ldots \prec w_r$. It is easy to see that if two edges, $w_i w_j$ $(i < j)$ and $w_k w_l$ $(k < l)$, cross each other, then we have $i < k < j < l$ or $k < i < l < j$. In other words, if two edges of this subgraph cross each other, the corresponding edges also cross in a drawing on the same vertex set, weakly isomorphic to the *convex* drawing C_r. Clearly, C_r contains a non-crossing copy of every tree with r vertices, so the same is true for G'. $\square$

In view of the last claim, it remains to prove that at least one of $G(100)$, $G(001)$, and G' has a complete subgraph of size $r = \lceil m^{1/3} \rceil$. Suppose, in order to obtain a contradiction, that this is not the case.

If some element $u \in U = \{u_1, u_2, \ldots, u_m\}$ had at least $r - 1$ larger neighbors in $G(001)$ with respect to the ordering $\prec$, then, by Claim 3.2 (ii), these neighbors together with u would induce a complete subgraph in $G(001)$, a contradiction.

Now we recursively construct a sequence $w_1 \prec w_2 \prec \ldots$ consisting of at least $m^{2/3}$ elements of U, which form an independent set in $G(001)$ (i.e., they induce a complete subgraph in $G(100) \cup G'$).

Let $W_0 := \emptyset$ and $U_0 := \{u_1, u_2, \ldots, u_m\}$. Suppose that, for some $p < m^{2/3}$, we have already found two subsets $W_p = \{w_1, w_2, \ldots, w_p\}$ and $U_p \subset \{u_1, u_2, \ldots, u_m\}$, such that

1. W_p is an independent set in $G(001)$,
2. every element of W_p precedes every element of U_p,
3. there is no edge between W_p and U_p,
4. $|U_p| \geq m - p(r - 1)$.

If $U_p \neq \emptyset$, let w_{p+1} be the smallest element of U_p with respect to the ordering $\prec$, and set $W_{p+1} := W_p \cup \{w_{p+1}\}$. Let U_{p+1} denote the set obtained from U_p by the deletion of w_{p+1} and its larger neighbors. Clearly, we have $|U_{p+1}| \geq |U_p| - r + 1$, so that this procedure can be repeated at least $\lceil m^{2/3} \rceil$ times.

Define the *rank* of any element $w \in W := \{w_1, w_2, \ldots\}$, as the number of vertices of the longest monotone path (with respect to $\prec$) which ends at w in the subgraph of $G(100)$ induced by W. There is no element whose rank is at least $m^{1/3}$, otherwise, by Claim 3.2 (i), the vertices of the corresponding path would induce a complete subgraph of size at least r in $G(100)$, contradicting our assumptions.

Therefore, we can suppose that at least $m^{1/3}$ elements of W have the same rank. According to the definitions, these elements form an independent set in $G(100)$ as well as in $G(001)$. Thus, they induce a complete subgraph in G', again a contradiction. This proves Theorem 1.

4   Concluding Remarks

I. It follows from the proof of Theorem 1 that Theorem 2 can be slightly strengthened.

Theorem 4.1. *Every complete topological graph with n vertices has a complete topological subgraph with $m \geq c \log^{1/6} n$ vertices, weakly isomorphic to a twisted complete graph T_m, or a complete topological subgraph with $p \geq c \log \log n$ vertices, weakly isomorphic to a convex complete graph C_p.*

II. The following statement is a direct corollary of the first result in [PSS96].

Theorem 4.2. *Every complete topological graph of n vertices contains at least $c \log n / \log \log n$ pairwise crossing edges.*

III. Both C_m and T_m, the convex and the twisted topological graphs with m vertices, respectively, determine precisely $\binom{m}{4}$ edge crossings. Therefore, the following theorem of Harborth, Mengersen, and Schelp [HMS95] is an immediate consequence of Theorem 2.

Corollary 4.3 *For any positive integer m, there exists a smallest number $n(m)$ such that every complete topological graph with at least $n(m)$ vertices has a complete subgraph with m vertices and with $\binom{m}{4}$ crossings between its edges.*

In fact, for large values of m, Theorem 2 implies a better bound on the function $n(m)$ than the proof given in [HMS95].

IV. Let F denote the graph obtained from a complete graph of 5 vertices by subdividing each of its edges with an extra vertex. Given a complete topological graph K_n of n vertices, define an abstract graph G. Let the vertex set of G consist of $\lfloor n/2 \rfloor$ edges of K_n, no two of which share an endpoint. Let two vertices, $e, e' \in E(K_n)$, be joined by an edge of G if and only if e and e' cross each other. It is easy to see that G does not contain F as an induced subgraph (see e.g. [EET76]).

It follows from a theorem of Erdős and Hajnal [EH89] that, if a graph with m vertices does not contain some fixed induced subgraph F, then it must have either an empty or a complete subgraph with at least $e^{c\sqrt{\log m}}$ vertices, where $c > 0$ is a constant depending on F. Putting these two facts together, we obtain

Corollary 4.4. *Any topological complete graph with n vertices has at least $e^{c\sqrt{\log n}}$ edges that are either pairwise disjoint or pairwise crossing.*

This suggests that the bounds in Theorems 1 and 4.2 are far from being optimal. We conjecture that both estimates can be replaced by n^δ, for some $\delta > 0$. As was pointed out in the Introduction, this holds for geometric graphs.

V. In the case of *geometric* graphs, one can introduce several partial orderings on the set of edges (cf. [PT94], [PA95]). This allows us to apply Dilworth's Theorem in place of Ramsey's Theorem, to find much larger homogeneous substructures.

References

[AH66]   S. Avital and H. Hanani: Graphs (Hebrew), *Gilyonot Lematematika* **3** (1966), 2–8.

[AR88]   D. Archdeacon and B. R. Richter: On the parity of crossing numbers, *J. Graph Theory* **12** (1988), 307–310.

[C81]   Ch. J. Colbourn: On drawings of complete graphs, *J. Combin. Inform. System Sci.* **6** (1981), 169–172.

[CN00]   G. Cairns and Y. Nikolayevsky: Bounds for generalized thrackles, *Discrete Comput. Geom.* **23** (2000), 191–206.

[EET76]   G. Ehrlich, S. Even, and R. E. Tarjan: Intersection graphs of curves in the plane, *J. Combinatorial Theory, Ser. B* **21** (1976), 8–20.

[EG73]   P. Erdős and R. K. Guy: Crossing number problems, *Amer. Math. Monthly* **80** (1973), 52–58.

[EH89]   P. Erdős and A. Hajnal: Ramsey-type theorems, *Discrete Appl. Math.* **25** (1989), 37–52.

[ES35]   P. Erdős and G. Szekeres: A combinatorial problem in geometry, *Compositio Mathematica* **2** (1935), 463–470.

[ES60]   P. Erdős and G. Szekeres: On some extremum problems in elementary geometry, *Ann. Universitatis Scientiarum Budapestinensis, Eötvös, Sectio Mathematica* **III–IV** (1960–61), 53–62.

[GJ85]   M. R. Garey and D. S. Johnson: Crossing number is NP-complete, *SIAM J. Algebraic Discrete Methods* **4** (1983), 312–316.

[GMPP91]   P. Gritzmann, B. Mohar, J. Pach, and R. Pollack: Embedding a planar triangulation with vertices at specified points, *Amer. Math. Monthly* **98** (1991), 165-166.

[GH90]   H.-D. Gronau and H. Harborth: Numbers of nonisomorphic drawings for small graphs, in: *Proceedings of the Twentieth Southeastern Conference on Combinatorics, Graph Theory, and Computing (Boca Raton, FL, 1989)*, *Congr. Numer.* **71** (1990), 105–114.

[G69]   R. K. Guy: The decline and fall of Zarankiewicz's theorem, in: *Proof Techniques in Graph Theory*, Academic Press, New York, 1969, 63–69.

[HM74]   H. Harborth and I. Mengersen: Edges without crossings in drawings of complete graphs, *J. Combinatorial Theory, Ser. B* **17** (1974), 299–311.

[HM90]   H. Harborth and I. Mengersen: Edges with at most one crossing in drawings of the complete graph, in: *Topics in Combinatorics and Graph Theory (Oberwolfach, 1990)*, Physica, Heidelberg, 1990, 757–763.

[HM92]   H. Harborth and I. Mengersen: Drawings of the complete graph with maximum number of crossings, in: *Proceedings of the Twenty-third Southeastern International Conference on Combinatorics, Graph Theory, and Computing (Boca Raton, FL, 1992)*, *Congr. Numer.* **88** (1992), 225–228.

[HMS95]   H. Harborth, I. Mengersen, and R. H. Schelp: The drawing Ramsey number Dr(K_n), *Australasian J. Combinatorics* **11** (1995), 151–156.

[HT94]   H. Harborth and Ch. Thürmann: Minimum number of edges with at most s crossings in drawings of the complete graph, in: *Proceedings of the Twenty-fifth Southeastern International Conference on Combinatorics, Graph Theory and Computing (Boca Raton, FL, 1994)*, *Congr. Numer.* **102** (1994), 83–90.

[K79]   Y. S. Kupitz: *Extremal Problems in Combinatorial Geometry, Lecture Notes Series* **53**, Aarhus Universitet, Matematisk Institut, Aarhus, 1979.

[LPS97]   L. Lovász, J. Pach, and M. Szegedy: On Conway's thrackle conjecture, *Discrete and Computational Geometry* **18** (1997), 369–376.

[M78]   I. Mengersen: Die Maximalzahl von kreuzungsfreien Kanten in Darstellungen von vollständigen n-geteilten Graphen (German), *Math. Nachr.* **85** (1978), 131–139.

[P99]   J. Pach: Geometric graph theory, in: *Surveys in Combinatorics, 1999 (Canterbury)*, *London Math. Soc. Lecture Note Ser.* **267**, Cambridge Univ. Press, Cambridge, 1999, 167–200.

[PA95]   J. Pach and P. K. Agarwal: *Combinatorial Geometry*, Wiley-Interscience, New York, 1995.

[PSS96]   J. Pach, F. Shahrokhi, and M. Szegedy: Applications of crossing numbers, *Algorithmica* **16** (1996), 111–117.

[PT94]   J. Pach and J. Törőcsik: Some geometric applications of Dilworth's theorem, *Discrete and Computational Geometry* **12** (1994), 1–7.

[PT98]   J. Pach and G. Tóth: Which crossing number is it, anyway? *Proceedings of the 39th Annual Symposium on Foundations of Computer Science,* 1998, 617–626.

[R64]   G. Ringel: Extremal problems in the theory of graphs, in: *Theory of Graphs and its Applications (Proc. Sympos. Smolenice, 1963)*, Publ. House Czechoslovak Acad. Sci., Prague, 1964, 85–90.

[TV98]   G. Tóth and P. Valtr: Note on the Erdős-Szekeres theorem, *Discrete Comput. Geom.* **19** (1998), 457–459.

[T77]   P. Turán: A note of welcome, *J. Graph Theory* **1** (1977), 7–9.

[W71]   D. R. Woodall: Thrackles and deadlock, in: *Combinatorial Mathematics and its Applications (Proc. Conf., Oxford, 1969)*, Academic Press, London, 1971, 335–347.

[WB78]   A. T. White and L. W. Beineke: Topological graph theory, in: *Selected Topics in Graph Theory (L. W. Beineke and R. J. Wilson., eds.)*, Academic Press, Inc. [Harcourt Brace Jovanovich, Publishers], London-New York, 1983, 15–49.

Minimum Weight Drawings of Maximal Triangulations*
(Extended Abstract)

William Lenhart[1] and Giuseppe Liotta[2]

[1] Dept. of Computer Science, Williams College
`lenhart@cs.williams.edu`
[2] Ingegneria Elettronica e dell'Informazione, Università degli Studi di Perugia
`liotta@diei.unipg.it`

Abstract. This paper studies the drawability problem for minimum weight triangulations, i.e. whether a triangulation can be drawn so that the resulting drawing is the minimum weight triangulations of the set of its vertices. We present a new approach to this problem that is based on an application of a well known matching theorem for geometric triangulations. By exploiting this approach we characterize new classes of minimum weight drawable triangulations in terms of their skeletons. The skeleton of a minimum weight triangulation is the subgraph induced by all vertices that do not belong to the external face. We show that all maximal triangulations whose skeleton is acyclic are minimum weight drawable, we present a recursive method for constructing infinitely many minimum weight drawable triangulations, and we prove that all maximal triangulations whose skeleton is a maximal outerplanar graph are minimum weight drawable.

1 Introduction

The study of the combinatorial properties of fundamental geometric graphs such as minimum spanning trees, Delaunay triangulations, proximity graphs, rectangle of influence graphs, maximum weight triangulations, and Voronoi trees is motivated not only by the theoretical appeal of the questions that this study raises, but also by the importance that such geometric structures have in different application areas including computer graphics, computer aided manufacturing, communication networks, and computational biology. Geometric graphs are straight-line drawings that satisfy some additional geometric constraints (for example pairs of adjacent vertices are deemed to be "close" according to some definition of proximity, while not adjacent vertices are far from each other in the drawing). Thus, the study of the combinatorial properties of a given type of geometric graph can be naturally turned into the following graph drawing question:

---

* Research supported in part by the project "Algorithms for Large Data Sets: Science and Engineering" of the Italian Ministry of University and Scientific and Technological Research (MURST 40%), and by the project "Geometria Computazionale Robusta con Applicazioni alla Grafica ed a CAD" of the Italian National Research Council (CNR).

J. Marks (Ed.): GD 2000, LNCS 1984, pp. 338–349, 2001.

What are those graphs admitting the given type of drawing?. This question has attracted increasing interest in both the graph drawing and the computational geometry communities and several papers have been published on the topic in recent years, including [2,4,7,8,9,10,14,15,16,17]. See also [6] for a survey.

The present paper is devoted to minimum weight triangulations. Despite the relevance of minimum weight triangulations in areas like numerical analysis and computational geometry, the problem of computing these triangulations efficiently has not yet been solved and their basic combinatorial properties are still not well-understood. We provide new insight on the combinatorial properties of minimum weight triangulations by addressing the *minimum weight drawability problem*, i.e. the problem of determining whether a triangulation T admits a straight line drawing Γ that is a minimum weight triangulation of the set of its vertices; we call Γ a *minimum weight drawing* of T and we say that T is *minimum weight drawable*.

The minimum weight drawability problem was first studied in [11,12] where it was proved that all maximal outerplanar graphs are minimum weight drawable and a linear time drawing algorithm was presented. As a side effect of the combinatorial characterization in [11,12], a linear time algorithm for computing the minimum weight triangulation of a set of points that are the vertices of a regular polygon was shown. These results have motivated the investigation of minimum weight drawable triangulations such that not all the vertices belong to the outer face, which is the subject of [13] and of a recent paper by Wang, Chin, and Yang [20].

In [13] families of minimum weight drawable triangulations are characterized in terms of their *skeleton*, i.e. the subgraph induced by their interior vertices. Classes of drawable triangulations whose skeleton is either acyclic or maximal are described and it is proved that the skeleton of a minimum weight triangulation can be any forest. In the same paper, the relationship between minimum weight drawability and Delaunay drawability is investigated. A triangulation is *Delaunay drawable* if it admits a drawing that is the Delaunay triangulation of the set of its vertices; characterizations of Delaunay drawable triangulations can be found in the works by Dillencourt [8,7,9]. In [13] an infinite family of minimum weight drawable, but non-Delaunay-drawable, triangulations are constructed, each of which as an acyclic skeleton.

Wang, Chin, and Yang [20] focus on the minimum drawability of triangulations with acyclic skeletons and show examples of triangulations of this type that do not admit a minimum weight drawing, thus solving one of the open problems in [13]. Wang, Chin, and Yang also provide a partial characterization of minimum weight drawable triangulations having acyclic skeletons, by showing that all triangulations whose skeleton is a regular star graph admit a minimum weight drawing.

In this paper we look at the minimum weight drawability problem from a new perspective. Namely, we present a new technique for proving that a straight-line drawing is a minimum weight triangulation of the set of its vertices. The technique compares distances between adjacent vertices against distances between any possible pairs of vertices in the drawing and is based on an application of a matching theorem by Aichholzer et al. [1] which establishes a correspondence between the edges of any two triangulations computed on the same point set. By using this technique we can prove the correctness of new drawing algorithms for new classes of minimum weight drawable triangulations. The main results that we establish in this paper can be listed as follows.

- We characterize those maximal triangulations with acyclic skeleton that admit a minimum weight drawing. In [13] only a partial characterization was presented.

- We devise a method for recursively constructing minimum weight drawable triangulations whose skeleton is a maximal triangulation. As an application of this method, we show minimum weight drawable triangulations with maximal skeletons that are *not* Delaunay drawable. The other previously known members of this family of triangulations all had acyclic skeletons [13].

- We show that all maximal triangulations whose skeleton is a maximal outerplanar graph are minimum weight drawable. This extends the result of [11] where the minimum weight drawability of maximal outerplanar graphs is proved.

2   Preliminaries

We assume familiarity with basic computational geometry, graph drawing and graph theory concepts. For further details see [3,5,18].

Theorem 1. [1] *Let P be a finite set of points in the plane and consider two triangulations T and T' of P. There exists a perfect matching between T and T' with the property that matched edges either cross or are identical.*

The *skeleton* $S(T)$ of a triangulation T is the graph induced by the set of its internal vertices. For example, the skeleton of a maximal outerplanar graph is the empty graph, the skeleton of a wheel graph consists of just one vertex, namely, the center of the wheel.

We will often be concerned with graphs G each of whose edges has an associated weight, namely the length of the corresponding segment in some straight-line drawing of G. We denote the weight of an edge e by $w(e)$. The weight of a set E of edges refers to the sum of the weights of the edges in E and is denoted by $w(E)$, as is the weight of a graph G or a drawing Γ.

3   Feasible and Forced Edges

Let P be a finite set of points in the plane. We denote by $Seg(P)$ the set of all segments having both endpoints in P. A set $E \subseteq Seg(P)$ is *feasible* if E is contained in some minimum weight triangulation of P; E is *forced* if it is contained in every minimum weight triangulation of P. Edges of the convex hull of P are clearly forced, as is any segment which is not crossed by any other segment connecting two points of P.

Our algorithms compute drawings where some edges are forced and the other edges are feasible. In order to show the correctness of our constructions, we rely on the following lemma, which gives a sufficient condition under which a set of edges is feasible. The lemma is an application of Theorem 1. In the lemma, for any $E \subseteq Seg(P)$, we denote by $I(E) \subseteq Seg(P)$ those edges which intersect at least one edge of E.

Lemma 1. Let $E \subseteq Seg(P)$ be such that

1. E is planar.

2. Every edge of E is light, that is, for every $e \in E$, every edge crossing e is at least as long as e.

3. For all $s\prime \in I(E)$ and all $s \in I(I(E)) - E - I(E)$ such that s crosses $s\prime$, $|s\prime| \geq |s|$.

Then E is feasible.

Sketch of Proof. Let T be a triangulation of P such that $E \not\subseteq T$. We will show how to modify T to obtain a triangulation containing E and having weight at most $w(T)$. Let $E' = E - T$, and let $G = T - I(E')$. Note that G is planar and that no edge of G intersects any edge of the (planar) set E', and so $G \cup E'$ is planar. Since any planar set of edges can be extended to a triangulation by the addition of zero or more edges, we can extend $G \cup E'$ to a triangulation T'. Hence, by Theorem 1, there exists a matching between T' and T in which every edge of T' is matched either with itself or with an edge of T that crosses it. Since no edge of E' is in T, E' is matched to a subset M of T so that every edge in E' crosses the edge in M that it matches. Therefore $M \subseteq T \cap I(E')$. Then $w(M) \geq w(E')$, since each edge of E' is light. So, $w(T) = w(G) + w(M) + w((T \cap I(E')) - M) \geq w(G) + w(E') + w((T \cap I(E')) - M) = w(G \cup E') + w((T \cap I(E')) - M)$.

Now, we extend $G \cup E'$ to a triangulation by adding $|(T \cap I(E')) - M|$ edges. Notice that each of the edges we add must be in $I(I(E)) - E - I(E)$, since $E \subseteq G \cup E'$, and edges in $I(E)$ would cross edges in $G \cup E'$. Thus, by Condition 3 of the lemma, each of the added edges can be at most as long as any edge in $I(E)$ that it crosses. Because $T \cap I(E')) - M \subset I(E)$, we have that the weight of the added edges is at most $w(T \cap I(E')) - M$. Therefore, by the inequality above we conclude that the new triangulation has weight at most $w(T)$.

In the next sections we present several applications of Lemma 1 to the minimum weight drawability problem.

4 Acyclic Skeletons

In [20] it is shown that not all triangulations with acyclic skeleton are minimum weight drawable. On the positive side, in [13] the following partial characterization of maximal triangulations with acyclic skeleton is proved.

Lemma 2. Let T be a maximal triangulation and let $S(T)$ be the skeleton of T. If $S(T)$ is a path, then T is minimum weight drawable and a minimum weight *drawing of T can be computed in linear time with the real RAM model of computation.*

In this section we complete the characterization of those maximal triangulations with acyclic skeleton that are minimum weight drawable. We start by characterizing acyclic skeletons of maximal triangulations.

Lemma 3. *Let T be a maximal triangulation and let $S(T)$ be the skeleton of T. If $S(T)$ is acyclic, then it is a tree with at most three leaves.*

Proof. Proof omitted in extended abstract.

Let T be a maximal triangulation whose skeleton is acyclic. By Lemma 3, two cases are possible: Either the skeleton of T is a tree with exactly one vertex of degree 3 or it is a path. If it is a path, the minimum weight drawability of T is guaranteed by Lemma 2. The next lemma studies the remaining case.

Lemma 4. *Every maximal triangulation whose skeleton is a tree with three leaves is minimum weight drawable.*

Proof. Proof omitted in extended abstract.

We can summarize the results of this section as follows.

Theorem 2. *Let T be a maximal triangulation with n vertices and let $S(T)$ be the skeleton of T. If $S(T)$ is acyclic, then T is minimum weight drawable, and a minimum weight drawing of T can be computed in $O(n)$ time in the real RAM model of computation.*

5   Maximal Skeletons

In this section we study the minimum weight drawability of triangulations whose skeleton is maximal. In Subsection 5.1 we show a recursive method to construct minimum weight drawable triangulations each having a skeleton that is a maximal triangulation. In subsection 5.2 we study the minimum weight drawability when the skeleton is a maximal outerplanar graph.

5.1   Skeletons That Are Maximal Triangulations

The next theorem allows us to describe a recursive method for drawing certain triangulations as minimum weight triangulations

Theorem 3. *Let T be a maximal triangulation and let $S(T)$ be its skeleton. If $S(T)$ is such that:*

    1. $S(T)$ is a maximal triangulation, and

    2. $S(T)$ is minimum weight drawable,

then T is minimum weight drawable.

Sketch of Proof. Let a_0, a_1, and a_2 be the three vertices of the outer face of T. Let v_0, v_1, and v_2 be the three vertices of the outer face of $S(T)$ and let Γ' be a minimum weight drawing of $S(T)$. We show how to construct a minimum weight drawing Γ of T from Γ' by adding vertices a_0, a_1, and a_2 and their incident edges. There are two cases to consider: Either (i) each a_i $(i = 0, 1, 2)$ is adjacent to two vertices of the outer face of $S(T)$, or (ii) there exists a vertex of the outer face of T adjacent to v_0, v_1, and v_2.

For Case (i), suppose a_i is adjacent to v_i and to v_{i+1} (in the rest of the proof we always assume $i = 0, 1, 2$ and all subscripts taken mod 3). Each a_i is represented as a point in the plane so that the following geometric constraints are satisfied (see Figure 1(a)):

Constraint 1: The coordinates of the vertices are chosen so that a_1, a_2, and a_3 form the convex hull of the new set of points.

Constraint 2: Vertex a_i is connected to v_i and to v_{i+1} by segments that do not intersect any edge of Γ'. The segment connecting a_i to v_{i+2} intersects Γ'.

Constraint 3: The distance between a_i and each vertex of Γ' is larger than length of the longest edge of the outer face of Γ'.

Let Γ be the resulting drawing and let P be the set of vertices of Γ. We prove that Γ is a minimum weight triangulation of P. All edges of the type $a_i a_{i+1}$ are forced because of *Constraint 1*. Similarly, the edges connecting a_0, a_1 and a_2 to v_0, v_1, and v_2 are forced because of *Constraint 2*.

Let E be the edges of the type $v_i v_{i+1}$. Clearly E is planar. $I(E)$ consists of segments connecting a vertex of Γ' to a vertex of the convex hull of P. Thus, by Constraint 3 every edge of E is light. Also notice that $I(I(E)) - E - I(E)$ consists of segments connecting pairs of vertices of Γ', which by *Constraint 3* are shorter than any element of $I(E)$. Therefore, by Lemma 1 we conclude that E is feasible.

Now, since E is feasible and is a (convex) triangle, all edges of the type $a_i a_{i+1}$ are forced. Thus, since Γ' is a minimum weight drawing, we conclude that Γ is a minimum weight triangulation of P.

We now consider Case (ii). Suppose that a_0 is adjacent to v_0, v_1, and v_2, that a_1 is adjacent to both v_1 and v_2, and that a_2 is adjacent only to v_2. A drawing of T is constructed in three steps: first a_0 and its incident edges are added to Γ', then a_1 is added, and finally a_2 is added. The coordinates of the vertices is chosen so that a_1, a_2, and a_0 form the convex hull of the new set of points. Also, the following constraints are satisfied (see Figure 1(b)):

Constraint a: Vertex a_0 is connected to v_0, v_1, and v_2 by segments that do not intersect any edge of Γ'. The distance between a_0 and each vertex of Γ' is larger than the length of the longest edge of the outer face of Γ'.

Constraint b: Vertex a_1 is connected to v_1, and v_2 by segments that do not intersect any edge of Γ'. A segment connecting a_1 to v_0 crosses only shorter segments.

Constraint c: Vertex a_2 is connected to v_2, by segments that do not intersect any edge of Γ'. Segments connecting a_2 to either v_0 or v_1 cross only shorter segments.

Observe that *Constraints a, b,* and *c* imply that the distance between a_i and each vertex of Γ' is larger than length of the longest edge of the outer face of Γ'. The reasoning to prove that the drawing defined with this construction is minimum weight is similar to that described for Case (i). We omit details for brevity.

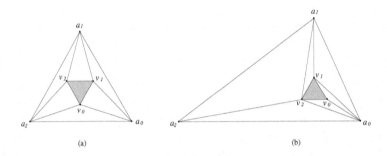

(a)   (b)

Fig. 1. Illustration for Theorem 3. The shaded grey regions represents Γ'. (a) A minimum weight drawing in which each a_i $(i = 0, 1, 2)$ is adjacent to two vertices of the outer face of Γ'. (b) A minimum weight drawing in which a vertex of the outer face of T adjacent to three vertices of the outer face of Γ'.

Theorem 3 provides a basic tool for constructing minimum weight drawable triangulations. One such triangulation, obtained after one step of the recursion, is depicted in Figure 2: Theorems 3 and 2 imply that the triangulation is minimum weight drawable. In the figure, some vertices of the triangulation are drawn as white circles: Removing the white vertices breaks the graph into four disconnected components. This means that the graph violates one of necessary conditions that all Delaunay drawable triangulations must satisfy [8]. Similarly, it can be verified that none of the triangulations recursively drawn by the above procedure are Delaunay drawable.

Lemma 5. *There exists an infinite family of triangulations that admit a minimum weight drawing, are not Delaunay drawable, and have skeletons that are maximal triangulations.*

We remark that the only families of minimum weight but not Delaunay drawable triangulations known so far had a considerably simpler combinatorial structure, since their skeletons were forced to be acyclic [13].

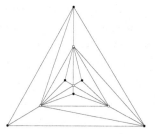

Fig. 2. A minimum weight drawable triangulation constructed by the recursive procedure of Theorem 3. The skeleton of the triangulation is a maximal triangulation. The triangulation is not Delaunay drawable.

5.2   Skeletons That Are Maximal Outerplanar Graphs

In [11,12] it is proved that all maximal outerplanar graphs admit minimum weight drawings and a linear time algorithm to compute these drawings is presented. In this section we prove that every maximal triangulation whose skeleton is a maximal outerplanar graph is minimum weight drawable. Our proof relies on the following approach:

- A special type of minimum weight drawing of $S(T)$ is computed by means of a variant of the algorithm in [11,12].
- Such a drawing of $S(T)$ is used as a building block to compute a minimum weight drawing of T.

Before giving more technical details, we briefly recall the basic idea behind the algorithm of [11,12]. Let G be a maximal outerplanar graph and let $D(G)$ be its dual; observe that $D(G)$ is a tree such that all non-leaf vertices have degree three. The vertices of G are drawn as cocircular points chosen to be a subset of the vertices of regular polygon Π. Π is defined so that the dual graph of its minimum weight triangulation is a complete tree having $D(G)$ as its subtree (in [11,12] it is shown that the minimum weight triangulation of a regular polygon coincides with its greedy triangulation). The minimum weight triangulation of G can be obtained by deleting vertices of degree 2 from the minimum weight triangulation of Π, until the dual of the remaining triangulation becomes identical to $D(G)$ (deleting a vertex of degree 2 and its incident edges from the minimum weight triangulation of Π corresponds to deleting a leaf from its dual tree). Since Π is defined in such a way that it may have exponentially many more vertices than G, additional tools are devised [11,12] by which a time complexity proportional to the number of vertices of G is achieved. Intuitively, the algorithm does not explicitly construct the minimum weight drawing of Π, but it uses the knowledge of the topology of its dual tree to directly compute the coordinates of the vertices of the minimum weight drawing of G.

We are now ready to prove the main result of this section. Let T be a maximal triangulation whose skeleton $S(T)$ is a maximal outerplanar graph. Let a_0, a_1, and a_2 be the three vertices of the outer face of T. We distinguish between those vertices of $S(T)$ that are adjacent to exactly one of a_0, a_1, and a_2 and those vertices of $S(T)$ that are adjacent to at least two of a_0, a_1, and a_2. Vertices of $S(T)$ of this second type are called *transition vertices*. Since T is a maximal triangulation, $S(T)$ has either two or three transition vertices. An example where $S(T)$ has two transition vertices is given in Figure 3 (a); an example where $S(T)$ has three transition vertices is given in Figure 3 (b).

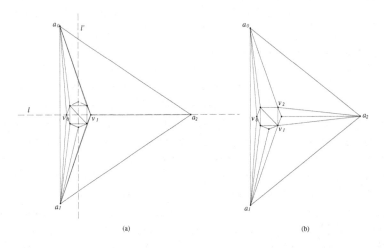

(a)   (b)

Fig. 3. (a) A minimum weight drawing of a triangulation whose skeleton is a maximal outerplanar graph with two transition vertices. (b) A minimum weight drawing of a triangulation whose skeleton is a maximal outerplanar graph with three transition vertices. Figure (a) also shows lines l and l' and the minimum weight triangulation of a regular polygon Π that are used by the drawing algorithm of Lemma 6.

Lemma 6. *It T is a maximal triangulation whose skeleton $S(T)$ is a maximal outerplanar graph with exactly two transition vertices, then T is minimum weight drawable.*

Sketch of Proof. Since $S(T)$ has exactly two transition vertices, one of them is adjacent to all of the outer vertices a_0, a_1, and a_2 of T, while the other is adjacent to only two of them. Let us label the two transition vertices as v_0 and v_1 and assume that v_0 is adjacent only to to a_0 and a_1. Let n_0 be the number of non-transition vertices of $S(T)$ that are adjacent to a_0, let n_1 be the number of non-transition vertices of $S(T)$ that are adjacent to a_1 and let $n = \max\{n_1, n_2\}$.

We modify the drawing algorithm of [11,12] as follows. A polygon Π is defined such that: (i) The dual tree of a minimum weight triangulation of Π is a complete tree having the dual of $S(T)$ as its subtree (we recall that the minimum weight triangulation of a regular polygon coincides with its greedy triangulation); (ii) Π has at least $2n+2$ vertices. A minimum weight drawing of $S(T)$ is now computed by vertex deletion from the minimum weight triangulation of Π. Vertices v_0 and v_1 are chosen to be antipodal points of the polygon; therefore, there are n vertices on both sides of the polygon between v_0 and v_1. This guarantees that we can delete vertices of degree 2 from the minimum weight triangulation of Π in such a way that: (i)the dual graph of the resulting triangulation after the deletions coincides with the dual graph of $S(T)$ (i.e. the resulting triangulation is a drawing of $S(T)$), and (ii) the two paths from v_0 to v_1 along the drawing of $S(T)$ consist of n_0 and of n_1 vertices, respectively. Let Γ' be the drawing of $S(T)$ obtained by this procedure. The fact that Γ' is a minimum weight drawing of $S(T)$ is a consequence of the property that deleting a vertex of degree 2 and its incident edges from a minimum weight triangulation of a set of cocircular points gives as a result a minimum weight triangulation of the remaining points.

In order to construct a minimum weight drawing of T we now add to Γ' vertices a_0, a_1, a_2, and their incident edges. Refer to Figure 3 (a). Let ℓ be the line through v_0 and v_1 and let ℓ' be the perpendicular bisector of the segment having v_0 and v_1 as its endpoints. We assume that by construction Γ' has n_0 vertices in the half-plane above ℓ and thus there are n_1 vertices in the half-plane below ℓ and that ℓ is horizontal. The coordinates of the vertices are chosen so that a_1, a_2, and a_0 form the convex hull of the new set of points. The following additional constraints are satisfied

Constraint 1: Vertex a_0 is drawn on the left-hand side of line ℓ' and above line ℓ. Vertex a_0 is connected to v_0, v_1 and to the n_0 non-transition vertices above ℓ by segments that do not intersect any edge of Γ'. The distance between a_0 and each vertex of Γ' is larger than the length of the longest edge of Γ'.

Constraint 2: Vertex a_1 is drawn on the left-hand side of line ℓ' and below line ℓ. Vertex a_1 is connected to v_0, v_1 and to the n_1 non-transition vertices below ℓ by segments that do not intersect any edge of Γ'. The distance between a_1 and each vertex of Γ' is larger than the length of the longest edge of Γ'.

Constraint 3: Vertex a_2 is drawn on the right-hand side of ℓ' and on line ℓ. Vertex a_2 is connected to v_1 by a segment longer than those segments connecting a_0 and a_1 to vertices of $S(T)$.

Let Γ be the resulting drawing. Observe that the edges of the outer face and edge a_2v_1 are forced. Let E be the set consisting of the edges of Γ' and of the segments connecting a_0 and a_1 to $S(T)$. Clearly E is planar. Also, $I(E)$ consists of segments connecting either vertices of Γ' to vertices of the convex hull or pairs of vertices of Γ'. By Constraints 1, 2, and 3 and since Γ' is a minimum weight drawing, we have that each segment e of E is crossed by segments of $I(E)$ that are no shorter than e; hence E is light. Now, since $I(I(E)) - I(E) - E = \emptyset$, it follows by Lemma 1 that E is feasible. Therefore Γ is a minimum weight drawing of T.

Lemma 7. *If T is a maximal triangulation whose skeleton $S(T)$ is a maximal outerplanar graph with exactly two transition vertices, then T is minimum weight drawable.*

Sketch of Proof. A minimum weight drawing Γ of T is computed by a variant of the algorithm in the proof of Lemma 6. Namely, Γ' is computed in the same way as in the proof of the above lemma, and vertices a_0 and a_1 are added by the same strategy. The coordinates of vertex a_2 are now chosen so that: (i) all segments connecting a_2 to the vertices of Γ' are shorter that all segments that can possibly cross them and (ii) all segments starting at a_2 and that cross remaining edges of the drawing are longer than these edges. An example of a drawing computed by this strategy is given in Figure 3 (b). The proof that Γ is a minimum weight drawing relies on Lemma 1 and is similar to that of Lemma 6.

Theorem 4. *Let T be a maximal triangulation and let $S(T)$ be its skeleton. If $S(T)$ is a maximal outerplanar triangulation, then T is minimum weight drawable and its minimum weight drawing can be computed in linear time in the real RAM model of computation.*

6 Conclusions and Open Problems

In this paper we have presented new results on the minimum weight drawability problem by characterizing new families of maximal triangulations that admit a minimum weight drawing. The new results are based on a sufficient geometric condition for a set of line segments to be part of a minimum weight triangulation.

The general problem of determining which triangulations are minimum weight drawable is still far from solved. As intermediate steps toward solving the problem, we might suggest pursuing the following:

1. Devise other sufficient conditions of the type given in Lemma 1 that allow us to better understand the geometric properties of minimum weight triangulations of given sets of points.

2. Study the minimum weight drawability of k-outerplanar graphs (a graph is k-outerplanar when it has a planar embedding such that all vertices are on disjoint cycles properly nested at most k deep). The proof techniques of Lemmas 6 and 7 may be a good starting point.

3. Further investigate the relationship between Delaunay drawability and minimum weight drawability. An interesting class to study seems to be the set of 4-connected triangulations, for which a characterization in terms of Delaunay drawability is known.

References

1. Oswin Aichholzer, Franz Aurenhammer, Siu-Wing Chen, Na oki Katoh, Michael Taschwer, Günter Rote, and Yin-Feng Xu. Triangulations intersect nicely. *Discrete Comput. Geom.*, 16:339–359, 1996.
2. T. Biedl, A. Bretscher, and H. Meijer. Drawings of graphs without filled 3-cycles. In *Graph Drawing (Proc. GD '99)*, volume 1731 of *Lecture Notes Comput. Sci.*, pages 359–368, 2000.
3. J. A. Bondy and U. S. R. Murty. *Graph Theory with Applications*. Macmillan, London, 1976.

4. P. Bose, W. Lenhart, and G. Liotta. Characterizing proximity trees. *Algorithmica*, 16:83–110, 1996.

5. G. Di Battista, P. Eades, R. Tamassia, and I. G. Tollis. *Graph Drawing*. Prentice Hall, Upper Saddle River, NJ, 1999.

6. G. Di Battista, W. Lenhart, and G. Liotta. Proximity drawability: a survey. In R. Tamassia and I. G. Tollis, editors, *Graph Drawing (Proc. GD '94)*, volume 894 of *Lecture Notes Comput. Sci.*, pages 328–339, 1995.

7. M. B. Dillencourt. Realizability of Delaunay triangulations. *Inform. Process. Lett.*, 33(6):283–287, February 1990.

8. M. B. Dillencourt. Toughness and Delaunay triangulations. *Discrete Comput. Geom.*, 5, 1990.

9. M. B. Dillencourt and W. D. Smith. Graph-theoretical conditions for inscribability and Delaunayr ealizability. In *Proc. 6th Canad. Conf. Comput. Geom.*, pages 287–292, 1994.

10. P. Eades and S. Whitesides. The realization problem for Euclidean minimum spanning trees is NP-hard. *Algorithmica*, 16:60–82, 1996.

11. W. Lenhart and G. Liotta. Drawing outerplanar minimum weight triangulations. *Inform. Process. Lett.*, 57(5):253–260, 1996.

12. W. Lenhart and G. Liotta. How to draw outerplanar minimum weight triangulations. In F. J. Brandenburg, editor, *Graph Drawing (Proc. GD '95)*, volume 1027 of *Lecture Notes Comput. Sci.*, pages 373–384. Springer-Verlag, 1996.

13. William Lenhart and Giuseppe Liotta. Drawable and forbidden minimum weight triangulations. In G. Di Battista, editor, *Graph Drawing (Proc. GD '97)*, volume 1353 of *Lecture Notes Comput. Sci.*, pages 1–12. Springer-Verlag, 1998.

14. G. Liotta, A. Lubiw, H. Meijer, and S. H. Whitesides. The rectangle of influence drawability problem. *Comput. Geom. Theory Appl.*, 10(1):1–22, 1998.

15. G. Liotta and H. Meijer. Voronoi drawings of trees. In *Graph Drawing (Proc. GD '99)*, volume 894 of *Lecture Notes Comput. Sci.*, pages 328–339, 2000.

16. A. Lubiw and N. Sleumer. Maximal outerplanar graphs are relative neighborhood graphs. In *Proc. 5th Canad. Conf. Comput. Geom.*, pages 198–203, 1993.

17. C. Monma and Subhash Suri. Transitions in geometric minimum spanning trees. *Discrete Comput. Geom.*, 8:265–293, 1992.

18. F. P. Preparata and M. I. Shamos. *Computational Geometry: An Introduction*. Springer-Verlag, 3rd edition, October 1990.

19. Cao An Wang, Francis Y. Chin, and Boting Yang. Maximum weight triangulation and graph drawing. *Inform. Process. Lett.*, 70(1):17–22, 1999.

20. Cao An Wang, Francis Y. Chin, and Boting Yang. Triangulations without minimum weight drawing. In *Algorithms and Complexity (Proc. CIAC 2000)*, volume 1767 of *Lecture Notes Comput. Sci.*, pages 163–173. Springer-Verlag, 2000.

A Layout Algorithm for
Bar-Visibility Graphs on the Möbius Band

Alice M. Dean

Department of Mathematics and Computer Science
Skidmore College, Saratoga Springs, NY 12866
adean@skidmore.edu, www.skidmore.edu/~adean

Abstract. We characterize two types of bar-visibility graphs on the
Möbius band (abbreviated "BVGMs"), in which vertices correspond to
intervals that are parallel or orthogonal to the axis of the band, depending on type, and in which adjacency corresponds to orthogonal visibility
of intervals. BVGMs with intervals orthogonal to the axis are shown to
be equivalent to the "polar visibility graphs" studied by Hutchinson [7].
BVGMs with intervals parallel to the axis are characterized as those
graphs G which satisfy the following conditions: G is embedded on the
Möbius band; the block-cutpoint tree of G is a caterpillar in which all
but at most one block is planar; and the non-planar block, if it exists, is
at the "head" of the caterpillar.

1 Introduction

A *bar-visibility graph* in the plane (abbreviated *BVGP*) is a graph having a representation in which vertices correspond to disjoint, horizontal intervals ("bars")
in the plane, with two vertices adjacent if and only if there is a vertical band
of visibility between their bars, i.e., a non-degenerate rectangle with upper and
lower sides contained in the bars representing the two vertices, and intersecting no other interval representing a vertex. Wismath [14], and independently,
Tamassia and Tollis [10], characterized the graphs which can be represented as
bar-visibility graphs in the plane as those graphs having a plane embedding with
all cutpoints on a single face, and they gave linear-time algorithms to produce
a layout of a bar-visibility graph.

Characterizations and algorithms have since been given for bar-visibility on
other surfaces. In [12] BVGs on the *cylinder* (bars parallel to the axis of the
cylinder) are considered, and in [11] *spherical* BVGs (which are equivalent to
cylindrical BVGs with bars orthogonal to the axis) are considered. BVGs on the
torus are considered in [9], and BVGs on the projective plane and the Klein
bottle are studied in [7]. In this paper we consider BVGs on the Möbius band,
primarily with bars parallel to the axis of the band. We generalize the results
of [12] to obtain a characterization and a linear-time layout algorithm for these
graphs.

J. Marks (Ed.): GD 2000, LNCS 1984, pp. 350–359, 2001.

2 Parallel BVGs on the Cylinder

In [12] Tamassia and Tollis characterize graphs having a visibility layout on a cylinder, with bars parallel to the axis of the cylinder, as those graphs which are planar and whose block-cutpoint tree is a caterpillar. We call such a layout a *BVGCP* layout. A fundamental feature of their layout is that the underlying rectangle, whose top and bottom sides are identified to form a cylinder, is divided into rows and columns. There is one row for each vertex, and one column for each face of the embedding. Each vertex is represented by a horizontal interval in the appropriate row, and each face by a vertical interval in the appropriate column; edges are non-degenerate rectangles whose upper and lower borders are contained in the intervals corresponding to their endpoints, and whose left and right borders are contained in the intervals corresponding to their incident faces.

Cylindrical embeddings and their BVGCP layouts are easily seen to induce BVGMP embeddings and layouts, as indicated in the following proposition. An example of the induced Möbius embedding and layout is shown in Fig. 1.

Proposition 1. *Let G be a graph with an embedding on the cylinder. Such an embedding induces an embedding of G on the Möbius band. Furthermore, if the cylindrical embedding has a corresponding BVGCP layout (i.e., if the block-cutpoint tree of G is a caterpillar), then that layout induces a BVGMP layout of G.*

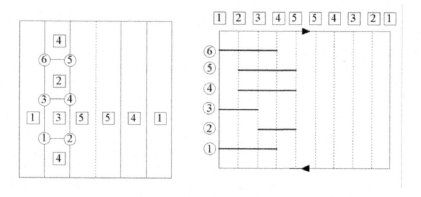

Fig. 1. A Möbius embedding and layout induced by a BVGCP

3 Layout Algorithms for Parallel BVGMs

In this section we generalize the algorithm of [12] to give a BVGMP layout for an arbitrary 2-connected graph embedded on the Möbius band. Our layout has *two* columns for each face of the embedding, arranged symmetrically around the vertical center line of a base rectangle for the Möbius band. The algorithm is somewhat more complex in this setting and requires some additional terminology.

Definition 1. 1. $R_E(G)$ denotes an embedding of a graph G on a base rectangle R_E for the Möbius band M, formed by identifying top and bottom borders of the rectangle R_E, in opposite directions.

2. A partial edge of $R_E(G)$ that intersects the top or bottom border of R_E is called a "split edge." A split edge that extends from the top to the bottom of R_E is called a "full-length" edge.

3. The base graph of $R_E(G)$, denoted G_b, is the plane graph induced by deleting all split edges.

4. If $R_E(G)$ has any split edges, then before identification of sides, there are distinct left and right faces, denoted L and R. In the cylindrical identification of [12], the faces L and R remain distinct after identification, but in the Möbius identification they merge into a single face adjacent to the boundary of the band.

5. Two Möbius embeddings of a graph, $M_1(G)$ and $M_2(G)$, are called equivalent if they have equivalent base rectangle embeddings (i.e., embeddings with identical vertex rotation schemes, cf. [13]). In particular, the faces of $M_1(G)$ and $M_2(G)$ that border the boundary of M are identical.

6. A Möbius embedding $M(G)$ is called cylindrical if it is equivalent to the Möbius embedding induced by a cylindrical embedding $C(G)$ (cf. the proof of Prop. 1, cf. [2]); otherwise it is called non-cylindrical. Note that a planar graph may have a non-cylindrical embedding (for example, it is simple to provide such a drawing for K_4), but a non-planar graph does not have a cylindrical drawing, since all cylindrical graphs are planar.

Remark 1. We present the algorithm to obtain a BVGMP layout from an embedding $R_E(G)$ in four stages. For more details, see [2].

- Obtain a layout for a 2-connected graph G with non-cylindrical embedding $R_E(G)$ such that G_B is also 2-connected, and such that the border of the face L contains 2 or more vertices;
- Obtain a layout for a 2-connected graph G with non-cylindrical embedding $R_E(G)$ such that G_B is also 2-connected, and such that the border of the face R contains 2 or more vertices;
- Obtain a layout for a 2-connected graph G with non-cylindrical embedding $R_E(G)$ such that G_B is 1-connected but not 2-connected;
- Obtain a layout for an arbitrary graph G with embedding $R_E(G)$.

Lemma 1. Let P be a 2-connected plane graph, and let $S = < s_1, \ldots, s_i >$ and $T = < t_1, \ldots, t_j >$ be disjoint counterclockwise paths on the outer boundary cycle of P (we think of S as containing the "bottom" vertices of the cycle, and T the "top" vertices). Given a sequence of n numbers, $v_1 < v_2 < \ldots < v_n$, P has an st-numbering (cf. [6]), with $s = s_1$ and $t = t_j$, in which the vertices of S are numbered consecutively upward starting with v_1, and the vertices of T are numbered consecutively upward ending with v_n. Furthermore, the numbers of the "left" vertices, namely those (if any) on the clockwise path from s_1 to t_j, have increasing numbers. Likewise, the "right" vertices, those on the counterclockwise path from s_i to t_1 also have increasing numbers.

Definition 2. *A numbering of a plane graph as described in Lemma 1, is called a* Möbius numbering *of the graph. The base graph on the left in Fig. 2 gives an example of a Möbius numbering, in which* $S = \{1, 2, 3, 4\}$ *and* $T = \{7, 8, 9, 10\}$.

Algorithm 1. *Input:* A non-cylindrical embedding $R_E(G)$ of a graph G on a base rectangle R_E for the Möbius band M, such that the base graph G_b is 2-connected (so in particular, G is 2-connected), and such that the pre-identification face L has at least two vertices on it. *Output:* A parallel BVGM layout $R_B(G)$ for G on a base rectangle R_B for M.

1. Construct a Möbius numbering for G_b such that the top (resp. bottom) vertices of the numbering are those that are incident with the upper (resp. lower) half of a split edge. Then the left (resp. right) vertices of the numbering lie on the border of the face L (resp. R);

2. Give the edges of G_b the orientation induced by the Möbius numbering of its vertices. G_b is an acyclic digraph with the bottom and top vertices of L as its unique source and sink. Complete this to an acyclic orientation of $R_E(G)$ by directing all split edges from bottom to top. Thus the borders of L and R are both oriented bottom to top;

3. Let D^* be the digraph whose vertices correspond to the faces of $R_E(G)$, including the internal faces, the faces L and R, and the upper half F_T of each split face F. (Note that we may assume that each split face has exactly one upper and one lower half, and that the two halves are distinct, cf. [2].) The edges of D^* are the dual edges of $R_E(G)$, including upper split edges but not lower split edges, and are directed from left to right across the oriented edges of $R_E(G)$. D^* is an acyclic digraph with the faces L and R as its unique source and sink;

4. The vertices of D^* are given a topological numbering corresponding to the orientation of D^*, and the lower half F_B of each split face is given the same number as its upper half. L receives the lowest number, R the highest, and the upper half-faces and R are numbered sequentially left to right. Each number is also given a subscript: subscript I (for "Alg. 1") for L, R, and all internal faces; T (for "top") for all upper half-faces; and B (for "bottom") for all lower half-faces;

5. Construct a base rectangle R_B for the BVGMP layout, with $n + 2$ rows and $2m$ columns, where n is the number of vertices of G, and m is the number of faces in the embedding $M(G)$ on the Möbius band M that is induced by $R_E(G)$. The bottom and top rows are not used for vertices; the remaining rows are numbered bottom to top in increasing order, using the Möbius numbering of G_b. The first m columns are numbered left to right in increasing order, using the subscripted, topological numbering of the left, internal, and upper faces; the next $m - 1$ columns are numbered left to right in *decreasing* order, using the lower and internal faces, but the subscript I

for each internal face is *changed* to II (for "Alg. 2", which makes use of these columns); the last column is given the I-subscripted number of R. Thus each vertex corresponds to one row, and each face to two columns, symmetrically arranged around the centerline of R_B, and no two columns have the same subscripted number.

6. Construct an interval to represent each $v \in G$. Let $Left(v)$ (resp. $Right(v)$) denote the number of the face in $R_E(G)$ to the left of the leftmost incoming and outgoing edges (resp. to the right of the rightmost incoming and outgoing edges). The interval for v is placed in the row having the number of v, and it extends from column $Left(v)$ to column $Right(v)$.

The resulting layout on R_B is equivalent to the embedding $R_E(G)$.

Example 1. The left-hand diagram in Fig. 2 shows an embedding of the Petersen graph on a base rectangle for the Möbius band, and the right-hand diagram shows the results of applying Algorithm 1; to help readability, the dual edges are not shown. The BVGMP layout that results is shown on the right in Fig. 2.

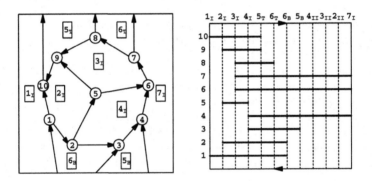

Fig. 2. Application of Algorithm 1 to the Petersen graph

We assume without loss of generality that both half-faces L and R contain one or more vertices, and that one or the other contains two or more vertices (cf. [2]). Algorithm 1 requires that the left-hand face L contain two or more vertices, so that the source and sink vertices are, respectively, at the bottom and top of L. If L does not have two vertices on its border, but R does, it is a simple matter to first reflect the embedding over the vertical center line of the base rectangle, exchanging the roles of L and R. However, when we extend the algorithm to the case in which the base graph G_b is not 2-connected, we want to use the same labels on all faces, regardless of which case each block falls into. Algorithm 1 can easily be modified so that a reflection of the embedding is not required. Details of the algorithm are omitted, but the essential idea is to lay out each vertex interval from right to left, using the column labels with subscript II rather than

I, and with $1_I = (m + 1)_{II}$ and $(m + 1)_I = 1_{II}$, where $m + 1$ is the number of faces on M.

Algorithm 2. *Input:* A non-cylindrical embedding $R_E(G)$ of a graph G on a base rectangle R_E for the Möbius band M, such that the base graph G_b is 2-connected, and such that the pre-identification face R has at least two vertices on it. *Output:* A parallel BVGM layout $R_B(G)$ for G on a base rectangle R_B for M.

Algorithms 1 and 2 give a BVGMP layout for a graph embedded on the Möbius band, if the base graph G_b is 2-connected. In the next algorithm we handle the case when G is 2-connected but G_b is not. We make the layout by filling in the bars one block at a time. First we need another definition.

Definition 3. *Suppose that $R_E(G)$ is an embedding of the 2-connected graph G on a base rectangle for Möbius band, with base graph G_b that is connected but not 2-connected, and let B be a block of G_b. We order the external faces (i.e., the upper and lower split faces, and L and R) from left to right and top to bottom, so that $L <$ any upper split face $<$ any lower split face $< R$. We denote by $Left(B)$ (resp., $Right(B)$) the first (resp., last) external face, in this ordering, that contains on its border some edge or split edge incident with B (such an edge may connect a vertex in B to another not in B).*

As an example, in the embedding at the upper left of Fig. 3, consider the G_b-block B containing the vertices v and y. $Left(B) = L$, and $Right(B)$ is the middle lower split face. A key observation is that, when we lay out each block of G_b, no adjacency involving an edge of B is represented by visibilities outside the columns from $Left(B)$ to $Right(B)$. If a vertex is a cutpoint of G, then its bar is a composite of the bars from the blocks to which it belongs.

Algorithm 3. *Input:* A non-cylindrical embedding $R_E(G)$ of a graph G on a base rectangle R_E for the Möbius band M, such that the base graph G_b is connected but not 2-connected, and such that L and R are the only external faces touching both the top and bottom sides of the base rectangle. *Output:* A parallel BVGM layout $R_B(G)$ for G on a base rectangle R_B for M.

1. The layout is done on a rectangle $R_B(G)$ with $2m$ columns and $n + 2$ rows, as in Algs. 1 and 2. The column labels are the same as before (with the first and the last columns doubly labeled for use in either Alg. 1 or Alg. 2). The row-labels are determined in subsequent steps, as each block is laid out.
2. The external faces of $R_E(G)$ are numbered as in Algs. 1 and 2. The numbers on the internal faces are assigned as each block is laid out. The blocks are laid out one by one, in the order of a *breadth-first traversal* of the block-cutpoint tree of G_b, by executing steps 3 to 6 for each block B.
3. Determine the values of $Left(B)$ and $Right(B)$; the bars of B are contained between the corresponding columns in the layout.
4. Contract the vertices of all blocks other than B, so that only the vertices of B remain; call the resulting embedded graph $E(B)$. At this point there may be multiple edges or loops at vertices of B that were cutpoints of G. If

neither face L nor R of $E(B)$ has two or more vertices on the border, then delete multiple split edges or split loops until that is the case. These edges are handled in Step 6.

5. Use either Alg. 1 or Alg. 2 to number and orient $E(B)$ and its dual. If a cut-point of B was numbered in a previous step, the *same number* must be used now as well. For non-cutpoints, use numbers that have not yet been used, chosen so that a Möbius-numbering results. For internal faces of $E(B)$, use the lowest possible face-numbers that have not yet been used. (The reason that the block-cutpoint tree is traversed in breadth-first order is to avoid vertex-numbering conflicts in this step; it doesn't matter what particular numbers are used, so long as they obey the rules for Möbius numbering.)

6. If any edges were deleted in Step 4, put them back now and orient them from bottom to top. For each vertex v in B, let $Left_B(v)$ be either $Left(v)$ in $E(B)$ or $Left(B)$, whichever is further right. Similarly, $Right_B(v)$ is either $Right(v)$ in $E(B)$ or $Right(B)$, whichever is further left. In other words, we "trim" the bar for v so that it is contained in the region from $Left(B)$ to $Right(B)$.

7. Once all the blocks of G have been labeled, we draw the bars for the layout of G. There is one row for each vertex, in increasing order of vertex numbers. The bar for a vertex v is the union of the bars from $Left_B(v)$ to $Right_B(v)$, for each block B containing v. The final result is a BVGMP layout for G.

Example 2. Figure 3 illustrates several steps of Algorithm 3, as well as the final layout. Blocks B_1 and B_2 are shown with other blocks contracted, and captions indicate which of Algorithms 1 and 2 is used, as well as the values of $Left(B_i)$ and $Right(B_i)$ (these values are computed *before* contracting the other blocks). Partial edges that were deleted in Step 4, and then put back in Step 6, are indicated by dotted lines.

Note that if an embedding $R_E(G)$ has G_b disconnected, then it is easy to construct a cylindrical embedding for G. Hence, if G is non-cylindrical and 2-connected, then G_b is connected. This, together with Algorithms 1, 2, and 3, gives the following theorem.

Theorem 1. *If G is a 2-connected graph embedded on the Möbius band, then there is an equivalent parallel BVGM layout of G.*

4   Characterizations of Parallel and Orthogonal BVGMs

In [12] Tamassia and Tollis show that a planar graph G admits a parallel BVG layout on the cylinder if and only if its block-cutpoint tree is a caterpillar. In this section we give an analogue of that result for parallel BVG layouts on the Möbius band. Now that we have established an algorithm for 2-connected graphs, the remaining results of [12] generalize quite straightforwardly (cf [2]).

Definition 4. *If the block-cutpoint tree T of a graph G is a caterpillar, then it is easy to see that the caterpillar's central path can be taken to be of the form $P = c_0 - B_1 - c_1 - B_2 - \ldots - c_{(k-1)} - B_k - c_k$, where the c_i are all the cutpoints of G, and the B_i are the non-leaf blocks of T. Any additional blocks are leaf-blocks, each containing a single cutpoint. We call this central path P the* spine *of T, and we call any leaf-block incident with either c_0 or c_k* head-block *of T.*

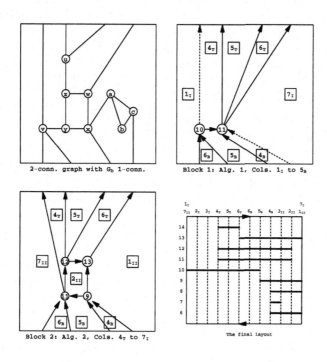

2-conn. graph with G_b 1-conn.

Block 1: Alg. 1, Cols. 1_I to 5_B

Block 2: Alg. 2, Cols. 4_T to 7_I

The final layout

Fig. 3. Applying Algorithm 3 to a 2-connected graph

Theorem 2. *A graph G has a parallel BVG layout on the Möbius band if and only if all the following conditions hold: (1) The block-cutpoint tree T of G is a caterpillar; (2) At most one block of G is non-planar; and (3) The non-planar block B, if it exists, must be a head-block in T.*

One can also consider BVGMs with bars *orthogonal* to the axis of the band. It is easy to show that these are equivalent to the *polar* and *circular* visibility graphs (abbreviated PVGs and CVGs) that are characterized by Hutchinson in [7]. These are graphs with vertices represented by circular arcs in the projective plane. In CVGs arcs are permitted to be complete circles, while in PVGs they are not. There is a reversible transformation taking CVGs and PVGs to orthogonal BVGMs, yielding the following equivalences. Characterizations of CVGs and BVGs, hence also of orthogonal BVGMs, can be found in [7].

Theorem 3. *A graph G is a CVG if and only if it is an orthogonal BVGM. G is a PVG if and only if it is an orthogonal BVGM in which bars may not be complete circles.*

5   Concluding Remarks and Acknowledgments

It would be interesting to find applications of visibility representations on the Möbius band in VLSI design or elsewhere, since the class of BVGMPs includes the class of BVGCPs. It should also be possible to extend these methods to higher-genus, non-orientable surfaces. Another visibility class that has been studied is *rectangle-visibility* graphs (RVGs), in which vertices are represented by rectangles, and edges by visibility in both the vertical and horizontal directions. While a number of papers have been published on RVGs in the plane [3,4,8], very little has been done on RVGs on higher-genus surfaces. Hutchinson and the author [5] have obtained edge-bounds for RVGs on the cylinder and torus.

This paper was written while the author was on sabbatical in England in 1999. She wishes to express her gratitude to the members of the School of Mathematical and Information Sciences at Coventry University for their warm welcome and generosity while she was a visitor there. She is also grateful to Joan Hutchinson and Tom Shermer for several helpful conversations.

References

1. G. Di Battista, P. Eades, R. Tamassia, and I. G. Tollis. **Graph Drawing: Algorithms for the visualization of graphs.** Prentice Hall, Upper Saddle River, NJ, 1999.
2. A. M. Dean. Bar-visibility graphs on the Möbius band. Manuscript, 1999.
3. A. M. Dean and J. P. Hutchinson. Rectangle-visibility representations of bipartite graphs. *Discrete Applied Math.*, 75:9–25, 1997.
4. A. M. Dean and J. P. Hutchinson. Rectangle-visibility of unions and products of trees. *J. Graph Algorithms and Applications*, 2:1–21, 1998.
5. A. M. Dean and J. P. Hutchinson. Rectangle-visibility graphs on surfaces. Manuscript, 1999.
6. S. Even and R. E. Tarjan. Computing an *st*-numbering. *Theoret. Comput. Sci.*, 2:339–344, 1976.
7. J. P. Hutchinson. On polar visibility representations of graphs. In: *Proc. Graph Drawing 2000*, J. Marks, ed., to appear.
8. J. P. Hutchinson, T. Shermer, and A. Vince. On representations of some thickness-two graphs (extended abstract). In: *Lecture Notes in Computer Science #1027*, F. Brandenburg, ed., Springer, Berlin, 159–166, 1995.
9. B. Mohar and P. Rosenstiehl. Tessellation and visibility representations of maps on the torus. *Discrete Comput. Geom.*, 19:249–263, 1998.
10. R. Tamassia and I. G. Tollis. A unified approach to visibility representations of planar graphs. *Discrete Comput. Geom.*, 1:321–341, 1986.
11. R. Tamassia and I. G. Tollis. Tessellation representations of planar graphs. In: *Proc. 27th Annual Allerton Conf.*, University of Illinois at Urbana-Champaign, 48–57, 1989.

12. R. Tamassia and I. G. Tollis. Representations of graphs on a cylinder. *SIAM J. Disc. Math.*, 4:139–149, 1991.
13. A. White. **Graphs, Groups and Surfaces (revised ed.).** North-Holland, Amsterdam, 1984.
14. S. K. Wismath. Characterizing bar line-of-sight graphs. In: *Proc. 1st ACM Symp. Comput. Geom.*, 147–152, 1985.

An Algorithm for Finding Three Dimensional Symmetry in Trees *

Seok-Hee Hong and Peter Eades

Basser Department of Computer Science, University of Sydney, Australia.
{shhong, peter}@cs.usyd.edu.au

Abstract. This paper presents a model for drawing trees symmetrically in three dimensions and a linear time algorithm for finding maximum number of three dimensional symmetries in trees.

1 Introduction

Symmetry is one of the most important aesthetic criteria for Graph Drawing. Drawings of graphs in Graph Theory textbooks are often symmetric, because the symmetry clearly reveals the structure of the graph.

However, previous work on symmetric graph drawing has only focused on two dimensions. The problem of determining whether a given graph can be drawn symmetrically is NP-complete in general [10]. Heuristics for symmetric drawings of general graphs have been suggested [4]. For restricted classes of graphs, there are polynomial time algorithms: Manning presents algorithms for constructing symmetric drawings of trees, outerplanar graphs and embedded planar graphs [8,9,10]; Hong gives algorithms for finding maximum number of symmetries in series-parallel digraphs and planar graphs [5,6].

In this paper, we extend symmetric graph drawing into three dimensions. Symmetry in three dimensions is much richer than that in two dimensions. For example, a maximal symmetric drawing of a tree in two dimensions is in Figure 1 (a), showing 12 symmetries. However, the maximal symmetric drawing of the same tree in three dimensions shows 48 symmetries, as in Figure 1 (b). In fact, using more complex examples, we can prove the following theorem.

Theorem 1. *For each integer $n \geq 10$ there is a tree T with n nodes such that T has no symmetric drawing in two dimensions, but has a drawing in three dimensions that displays $4\lfloor \frac{n-6}{8} \rfloor + 2$ symmetries.* [1]

This paper is organized as follows. In the next section, we give a model for drawing graphs symmetrically in three dimensions. The main results of the paper

---

* This research has been supported by a Postdoctoral Fellowship from the Korean Science and Engineering Foundation and a grant from the Australian Research Council. Animated drawings are available from S. Hong at http://www.cs.usyd.edu.au/@shhong/research3.htm.
[1] In this extended abstract, proofs are omitted.

J. Marks (Ed.): GD 2000, LNCS 1984, pp. 360–371, 2001.
© Springer-Verlag Berlin Heidelberg 2001

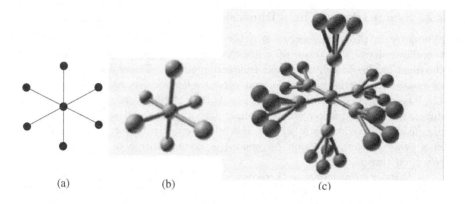

(a)   (b)   (c)

Fig. 1. *Symmetric drawings of trees in two and three dimensions.*

are in Section 3: here we present an algorithm for finding maximum number of symmetries of trees in three dimensions. A significant contribution of Section 3 is the introduction of a new data structure called the "Isomorphism Class Tree"; this structure is critical for the efficiency of the algorithm. A simple drawing algorithm is briefly described in Section 4. Section 5 concludes.

2   Symmetric Graph Drawing in Three Dimensions

In this section, we first review the model for symmetric graph drawing in two dimensions. To explain symmetry in three dimensions, we review some terminology from geometric symmetry and group theory [2,7,11]. Finally describe our new symmetry model in three dimensions.

2.1   Symmetric Graph Drawing in Two Dimensions

A symmetry of a two dimensional figure is an isometry of the plane that fixes the figure. There are two types of two dimensional symmetry, *rotational symmetry* and *reflectional (or axial) symmetry*. Rotational symmetry is a rotation about a *point* and reflectional symmetry is a reflection in an *axis*.

Symmetry in graph drawing is closely related to automorphisms of graphs: a symmetry of a graph drawing induces an automorphism of the graph. In this case, we say that the drawing *displays* the automorphism. If an automorphism is displayed as a symmetry in a drawing of the graph, then it is a *geometric* automorphism. The most critical part of the problem of drawing a graph symmetrically is to find a large group of geometric automorphisms. This formal model for symmetric drawing in two dimensions was introduced by a number of authors [3,5,6,8,9,10].

2.2   Symmetries in Three Dimensions

Symmetry in three dimensions is richer and more complex than symmetry in two dimensions. The types of symmetry in three dimensions can be roughly classified as *direct symmetry* and *indirect symmetry*. These are further refined as *rotation, reflection, inversion* and *rotary reflection (or rotary inversion)* [7, 11]. The difference from two dimensions is that a rotational symmetry in three dimensions is a rotation about an *axis* and a reflectional symmetry in three dimensions is a reflection in a *plane*. Inversion (or central inversion) is a reflection in a *point*. Rotary reflection (inversion) is a composition of a rotation and a reflection (inversion).

A *finite rotation group* in three dimensions is one of following three types. A *cyclic group (C_n)*, a *dihedral group (D_n)* and the rotation group of one of the *Platonic solids* [2,11]. There are only five regular Platonic solids, the *tetrahedron*, the *cube*, the *octahedron*, the *dodecahedron* and the *icosahedron*.

There are many types of *full symmetry groups* of a finite object in three dimensions. The complete list of all possible symmetry groups in three dimensions can be found in [2,7,11]. However, all are variations on just three types: pyramids, prisms, and Platonic solids. It can be shown that for the case of *trees*, a *maximum* size three dimensional symmetry group is one of three types: a regular pyramid configuration, a regular prism configuration, and the Platonic solids configuration.

A *regular pyramid* is a pyramid with a regular k-gon as its base. There is only one *k-fold rotation axis*, passing through the apex and the center of its base. There are k rotational symmetries, each of which is a rotation of $2\pi i/k$, $i = 0, 1, \ldots, k-1$. Also there are k reflectional symmetries in *reflection planes*, each containing the rotation axis. In total, the regular pyramid has $2k$ symmetries.

A *regular prism* has a regular k-gon as its top and bottom face. There are $k+1$ rotation axes and they can be divided into two classes. The first one, called the *principal* axis, is a k-fold rotation axis which passes through the centers of the two k-gon faces. The second class, of *secondary* axes, consists of k 2-fold rotation axes which lie in a plane perpendicular to the principal axis. The number of rotational symmetries is $2k$. Also, there are k reflection planes, each containing the principal axis, and another reflection plane perpendicular to the principal axis. Further it has $k-1$ rotary reflections. If k is even, then they are the same as rotary inversions including the central inversion. In total, the regular prism has $4k$ symmetries.

The tetrahedron has four 3-fold rotation axes and three 2-fold rotation axes. It has 12 rotational symmetries and in total 24 symmetries. The octahedron has three 4-fold rotation axes, four 3-fold axes, and six 2-fold rotation axes. It has 24 rotational symmetries and a full symmetry group of size 48. The icosahedron has six 5-fold rotation axes, ten 3-fold rotation axes, and fifteen 2-fold rotation axes. It has 60 rotational symmetries and a full symmetry group of size 120. Note that the cube and the octahedron are dual solids, and the dodecahedron and the icosahedron are dual. For details, see [2,7,11].

2.3   Symmetric Graph Drawing in Three Dimensions

A symmetry of a three dimensional graph drawing induces an automorphism of the graph, and this automorphism is *displayed* by the symmetry. We say that the automorphism is a *three dimensional symmetry* of the graph.

To draw a graph symmetrically in three dimensions, there are two steps: first find the three dimensional symmetries, then construct a drawing which displays these symmetries. The first step is the more difficult; given the three dimensional symmetries, the drawing is easy to construct. This paper concentrates on the first step.

For the purpose of drawing graphs symmetrically, we require a graph drawing to satisfy three *non-degeneracy* conditions: no two vertices are located at the same point, no two edges overlap (they may intersect at a point), and no vertex lies on an edge with which it is not incident.

Now we are ready to find three dimensional symmetry in trees. In the next section, we present an algorithm for finding maximum number of symmetries of trees in three dimensions.

3   Symmetry Finding Algorithm

In this section we describe an algorithm for finding three dimensional symmetry in trees.

The *center* of a tree is a vertex c such that the maximum distance between c and any leaf is minimized. The algorithm treats the input tree as a rooted tree, rooted at a center. Every tree has either one center or a set of two adjacent centers; however, for this algorithm, we need a single center and if there are two centers, then we add a new vertex on the edge joining the two centers to make one center. The center of a tree T is fixed by every automorphism of T and thus every symmetry of a drawing of T fixes the location of the center.

The basic idea of the symmetry finding algorithm is to construct all possible symmetric configurations and then find the configuration which has the maximum number of symmetries. To construct the symmetric configurations, we place the center of the input tree at the apex of a pyramid, or at the centroid of a prism and the Platonic solids. Then we place each subtree attached to the center to form a symmetric configuration.

The algorithm computes an auxiliary tree called the *Isomorphism Class Tree* (ICT). This tree is a fundamental structure defining the isomorphisms between subtrees in the input tree. Once the ICT has been computed, we use separate subroutines to find the symmetries in the pyramid, prism, and the Platonic solids configurations. These subroutines use the data stored at the root and the first level of the ICT. Thus, the overall algorithm can be divided into four steps, as follows.

```
Algorithm 3DSymmetry_Tree
Input: A tree T.
Output: A maximum size group of three dimensional symmetries of T.
```

1. Find the center of T and root T at the center.
2. Construct the *Isomorphism Class Tree* (ICT) of T.
3. Find symmetries of each type.
   a) Construct a pyramid configuration.
   b) Construct a prism configuration.
   c) Construct Platonic solids configuration.
4. Output the group of the configuration which has maximum size.

In Section 3.1, we define the ICT and describe an algorithm to construct it. Steps 3 (a), (b) and (c) are described in Section 3.2, 3.3 and 3.4. The following theorem summarises the results.

Theorem 2. Algorithm 3DSymmetry_Tree *computes a maximum size three dimensional symmetry group of a tree in linear time.*

3.1   The Isomorphism Class Tree

Roughly speaking, the *Isomorphism Class Tree (ICT)* of a tree T represents the isomorphism classes of subtrees of T, some relationships between these classes, the size of each class, and the sizes of the rotational symmetry groups of the subtrees. Before giving the formal definition of the ICT, first we explain why it is needed. As an example, we consider the pyramid configuration, because it is simplest.

Let c be the center of T. As mentioned above, we root T at c. Deleting c and all its incident edges results in a collection of rooted disjoint subtrees $T_1, T_2, \ldots, T_m$. Each T_i is rooted at the vertex c_i that was adjacent to the center. Using a rooted tree isomorphism algorithm [1] we can partition the T_i into rooted isomorphism classes $I_1, I_2, \ldots, I_k$. That is, if T_i and T_j are isomorphic subtrees, then they belong to the same isomorphism class. Let $n_i = \mid I_i \mid$ and g be the *greatest common divisor* (*gcd*) of all n_i. Then we can construct a pyramid with a g-fold rotation axis by placing the center at the apex and distributing the subtrees in the reflection planes, each containing a side edge of the pyramid (See Figure 3 (a)). It is clear that for each divisor j of g, there is a pyramid drawing which displays j rotational symmetries.

However, other symmetries are possible. We can choose one subtree T_j from I_i and place it such that the edge (c, c_j) is on the rotation axis as in Figure 3 (b). Note that in this case, the tree T_j must be fixed by a rotational symmetry. Thus the symmetry group of the pyramid is the intersection of the two symmetry groups: one of T_j, and the other which permutes the remaining subtrees $T_1, T_2, \ldots, T_{j-1}, T_{j+1}, \ldots, T_m$. The size of this group is a divisor of $gcd(e, n_1, n_2, \ldots, n_i - 1, \ldots, n_k)$ where e is the size of a rotational symmetry group of T_j.

Suppose that L_i is the set of sizes of rotational symmetry groups of T_j, and G_i is the set of divisors of $gcd(n_1, n_2, \ldots, n_i - 1, \ldots, n_k)$. Using this notation, we can compute the set L of sizes of the rotational symmetry groups (in the pyramid configuration) for T as follows.

Algorithm ComputeL
Input: The sizes $n_1, n_2, \ldots, n_k$ of the isomorphism classes of the subtrees at node u.
Output: The set L of sizes of rotational symmetries of the subtree rooted at u.

1. L = the set of divisors of $gcd(n_1, n_2, \ldots, n_k)$.
2. for $i = 1$ to k do
   a) Compute G_i = the set of divisors of $gcd(n_1, n_2, \ldots, n_i - 1, \ldots, n_k)$.
   b) Add $L_i \cap G_i$ to L.

This algorithm requires the computation of L_i for $i = 1, 2, \ldots, k$; in principle, this can be computed recursively. In practice it is more efficient to use the *Isomorphism Class Tree* (ICT), defined as follows.

Each node in the ICT represents an isomorphism class. Suppose that c is the center of a tree T and $I_1, I_2, \ldots, I_k$ are the rooted isomorphism classes of the rooted subtrees $T_1, T_2, \ldots, T_m$ of c. The root node r of the ICT represents the whole tree T. The children $u_1, u_2, \ldots, u_k$ of r represent the isomorphism classes $I_1, I_2, \ldots, I_k$ respectively. Suppose that $T_j \in I_i$ and c_j is the root of T_j. We can recursively decompose T_j into subtrees $T_{j_1}, T_{j_2}, \ldots, T_{j_p}$ by deleting c_j and then divide them into isomorphism classes $I_{i_1}, I_{i_2}, \ldots, I_{i_q}$. Then u_i has $u_{i_1}, u_{i_2}, \ldots, u_{i_q}$ as its children in the ICT. The ICT can be recursively constructed in this way until a subtree becomes a node. Figure 2 shows an example of the ICT.

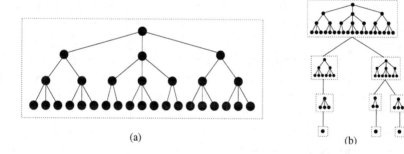

(a)   (b)

Fig. 2. *(a) a tree T (b) The Isomorphism Class Tree (ICT) of T.*

Suppose that v is a node in the ICT, representing an isomorphism class I_v and suppose that $T_j \in I_v$. We associate two values with v: the integer $n_v = |I_v|$ and the set L_v of sizes of the rotational symmetries of T_j. These two values are useful in finding the three dimensional symmetries of a tree.

Next we consider algorithms for constructing the ICT and its associated values. Using a rooted tree isomorphism algorithm [1], it is simple to construct the ICT and each n_v in linear time. However, we need an algorithm to compute L_v of each v. This can be computed by applying ComputeL in a bottom up approach on the ICT. However, a direct implementation of `ComputeL` may be expensive; we now show how to do it in linear time. Suppose that u is a node of T and $I_1, I_2, \ldots, I_k$ are the rooted isomorphism classes of the rooted subtrees $T_1, T_2, \ldots, T_m$ of u. We want to implement `ComputeL` at the node v corresponding to u in time $O(m)$. This can be done as follows.

We use a bit array to represent G_i, that is, $G_i[p] = 1$ if and only if p is a divisor of each of $n_1, n_2, \ldots, n_i - 1, \ldots$, and n_k. Also we represent L_i as a bit array: if $T_j \in I_i$ has a drawing which displays p rotations, then $L_i[p] = 1$ and $L_i[p] = 0$ otherwise. Note that if $L_i[p] = 1$, then $L_i[q] = 1$ for all divisors q of p. The output L of `ComputeL` can be represented in the same way. Then to compute $G_i \cap L_i$ and add it to L, we take the bitwise AND of L_i and G_i, and then the bitwise OR with L; this can be done in time $O(\min(\max(G_i), \max(L_i)))$.

Since $gcd(n_1, n_2, \ldots, n_k) \leq \min(n_1, n_2, \ldots, n_k)$, Step 1 of `ComputeL` can be implemented in time $O(k \min(n_1, n_2, \ldots, n_k))$, which is $O(m)$.

For Step 2, we consider two cases.

- For all p, $n_p > 1$. In this case,

$$gcd(n_1, n_2, \ldots, n_i - 1, \ldots, n_k) \leq \min(n_1, n_2, \ldots, n_i - 1, \ldots, n_k). \quad (1)$$

  Thus $\max(G_i) \leq \min(n_1, n_2, \ldots, n_k)$. It can be deduced that both parts of step 2 can be implemented in time $O(k \min(n_1, n_2, \ldots, n_k))$, which is $O(m)$.
- For some p, $n_p = 1$. In this case, the inequality (1) does not hold. However, for $i \neq p$, $G_i = \{1\}$; thus we only need to execute steps 2(a) and 2(b) for the single case $i = p$. Both Step 2(a) and Step 2(b) take time $O(\max(G_p))$, which is $O(\max(n_1, n_2, \ldots, n_{p-1}, n_{p+1}, \ldots, n_k))$, which is $O(m)$.

It follows that `ComputeL` can be implemented in time $O(m)$, and thus the whole of the ICT, including each L_v, can be computed in linear time.

The following sections describe steps 3(a), 3(b), and 3(c) of `3DSymmetry_Tree` in turn. These algorithms use the ICT; in fact, they only consider the root r and its children u_i (together with n_i and L_i) in the ICT.

3.2   Pyramid Configuration

In this section, we give an algorithm for finding all possible rotational symmetries in the pyramid configuration. The basic idea is to construct a pyramid-type drawing of a tree by placing the center of the tree at the apex of a pyramid, some fixed subtrees about the rotation axis, and k isomorphic subtrees in the reflection planes that contain the side edges of the pyramid. The resulting drawing has the same symmetry group of the k-gon based pyramid.

The set L_r at the root of the ICT contains the sizes of rotational symmetry groups which fix at most one subtree. However, there is another way of constructing symmetric drawings of trees by placing another subtree on the rotation axis: we can place *two* possibly different subtrees on the rotation axis, if the symmetry groups of the fixed subtrees are appropriate.

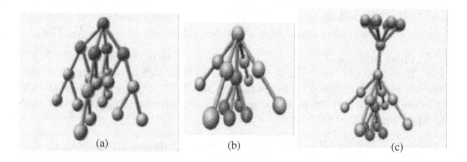

Fig. 3. *The pyramid configuration with different number of fixed subtrees.*

For example, Figure 3 (a) shows the pyramid with no fixed subtree and Figure 3 (b) shows the pyramid with one fixed subtree. Figure 3 (c) shows the pyramid with two fixed subtrees.

The pyramid algorithm mainly depends on the number of fixed subtrees on the rotation axis. This is at most two, as described in the following Lemma.

Lemma 1. *Suppose that T is a tree rooted at the center c of T. Let $T_1, T_2, \ldots, T_m$ be the rooted subtrees of c. In a pyramid configuration, there are at most two fixed subtrees T_i and T_j on the rotation axis.*

From Lemma 1 one can derive the following theorem, which forms the basis of the algorithm.

Theorem 3. *Suppose that T is a tree, r is the root node of the ICT of T, and $u_1, u_2, \ldots, u_k$ are the children of r in the ICT. Then:*

1. *If g is the size of a rotation group in the pyramid configuration, then g is a divisor of $\gcd(n_1, n_2, \ldots, n_k)$.*
2. *If there is one fixed subtree T_j, then the rotation group of the pyramid is the intersection of the rotation group of T_j and a rotation group of the remaining T_i $(i = 1, 2, \ldots, m, i \neq j)$. Suppose that u_p represents the isomorphism class I_p for which $T_j \in I_p$, and G_p is the set of divisors of $\gcd(n_1, n_2, \ldots, n_p - 1, \ldots, n_k)$. Then the intersection is $G_p \cap L_p$.*
3. *If there are two isomorphic fixed subtrees T_i and T_j, then the rotation group of the pyramid is the intersection of the rotation group of T_i (or T_j) and the rotation group of the remaining T_ℓ $(\ell = 1, 2, \ldots, m, \ell \neq i, j)$. Suppose that u_p represent the isomorphism class I_p to which T_i and T_j belong, and G'_p is the set of divisors of $\gcd(n_1, n_2, \ldots, n_p - 2, \ldots, n_k)$. Then the intersection is $G'_p \cap L_p$.*
4. *If there are two nonisomorphic fixed subtrees T_i and T_j, then the rotation group of the pyramid is the intersection of the rotation group of T_i, the rotation group of T_j and the rotation group of the remaining T_ℓ $(\ell = 1, 2, \ldots, m, \ell \neq i, j)$. Suppose that u_p (u_q) represents the isomorphism class*

I_p (I_q) to which T_i (T_j) belong, and G_{pq} is the set of divisors of $\gcd(n_1, n_2, \ldots, n_p - 1, \ldots, n_q - 1, \ldots, n_k)$. Then the intersection is $G_{pq} \cap L_p \cap L_q$.

A direct implementation based on Theorem 3 may be computationally expensive. To reduce the time complexity, we use an approach based on trial division. We test candidate sizes g of the rotation group. First we compute the remainder r_i of the division of n_i by g. There are four cases, corresponding to the four parts of Theorem 3:

1. If all $r_i = 0$ for all i, then g is the size of a rotation group, with no fixed subtree.
2. Suppose that there is only one p such that $r_p = 1$ and all the others are 0; further suppose that $g \in L_p$. Then there is a rotation of size g, with one fixed subtree.
3. Suppose that there is only one p such that $r_p = 2$ and all the others are 0; further suppose that $g \in L_p$. Then there is a rotation of size g, with two isomorphic fixed subtrees.
4. Finally, suppose that there are only two r_p and r_q which are 1, and all other are 0; further suppose that $g \in L_p \cap L_q$. Then there is a rotation of size g, with two nonisomorphic fixed subtrees.

If none of the four cases above hold, then g is not the size of a rotation group.

Next we need to compute the intersection of the possible candidate g and the rotation group of the fixed subtrees (L_p or L_q). We can use the same bit representation which was used for L_v in the ICT as in the previous section. Each candidate g is represented by a bit array G in a similar fashion. Again, the intersections are computed by a bitwise AND, but in this case there may be more than two vectors involved (when there are two candidates, as in item 4 above).

Now we present an algorithm to find the symmetries in a pyramid configuration.

Algorithm Pyramid
(1) For each candidate g do
    (1.1) Compute the remainder r_i of n_i divided by g, for $1 \leq i \leq k$.
    (1.2) Compute the set G of divisors of g.
    (1.3) Compute the set L of sizes of all rotational symmetry groups:
        (1.3.1) If $r_i = 0$ for $1 \leq i \leq k$, then add G to L.
        (1.3.2) If $r_p = 1$, and $r_i = 0$ for $i \neq p$, then add $G \cap L_p$ to L.
        (1.3.3) If $r_p = 2$ and $r_i = 0$ for $i \neq p$, then add $G \cap L_p$ to L.
        (1.3.4) If $r_p = 1$, $r_q = 1$ and $r_i = 0$ for $i \neq p, q$, then add $G \cap L_p \cap L_q$ to L.
(2) Return the maximum element of L.

An analysis similar to that used in the previous section shows that **Pyramid** can be implemented in linear time. Note that the output of **Pyramid** is the maximum number of *rotational* symmetries in the pyramid configuration; the maximum size of a symmetry group in the pyramid configuration is twice as big.

3.3   Prism Configuration

A tree drawn in a prism configuration has at most three types of rotation axes for fixing subtrees, and the number of fixed subtrees can be larger than with the pyramid configuration. This is described in the following Lemma.

Lemma 2. *Suppose that T is a tree rooted at the center c of T. Let $T_1, T_2, \ldots,$ T_m be rooted subtrees of c. In a regular k-gon prism configuration, there are zero or two fixed isomorphic subtrees on the k-fold rotation axis. Also, there are at most two types of k fixed isomorphic subtrees on the secondary axes.*

For example, Figure 4 (a) shows the prism with no fixed subtree and Figure 4 (b) shows the prism with fixed subtrees on the principal axis. Figure 4 (c) shows the prism with fixed subtrees on the principal and the secondary axes.

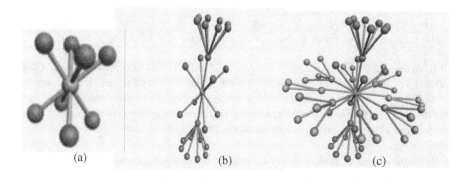

(a)   (b)   (c)

Fig. 4. *The prism configuration with different number of fixed subtrees.*

From Lemma 2 we can derive the following theorem, which forms the basis of the prism algorithm.

Theorem 4. *Suppose that T is a tree and the ICT is the Isomorphism Class Tree of T. Suppose that r is the root node of the ICT and $u_1, u_2, \ldots, u_k$ are the children of r in the ICT. Each u_i has n_i and L_i. Suppose that g is the size of the principal rotation axis of the prism configuration. Let r_i be the remainder of each n_i divided by g and m_i be the quotient of n_i divided by g. Then g satisfies both of the following conditions.*

1. *At most two m_i are odd and $2 \in L_i$.*
2. *At most one r_i is 2 and $g \in L_i$.*

Using theorem 4, one can construct a linear time algorithm, similar to **Pyramid**, to find the maximum number of rotational symmetries in a prism configuration. We omit this algorithm. Note that maximum number of symmetries in the prism configuration is four times as big as the maximum number of rotational symmetries in the prism configuration.

3.4   Platonic Solids Configuration

The Platonic solids have many rotation axes. However, the symmetry groups of the Platonic solids are fixed, and we only need to test whether we can construct a three dimensional drawing of a tree which has the same symmetry group as one of the Platonic solids. Using the similar method in the previous section, we can test this in a relatively simple way. For an example, we consider the cube. The number of the fixed subtrees on each axis is described in the following Lemma.

Lemma 3. *Suppose that T is a tree rooted at the center c of T. Let $T_1, T_2, \ldots, T_m$ be rooted subtrees of c. In the cube configuration, there are either zero or six fixed isomorphic subtrees on the 4-fold rotation axes. Also there are either zero or eight fixed isomorphic subtrees on the 3-fold rotation axes and either zero or twelve fixed isomorphic subtrees on the 2-fold rotation axes.*

For example, Figure 1 (c) shows the fixed subtree on the 4-fold axes of the cube configuration. From Lemma 3 we can derive the following theorem.

Theorem 5. *Suppose that T is a tree and the ICT is the Isomorphism Class Tree of T. Suppose that r is the root node of the ICT and $u_1, u_2, \ldots, u_k$ are the children of r in the ICT. Each u_i has n_i and L_i. Let r_i be the remainder of each n_i divided by 24. Then, if each r_i and L_i satisfies one of the following conditions, T has the same symmetry group as the cube.*

1. *At most one r_i is 6 and $4 \in L_i$ and at most one r_j is 8 and $3 \in L_j$ and at most one r_k is 12 and $2 \in L_k$.*
2. *At most one r_i is 14 and $4, 3 \in L_i$ and at most one r_j is 12 and $2 \in L_j$.*
3. *At most one r_i is 18 and $4, 2 \in L_i$ and at most one r_j is 8 and $3 \in L_j$.*
4. *At most one r_i is 20 and $3, 2 \in L_i$ and at most one r_j is 6 and $4 \in L_j$.*
5. *At most one r_i is 2 and $4, 3, 2 \in L_i$.*

Theorem 5 can be used to construct a linear time algorithm to test whether a tree has the same symmetry group as the cube. Similar results can be used to construct algorithms to test whether a tree has the same symmetry group as the icosahedron and the tetrahedron. We omit these algorithms.

4   Drawing Algorithm

Given a tree T, and a group Γ of three dimensional symmetries of T, it is straightforward to construct a straight-line drawing of T which displays Γ.

We draw the center of the tree at the origin. Non-fixed subtrees of the center are drawn in reflection planes through the origin, in such a way that the drawings of isomorphic subtrees are congruent. Fixed subtrees are drawn recursively. By assigning disjoint areas of the reflection planes to different subtrees, one can ensure planarity of the drawing.

5   Conclusion

This paper presents a linear time algorithm to construct maximally symmetric drawings of trees in three dimensions. These symmetries are symmetries of the *whole* drawing. It is also possible to extend this work to display symmetries of part of the drawing by providing a model and algorithms for such "partial symmetries" in three dimensions.

As further work, we would like to draw planar graphs symmetrically in three dimensions. Heuristics for drawing general graphs symmetrically in three dimensions remains as a challenge.

References

1. A. Aho, J. Hopcroft and J. Ullman, *The Design and Analysis of Computer Algorithms*, Addison-Wesley, 1974.
2. M. A. Armstrong, *Groups and Symmetry*, Springer-Verlag, 1988.
3. P. Eades and X. Lin, Spring Algorithms and Symmetry, *Computing and Combinatorics*, Springer Lecture Notes in Computer Science 1276, (Ed. Jiang and Lee), pp. 202-211.
4. H. Fraysseix, An Heuristic for Graph Symmetry Detection, *Graph Drawing'99*, Lecture Notes in Computer Science 1731, (Ed. J. Kratochvil), pp. 276-285, Springer Verlag, 1999.
5. S. Hong, P. Eades, A. Quigley and S. Lee, Drawing Algorithms for Series-Parallel Digraphs in Two and Three Dimensions, In S. Whitesides, editor, Graph Drawing (Proc. GD'98), vol. 1547 of Lecture Notes in Computer Science, pp. 198-209, Springer Verlag, 1998.
6. S. Hong, P. Eades and S. Lee, An Algorithm for Finding Geometric Automorphisms in Planar Graphs, *Algorithms and Computation*, Lecture Notes in Computer Science 1533, (Ed. Chwa and Ibarra), pp. 277-286, Springer Verlag, 1998.
7. E. H. Lockwood and R. H. Macmillan, *Geometric Symmetry*, Cambridge University Press, 1978.
8. J. Manning and M. J. Atallah, Fast Detection and Display of Symmetry in Trees, *Congressus Numerantium*, 64, pp. 159-169, 1988.
9. J. Manning and M. J. Atallah, Fast Detection and Display of Symmetry in Outerplanar Graphs, *Discrete Applied Mathematics*, 39, pp. 13-35, 1992.
10. J. Manning, Geometric Symmetry in Graphs, *Ph.D. Thesis, Purdue Univ.*, 1990.
11. G. E. Martin, *Transformation Geometry, an Introduction to Symmetry*, Springer, New York, 1982.

On Maximum Symmetric Subgraphs

Ho-Lin Chen[1], Hsueh-I. Lu[2], and Hsu-Chun Yen[1]

[1] Department of Electrical Engineering, National Taiwan University, Taipei 106, Taiwan, R.O.C. dnbcom@ms4.hinet.net, yen@cc.ee.ntu.edu.tw.
[2] Institute of Information Science, Academia Sinica, Taipei 115, Taiwan, R.O.C. hil@iis.sinica.edu.tw.

Abstract. Let G be an n-node graph. We address the problem of computing a maximum symmetric graph H from G by deleting nodes, deleting edges, and contracting edges. This NP-complete problem arises naturally from the objective of drawing G as symmetrically as possible. We show that its tractability for the special cases of G being a plane graph, an ordered tree, and an unordered tree, depends on the type of operations used to obtain H from G. Moreover, we give an $O(\log n)$-approximation algorithm for the intractable case that H is obtained from a tree G by contracting edges. As a by-product, we give an $O(\log n)$-approximation algorithm for an NP-complete edit-distance problem.

1 Introduction

As graphs are known to be one of the most important abstract models in various scientific and engineering areas, *graph drawing* has naturally emerged as a fast growing research topic in computer science. Among various aesthetics investigated in the literature, *symmetry* has received much attention recently [2,3,4, 5,6,9,10,16,15,17]. As a symmetric graph can be decomposed into a number of isomorphic subgraphs, only a portion of the graph, together with the symmetry information, is sufficient to define the original graph. In this way, symmetric graphs can often be represented in a more succinct fashion than their asymmetric counterparts. In reality, however, we often have to lower our expectations by allowing imperfection while considering the so-called 'nearly symmetric' drawings of graphs. To draw a graph in a nearly symmetric fashion, a good starting point might be to draw its symmetric subgraph as large as possible first, and then add the remaining nodes and edges to the drawing. Like many of the graph drawing problems, determining whether a graph has an axial or rotational symmetry is computationally intractable [15]. The maximum symmetric subtrees (i.e., subtrees that exhibit symmetric drawings), however, can be computed in polynomial time [4]. For series-parallel graphs, algorithms that display as much symmetry as possible can be found in [10]. Aside from the above algorithmic aspects of symmetric drawings, in [6] several types of symmetries, including axial symmetries and rotational symmetries, have been characterized in a unified way using geometric automorphism groups.

To formulate the notion of *near symmetry*, in this paper we propose a quantitative measure to capture the *degree of symmetry* in drawing graphs, and then

J. Marks (Ed.): GD 2000, LNCS 1984, pp. 372–383, 2001.

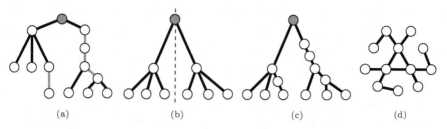

Fig. 1. (a) An asymmetric graph G rooted at the gray node. (b) A maximum axially symmetric graph H obtained from G by contracting edges. (c) A nearly symmetric drawing of G. (d) A 3-rotational symmetric graph.

investigate the complexity of computing such a measure for ordered trees, unordered trees, and plane graphs. An $O(\log n)$-approximation algorithm is given for an NP-complete case for unordered trees. Given a graph G, our symmetry measure of G is the maximum number of edges in a symmetric graph H that is obtained from G by applying a sequence of edge contractions, node deletions, or edge deletions. For example, the asymmetric graph G in Figure 1(a) can be turned into a symmetric graph H in Figure 1(b) by contracting the gray edges. One can verify that H is a maximum axially symmetric graph obtainable from G by contracting edges. Therefore, the degree of axial symmetry for G is 8. Visually, the graph in Figure 1(a) admits a nearly symmetric drawing as displayed in Figure 1(c). In light of the above, our symmetry measure can be thought of as a quantitative way of defining the notion of *near symmetry* in graph drawings. Graphs with higher degrees of symmetry have the tendency to exhibit better symmetric appearances visually. For more about nearly symmetric drawings, the reader is referred to [9].

As stated above, our main concern is to turn a graph into a symmetric graph with maximum number of edges through edge contractions, edge deletions, and node deletions. If the resulting graph is axially (respectively, rotationally) symmetric, we call the problem DAS (respectively, DRS), standing for degree of axial (respectively, rotational) symmetry. The tractability results are summarized in Tables 1 and 2. By allowing nodes to be drawn sufficiently close to each other (see Figure 1(c)), edge contractions, node deletions, and edge deletions seem to be rather natural to define near symmetry. In our setting, if a graph cannot be turned into a symmetric one, then the degree of symmetry is zero.

Our DAS problems are related to GRAPH ISOMORPHISM problems. Such problems include TREE INCLUSION [12] and EDIT DISTANCE [20], which have applications to analyzing molecular structures in biology. Given two labelled trees A and B, TREE INCLUSION is to determine whether A can be obtained from B by 'contracting' nodes, whereas EDIT DISTANCE is to determine the minimum number of 'changes,' 'contracting' (or its dual) needed to transform A into B. The main disparity between our DAS problem and those related to graph isomorphism is that the latter deal with two or more graphs from which some common substructures are extracted, whereas in the case of DAS, a single graph

Table 1. The tractability for maximum axially symmetric subgraph.

		node deletion	edge deletion	edge contraction
tree	ordered	P [Theorem 1]	P [4]	P [Theorem 1]
	unordered	P [Theorem 3]	P [4]	NPC [Theorem 2]
graph	plane	P [Theorem 4]	NPC [Theorem 5]	NPC [Theorem 5]
	general	NPC [15]	NPC [15]	NPC [15]

Table 2. The tractability for maximum rotationally symmetric subgraph with respect to unordered trees.

	node deletion	edge deletion	edge contraction
$O(1)$-degree trees	P [Theorem 8]	P [1]	Do not preserve degree bound
general trees	NPC [Theorem 9]	NPC [Theorem 9]	NPC [Theorem 9]

is considered. (Although, as we shall see later, a number of techniques commonly seen in the graph isomorphism research turn out to play a constructive role in our design and analysis.) As for trees, our DAS problem also differs from the TREE INCLUSION and EDIT DISTANCE in the following. First, our DAS problem is defined over graphs (while treating trees, both ordered and unordered, as a special case), as opposed to TREE INCLUSION and EDIT DISTANCE which are explicitly targeted for trees only. Second, instead of given two 'labelled' trees as inputs in the TREE INCLUSION and EDIT DISTANCE cases, in our symmetric drawing setting the input consists of a single graph (or tree). In addition, our graphs are not labelled. Finally, in addition to *contractions*, we also consider node deletions and edge deletions. An interesting by-product of our work is an $O(\log n)$-approximation algorithm for an NP-complete problem related to EDIT DISTANCE problems.

The rest of the paper is organized as follows. Section 2 defines the degree-of-symmetry problems. Section 3 studies the tractability of the problems for axial symmetry, and gives two approximation algorithms, one for a symmetry problem, the other for a problem related to edit distance. Section 4 studies the tractability of the problems for rotational symmetry with respect to unordered trees. Section 5 concludes the paper with some future research directions.

2   Preliminaries

Let G be a graph. Let $|G|$ denote the number of nodes in G. A *drawing* of G on the plane is a mapping D from the nodes of G to $\Re^2$, where $\Re$ is the set of real numbers. That is, each node v is placed at point $D(v)$ on the plane, and each edge (u, v) is displayed as a line segment connecting $D(u)$ and $D(v)$. A graph G has an *axial symmetry* if there exists a drawing D under which the image of G is symmetric with respect to a straight line on the plane. G has a *k-rotational symmetry* if there exists a drawing D such that D is unchanged if

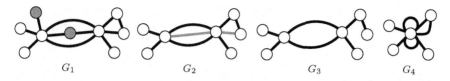

Fig. 2. Four graphs for illustrating the operations of deleting nodes, deleting edges, and contracting edges.

the plane is rotated at some point by $360/k$ degrees. For example, the drawing shown in Figure 1(b) has an axial symmetry; and the drawing in Figure 1(d) has a 3-rotational symmetry. Axial symmetry and rotational symmetry are two special kinds of *geometric automorphisms* defined by Hong, Eades, and Lee [8]. The reader is referred to [6,8,9,15] for more about symmetry in graph drawing.

For a connected G, the following basic graph operations are used throughout the paper.

- *Node deletion.* Only nodes of degree no more than two can be deleted from G. For any degree-one node w of G, deleting w is the operation of removing w and its incident edge from G. If the degree of w is two, where u and v are its neighbor in G, then deleting w is the operation of deleting w, (u, w), and (v, w) from G, and then adding a copy of (u, v) to G.
- *Edge deletion.* The operation can only be applied to an edge (u, v) whose removal does not disconnect G, or one of u and v is of degree one. Deleting (u, v) is the operation of removing edge (u, v) (or exactly one copy of the parallel edges incident to u and v). In case a degree-one node is involved, the node is removed as well.
- *Edge contraction.* For any edge (u, v) of G, contracting (u, v) is the operation of deleting the edge incident to u and v, and then merging u and v into a single node. Note that if G has m parallel edges incident to u and v, then all but one of those m parallel edges become self-loops incident to the merged node in the resulting graph.

Examples are shown in Figure 2: G_2 is obtained from G_1 by deleting the gray nodes; G_3 is obtained from G_2 by deleting the gray edges; and G_4 is obtained from G_2 by contracting the gray edges. Given graphs G and H, we write $G \overset{\text{ec}}{\rightarrow} H$, $G \overset{\text{nd}}{\rightarrow} H$, and $G \overset{\text{ed}}{\rightarrow} H$ to signify that H can be obtained from G through a sequence of edge contractions, node deletions, and edge deletions, respectively. Clearly, a node deletion can be viewed as an edge contraction. Hence, $G \overset{\text{nd}}{\rightarrow} H$ implies $G \overset{\text{ec}}{\rightarrow} H$. One can easily see that if H consists of a single node and no edges, then $G \overset{\text{ec}}{\rightarrow} H$ and $G \overset{\text{ed}}{\rightarrow} H$ hold for any nonempty G. However, $G \overset{\text{nd}}{\rightarrow} H$ does not necessarily hold, since parallel edges cannot be removed through node deletions.

Given a graph G, the *degree of axial* (respectively, *rotational*) *symmetry* of G is defined to be the size of a maximum axially (respectively, rotationally) symmetric graph H that can be derived from G through basic graph operations. In this paper, we investigate the tractability of determining the degree of axial symmetry (DAS) and the degree of rotational symmetry (DRS). Let DAS_{ec}, DAS_{nd}, and DAS_{ed} denote the DAS problem with respect to edge contraction, node deletion, and edge deletion, respectively. Let DRS_{ec}, DRS_{nd}, and DRS_{ed} be defined similarly. For any graphs G_1 and G_2, let $dist_{ec}(G_1, G_2)$, $dist_{ed}(G_1, G_2)$, and $dist_{nd}(G_1, G_2)$ denote the minimum number of edge contractions, edge deletions, and node deletions required to transform G_1 and G_2 into two identical graphs, respectively.

Fact 1 (see [15]) *For each operation* op $\in \{ec, ed, nd\}$, DAS_{op} *and* DRS_{op} *for graphs are NP-complete.*

3   Axial Symmetry

3.1   Ordered Trees

We begin by considering the DAS problems for ordered and unordered trees. Unless stated otherwise, trees are assumed to be rooted for the rest of the paper. For any ordered forest F, let reflection(F) be the ordered forest obtained from F by (a) reversing the order of the trees in F, and (b) reversing the order of the children of v for each node v in F. For any ordered tree T, let $T(i)$ be the subtree of T rooted at v_i, where $v_1, v_2, \ldots, v_n$ is the postordering of T.

Fact 2 (see [20,13]) *For any ordered forests F_1 and F_2, $dist_{ec}(F_1, F_2)$ can be computed in polynomial time.*

Theorem 1. DAS_{ec} *and* DAS_{nd} *for ordered trees can be solved in polynomial time.*

Proof. Clearly, one of the following two cases holds for any axially symmetric tree T.

- *Case 1*: The root of T has $2k + 1$ children for some integer k. Let v_i be the $(k + 1)$-st child of the root of T. Let F_1 be the subforest of T induced by the nodes v_j with $j < \ell$, where v_ℓ is the leftmost leaf node of $T(i)$. Let F_2 be the subforest of T induced by the nodes v_j with $i < j < n$. Clearly, F_1 is identical to reflection(F_2).
- *Case 2*: The root of T has $2k$ children for some integer k. If $k \geq 1$, then let v_i be the k-th child of the root of T. Let F_1 be the subforest of T induced by the nodes v_j with $1 \leq j \leq i$. Let F_2 be the subforest of T induced by the nodes v_j with $i < j < n$. Clearly, F_1 is identical to reflection(F_2).

We prove the statement for DAS_{ec} by giving a dynamic-programming algorithm. Let T be the given ordered tree, where $v_1, v_2, \ldots, v_{|T|}$ is the postordering of T. Clearly, $v_{|T|}$ is the root of T. By the above observations for any axially symmetric ordered tree, one can verify that the algorithm shown in Figure 3 correctly computes the minimum number of edge contractions required to turn T into an axially symmetric ordered tree. An example that illustrates **number-of-contraction** for v_7 is shown in Figure 4. The correctness of

```
function number-of-contraction(T) {
    for i = 1 to |T| − 1 do {
        let A_i be the set of ancestors of v_i in T_i;
        let F_1 be the subforest of T induced by {v_1, v_2, . . . , v_i};
        let F_2 be the subforest of T induced by {v_{i+1}, v_{i+2}, . . . , v_{|T|}} − A_i;
        let F_3 be obtained by removing T(i) from F_1;
        let c_i = min{dist_ec(F_1, reflection(F_2)),
                     dist_ec(F_3, reflection(F_2)) + number-of-contraction(T(i))};
    }
    return min_{1≤i≤|T|−1} |A_i| + c_i;
}
```

Fig. 3. An algorithm for computing the minimum number of edge contractions required to turn T into an axially symmetric graph.

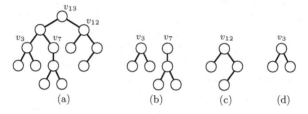

Fig. 4. (a) An ordered tree. (b) The F_1 for v_7. (c) The F_2 for v_7. (d) The F_3 for v_7.

number-of-contraction is immediate from the above two-case observation. By Fact 2, it is not difficult to see that number-of-contraction runs in polynomial time.

The statement for DAS$_{nd}$ can be proved similarly. □

3.2 Unordered Trees

Theorem 2. DAS$_{ec}$ *for unordered trees is NP-complete.*

Proof. By Fact 1, it suffices to show a reduction from the following variant of SATISFIABILITY, which remains NP-complete [12]: The input f is a set of m clauses $C_1, C_2, \ldots, C_m$ over variables $v_1, v_2, \ldots, v_n$, where each $\neg v_i$ appears in at most one clause of f. Let $\ell = (12m + 10)(m + \sum_{i=1}^{m} |C_i|) + n$, where $|C_i|$ is the number of literals in C_i. For each $1 \leq k \leq m$, let G_k be as shown in Figure 5(b). Clearly, the height of each G_i is $3m + 5$. The unordered tree instance T_f consists of two subtrees L and R, each begins with a long path of length ℓ. The length-ℓ path of L has subtrees $G_1, \ldots, G_m$. The length-ℓ path of R has subtrees $P_1, \ldots, P_n$, where P_i is defined as follows. If $\neg v_i$ does not appear in f, then P_i has a subtree G_j for each index j with $v_i \in C_j$. If $\neg v_i$ appears in a clause, say, C_r of f, then P_i is exactly G_r, whose leftmost leaf node has a subtree G_j for each index j with $v_i \in C_j$. An example of T_f is shown in Figure 5(a).

Now we prove that f is satisfiable if and only if $T_f \overset{ec}{\rightarrow} T$ holds for some axially symmetric tree T with $|T| \geq 1 + 2|L|$. Suppose f is satisfiable by a truth

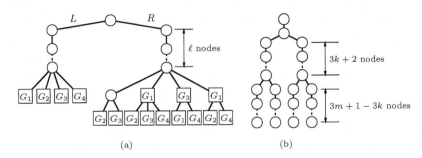

Fig. 5. (a) The T_f for $f = \{\neg v_2 \vee v_3 \vee \neg v_4,\ v_1 \vee v_2 \vee v_4,\ v_1 \vee v_2 \vee \neg v_3,\ v_2 \vee v_3 \vee v_4\}$. (b) The building block G_k.

assignment ϕ. We show how to obtain L by applying edge contractions to R. If $\phi(v_i) =$ false, then we remove all those G_j with $v_i \in C_j$ from P_i by edge contractions. If $\phi(v_i) =$ true, then we remove all but those G_j with $v_i \in C_j$ from P_i by edge contractions. Since each C_j, with $1 \le j \le m$, is satisfied by ϕ, it is clear that at least one copy of G_j remains in the resulting R. Thus, we can apply additional edge contractions on the resulting R to obtain L.

Now we assume that there is an axially symmetric tree T such that $T_f \overset{\text{ec}}{\leadsto} T$ and $|T| \ge 1 + 2|L|$ hold. One can easily verify that L and the length-ℓ path in R must stay intact in T. In the symmetric drawing, if block G_i in L is mapped into block G_j in R, then $i = j$. Note that the depth of each leaf node in each G_i, with $1 \le i \le m$, is exactly $m + 6$. Therefore, if $i \ne j$, then $G_i \overset{\text{ec}}{\leadsto} G_j$ does not hold. It follows that each G_i in the right subtree of T comes from a P_j of R. If P_j was composed of G_i whose leftmost leaf node has some subtrees, then let $\phi(v_i) =$ false. Otherwise, let $\phi(v_i) =$ true. Hence, ϕ indeed satisfies f. $\square$

Theorem 3. DAS$_{\text{nd}}$ *for unordered trees can be solved in polynomial time.*

Proof. Let $v_1, v_2, \dots, v_n$ be a postordering of the input tree T. If in the outcome of the conversion between $T(i)$ and $T(j)$, v_i and v_j remain to be the roots of $T(i)$ and $T(j)$, then the minimum number of node deletions required is written as $d(T(i), T(j))$. (Note that in the resulting symmetric mapping, the two roots need not be v_i and v_j.) Let das$(T(i))$ be the number of node deletions required to turn the $T(i)$ into a symmetric one.

For each i, let V_i consist of the children of v_i in T. We first show how to compute $d(T(i), T(j))$ in polynomial time. Construct a complete bipartite graph $B = (V_i \cup V_j, E)$ as follows: For each edge (v_p, v_q) with $v_p \in V_i$ and $v_q \in V_j$, let the edge weight $w(v_p, v_q)$ be $|T(p)| + |T(q)| - \text{dist}_{\text{nd}}(T(p), T(q))$, and compute the maximum matching of graph G. Since v_i and v_j must be mapped into each other for computing $d(T(i), T(j))$, and the numbers of children of v_i and v_j do not increase under node deletions, we know $d(T(i), T(j)) = |T(i)| + |T(j)| -$ matching(B), where matching(B) is the weight of the maximum matching of B. Since maximum matching can be computed in polynomial time [14], so can $d(T(i), T(j))$.

It remains to show that each $\mathrm{das}(T(i))$ can be computed in polynomial time. Construct a weighted graph $G_i = (V_i, E_i)$. For each pair of nodes v_p and v_q in V_i, let the weight of edge (v_p, v_q) be $|T(p)| + |T(q)| - \mathrm{dist}_{\mathrm{nd}}(T(p), T(q))$. We then do the following:

Step 1. The number of operations required to let $T(i)$ become symmetric without any nodes on the symmetry axis is $|T(i)| - \mathrm{matching}(G_i)$.

Step 2. Delete one node v_j of G_i, and then compute the maximum matching of the resulting graph. The number of operations required to let $T(j)$ become symmetric with v_j on the symmetry axis is $|T(j)| -$ the sum of edge weights found $- \mathrm{das}(T(j))$.

Step 3. Repeat Step 2 for each node in G.

Step 4. The minimum number of required operations is the minimum value found in Steps 1 and 3.

Clearly, the solution of a maximum matching contains edge (v_i, v_j) of weight w if and only if $T(i)$ and $T(j)$ can be made identical by deleting w nodes. Thus, the above procedure computes $\mathrm{das}(T(i))$ correctly, proving the theorem. □

3.3 Plane Graphs

In this subsection we study the DAS problems for planarly embedded graphs.

Theorem 4. DAS$_{\mathrm{nd}}$ *for plane graphs can be solved in polynomial time.*

Proof. The proof is a slight extension of a result in [15]. First we apply node deletions to all the degree-1 and degree-2 nodes of G. Let the resulting graph be H. Hence, if after finite node deletion operations on G, we get a symmetric graph G_1, and suppose P is a geometric automorphism on G_1, then P must also be a geometric automorphism on H. Using the algorithm that finds automorphism for plane graphs [11], all the geometric automorphisms of H can be obtained in polynomial time. And for each automorphism P, we then compare the numbers of degree-1 and degree-2 nodes of G on the corresponding edges of P by Theorem 1. Thus, the theorem is proved. □

Theorem 5. DAS$_{\mathrm{ed}}$ *and* DAS$_{\mathrm{ec}}$ *for plane graphs are NP-complete.*

Proof. By Fact 1, it suffices to ensure the NP-hardness of DAS$_{\mathrm{ed}}$ and DAS$_{\mathrm{ec}}$ for plane graphs. The NP-hardness of DAS$_{\mathrm{ed}}$ can be obtained by a reduction from the NP-complete Hamiltonian cycle problem on a plane graph G [7]. For a given n-node m-edge plane graph G, we construct a plane graph H by connecting G, an $(m+2)$-node path P_{m+2}, and an n-node cycle C_n. One can verify that there exists an axially symmetric plane graph H' whose edge number is greater than or equal to $2n + m + 3$ and $H \overset{\mathrm{ed}}{\to} H'$ if and only if G admits a Hamiltonian cycle.

Now we prove the NP-hardness of DAS$_{\mathrm{ec}}$. Let G^* be the dual graph of the input plane graph G with the node (and its incident edges) associated with the external face in G removed. Clearly, G admits a symmetric drawing if and only if G^* admits a symmetric drawing. Also, an edge contraction on G corresponds to an edge deletion on G^*. Therefore the NP-hardness of DAS$_{\mathrm{ec}}$ for G follows from that of DAS$_{\mathrm{ed}}$ for G^*. □

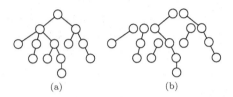

Fig. 6. (a) An unordered tree T. (b) A path decomposition of T.

3.4   Approximation Algorithms

As shown in Theorem 2, DAS$_{ec}$ for unordered trees is NP-complete. In the following, we give an approximation algorithm for DAS$_{ec}$. We shall see later that the technique of our approximation algorithm for DAS$_{ec}$ has an application to the problem of finding the maximum isomorphic subtrees (under contraction) between two unordered unlabeled trees. A related result in the literature is that the edit-distance problem for unordered labeled trees is MAX SNP-hard [19].

Let decomp(T, r) be the following procedure of decomposing a tree T with root r into k paths, where k is the number of leaves. The leaves are labeled from 1 to k such that a leaf with a smaller depth is assigned a smaller label. The k paths are constructed as follows. An edge $e = (v, w)$, where v is closer to r than w, belongs to path $P_i, 1 \le i \le k$, if i is the largest label among all the leaves of the subtree rooted at w. An example is shown in Figure 6. Let e be the base of the natural logarithm.

Theorem 6. *For any n-edge rooted tree T, there is a polynomial-time algorithm that outputs a symmetric tree T' such that $T \overset{ec}{\to} T'$ and T' contains at least $\frac{n}{e \ln n}$ edges.*

Proof. Let r be the root of T. We first perform decomp(T, r), and let $R = \{P_1, \ldots, P_k\}$, where k is the number of leaves of T, be the resulting set of paths. We then decompose R into $R_1, \ldots, R_{\lceil \ln n \rceil}$ such that any path in R_i has length between e^{i-1} and e^i, for each $1 \le i \le \lceil \ln n \rceil$. Then, find the set $R_j, 1 \le j \le \lceil \ln n \rceil$, with the maximum number of edges among all R_i. Now $\forall i, i \ne j$, contract all the edges in R_i and all the edges $e = (u, v)$ in R_j with the distance from v to the leaf of the path (in R) containing e greater than or equal to e^{j-1}. The resulting graph is a tree with paths of equal length $\lceil e^{j-1} \rceil$ directly attached to the root r, and hence, the graph is symmetric. For those paths in R_j, at least $1/e$ of the total number of edges of R_j is left after the mentioned contraction operations. (Recall that the length of each path in R_j has lower and upper bound of $\lceil e^{j-1} \rceil$ and $\lceil e^j \rceil$, respectively.) In view of the above, the total number of edges left in the resulting symmetric tree is at least $\frac{n}{e \ln n}$. □

By Theorem 6, we are able to find an approximation algorithm of guaranteed performance for computing, given two unordered and unlabeled trees T_1 and T_2, the maximum tree T such that both $T_1 \overset{ec}{\to} T$ and $T_2 \overset{ec}{\to} T$.

Theorem 7. *There is a polynomial-time algorithm which, given two unordered unlabeled trees T_1 and T_2, outputs a tree T having at least $\frac{4 \cdot opt}{e^2 \ln n_1 \ln n_2}$ edges such* *that $T_1 \overset{ec}{\to} T$ and $T_2 \overset{ec}{\to} T$, where n_1 and n_2 are the numbers of edges in T_1 and T_2, respectively, and opt is the number of edges in an optimal tree.*

Proof. For any given trees T_1 and T_2, we run decomp(T_1, r_1) and decomp(T_2, r_2). Then we separate the edges in T_1 and T_2 into $\frac{\ln n_1}{2}$ and $\frac{\ln n_2}{2}$ subsets, respectively, using the same strategy stated in the proof of Theorem 6. The length of a path in the i-th set is between e^{2i-2} and e^{2i}. For the i-th subset of T_1 and the j-th subset of T_2 (assuming that $i \geq j$), since each path in the j-th subset of T_2 has length at most e^{2j}, the maximum number of edges that can be matched is at most $e^{2j} \cdot \mu(i,j)$, where $\mu(i,j)$ is the minimum of (a) the number of paths in the i-th subset of T_1, and (b) the number of paths in the j-th subset of T_2. However, according to the contraction method used in Theorem 6, it is easy to see that the number of edges matched is at least $\mu(i,j) \cdot e^{2j-2}$ (i.e., $\frac{1}{e^2}$ of the maximum number of edges matched). Our contraction procedure is similar to that of Theorem 6. In the partitions of T_1 and T_2 mentioned above, we mark those edges belonging to the optimal solution opt. Since there are only $\frac{\ln n_1}{2}$ subsets of T_1 and $\frac{\ln n_2}{2}$ subsets of T_2, there exists a pair (i,j) such that the i-th subset of T_1 and the j-th subset of T_2 have more than $\frac{4 \cdot \text{opt}}{\ln n_1 \ln n_2}$ edges matched in opt. Our algorithm generates a tree of at least $\frac{1}{e^2} \cdot \frac{4 \cdot \text{opt}}{\ln n_1 \ln n_2}$ edges. The subtree achieving the above approximation ratio can be obtained by considering all possible combinations of i and j. □

4   Rotational Symmetry

Theorem 8. *For any constant $k \geq 3$, k-DRS$_{nd}$ for unordered trees with $O(1)$ degree can be solved in polynomial time.*

Proof. Let T be a given tree, whose maximum number of children of each node is at most δ. For brevity, we may assume that each non-leaf node of T has exactly δ children. This assumption can be removed without too much effort. Let $v_1, v_2, \ldots, v_n$ be the postordering of T. Let $v_{c(i,j)}$ be the j-th child of v_i. Let $d(i_1, i_2, \ldots, i_k)$ be the minimum number of node deletions required to transform $T(i_1), T(i_2), \ldots, T(i_k)$ into k identical trees. Given a set $S = \{s_1, s_2, \ldots, s_k\}$, a mapping $\sigma : S \to S$ is called a *permutation* of S if $\{\sigma(s_1), \sigma(s_2), \ldots, \sigma(s_k)\} = S$. It is not difficult to see that $d(i_1, i_2, \ldots, i_k)$ can be computed by the following recursive procedure: If $\{v_{i_1}, \ldots, v_{i_k}\}$ contains a leaf of T, then let $d(i_1, i_2, \ldots, i_k)$ be the sum of $|T(i_j)|$ over all leaves v_{i_j} of T. Otherwise, $d(i_1, \ldots, i_k)$ is the minimum of (a) $\min_{1 \leq p \leq k, 1 \leq q \leq \delta} d(i_1, \ldots, i_{p-1}, c(i_p, q), i_{p+1}, \ldots, i_k) + |T(i_p)| - |T(c(i_p, q))|$ and (b) $d_2 = \min_{\sigma_1, \ldots, \sigma_k \in \Sigma} \sum_{1 \leq t \leq \delta} d(c(i_1, \sigma_1(t)), \ldots, c(i_k, \sigma_k(t)))$, where Σ consists of all permutations of $\{1, \ldots, \delta\}$. Since $k = O(1)$ and $\delta = O(1)$, we know that $d(i_1, i_2, \ldots, i_k)$ can be computed in polynomial time. If the number of children of the root is no more than k, then the degree of k-rotational symmetry of T is zero. Otherwise, the degree of k-rotational symmetry of T is exactly $n - 1$ minus the minimum of $\sum_{i=1}^{\lfloor \frac{\delta}{k} \rfloor} d(r_{i,1}, r_{i,2}, \ldots, r_{i,k}) + \sum \{|T(j)| : v_j$ is a child of the root, $j \notin \{r_{i,j} \mid 1 \leq i \leq \lfloor \frac{\delta}{k} \rfloor, 1 \leq j \leq k\}\}$, where the minimum is taken over all choices of distinct children $v_{r_{i,j}}$ ($1 \leq i \leq \lfloor \frac{\delta}{k} \rfloor, 1 \leq j \leq k$) of the root. □

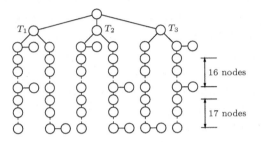

Fig. 7. The tree T_R constructed for the problem instance R of TRIPARTITE MATCHING, where $R = \{\{b_{1,1}, b_{2,1}, b_{3,2}\}, \{b_{1,1}, b_{2,2}, b_{3,2}\}, \{b_{1,2}, b_{2,2}, b_{3,1}\}\}$.

Theorem 9. *For any constant $k \geq 3$, k-DRS$_{nd}$, k-DRS$_{ed}$ and k-DRS$_{ec}$ for unordered trees are NP-complete in general.*

Proof. By Fact 1, a reduction from TRIPARTITE MATCHING [18] suffices. Let R be a given instance for TRIPARTITE MATCHING, which is a ternary relation over $B_1 \times B_2 \times B_3$, where $B_i = \{b_{i,1}, b_{i,2}, \ldots, b_{i,n}\}$ for each $i \in \{1, 2, 3\}$. We define the tree T_R for the DRS problems as follows. Define a function $f(k)$ by letting $f(1) = 1$ and $f(k) = f(k-1) + n^4 + k - 2$ for each k with $2 \leq k \leq |R|$. Let P be a path of $f(|R|)$ nodes rooted at an endpoint of P. For each $i \in \{1, 2, 3\}$ and $j \in \{1, 2, \ldots, n\}$, let $T_{i,j}$ be a rooted binary tree obtained from P as follows: the node of depth $f(k)$ is attached by a node if and only if the k-th relation in R contains $b_{i,j}$. Let T_R be a rooted tree consisting of the subtrees T_1, T_2, and T_3, where each T_i is a rooted tree consisting of the subtrees $T_{i,j}$ with $1 \leq j \leq n$. An example is shown in Figure 7. One can verify that R admits a size-n tripartite matching if and only if the degree of 3-rotational symmetry of T_R is at least $3|R| \cdot f(|R|) + 3n + 4$. $\square$

5   Conclusions

From a computational complexity viewpoint, we have investigated the problem of transforming a graph into a symmetric one using a number of graph operations including edge contractions, edge deletions, and node deletions. For a given graph, our goal is to find the maximum symmetric graph that can be derived from the original graph. An approximation algorithm with a guaranteed performance has been given to one of the NP-complete problems. Our study here can be viewed as a step towards a formal treatment of the notion of near symmetry in graph drawings. An interesting future research would be to find a sequence of operations that leads to the symmetric graph, and see how such information can be used to give a nearly symmetric display of an originally asymmetric graph (perhaps, by adjusting edge lengths and levels).

References

1. T. Akutsu, An RNC algorithm for finding a largest common subtree of two trees, *IEICE Transactions on Information Systems*, E75-D, pp. 95–101, 1992.
2. S. Bachl, Isomorphic Subgraphs, *International Symposium on Graph Drawing* (GD'99), LNCS 1731, pp. 286–296, 1999.
3. G. Di Battista, P. Eades, and R. Tamassia and I. Tollis, *Graph Drawing: Algorithms for the Visualization of Graphs*, Prentice-Hall, 1999.
4. K. Chin and H. Yen, The Symmetry Number Problem for Trees, manuscript, 1998.
5. F. de Fraysseix, A Heuristic for Graph Symmetry Detection, *International Symposium on Graph Drawing* (GD'99), LNCS 1731, pp. 276–285, 1999.
6. P. Eades and X. Lin, Spring Algorithms and Symmetry, *Computing and Combinatorics* (COCOON'97), LNCS 1276, pp. 202–211, 1997.
7. M. Garey and D. Johnson, *Computers and Intractability: A Guide to the Theory of NP-Completeness*. W. H. Freeman, New York, 1979.
8. S. Hong, P. Eades, and S. Lee, Finding Planar Geometric Automorphisms in Planar Graphs, *9th International Symposium on Algorithms and Computation* (ISAAC'98) LNCS 1533, Springer, pp. 277–286, 1998.
9. S. Hong, P. Eades, A. Quigley, and S. Lee, Drawing Algorithms for Series-Parallel Digraphs in Three Dimensions, *International Symposium on Graph Drawing* (GD'98), LNCS 1547, pp. 198–209, 1998.
10. S. Hong, P. Eades, A. Quigley, and S. Lee, Drawing Series-Parallel Digraphs Symmetrically, manuscript, 1999.
11. J. Hopcroft and R. Tarjan, A V^2 Algorithm for Determining Isomorphism of Planar Graphs, *Information Processing Letters*, 1(1):32–34, 1971.
12. P. Kilpelainen, and H. Mannila, Ordered and Unordered Tree Inclusion, *SIAM Journal on Computing* 24(2):340–356, 1995.
13. P. Klein, Computing the Edit Distance Between Unrooted Ordered Trees, *6th European Symposium on Algorithms* (ESA'98), LNCS 1461, 91–102, 1998.
14. E. Lawer, *Combinatorial Optimization: Networks and Matroids*, New York: Holt, Rinehart & Winston, 1976.
15. J. Manning, *Geometric Symmetry in Graphs*, Ph.D. Dissertation, Department of Computer Science, Purdue University, 1990.
16. J. Manning and M. Atallah. Fast Detection and Display of Symmetry in Trees, *Congressus Numerantium*, 64:159–169, 1988.
17. J. Manning, M. Atallah, K. Cudjoe, J. Lozito, and R. Pacheco. A System for Drawing Graphs with Geometric Symmetry, *International Symposium on Graph Drawing* (GD'95), LNCS 894, pp. 262–265, 1995.
18. C. Papadimitriou, *Computational Complexity*, Addison-Wesley, 1994.
19. K. Zhang, and T. Jiang, Some MAX SNP-hard Results Concerning Unordered Labeled Trees, *Information Processing Letters* 49(5):249–254, 1994.
20. K. Zhang, and D. Shasha, Simple Fast Algorithms for the Editing Distance between Trees and Related Problems, *SIAM Journal on Computing* 18(6):1245–1262, 1989.

Clan-Based Incremental Drawing

Fwu-Shan Shieh[1] and Carolyn L. McCreary[2]

[1]Minolta-QMS, Inc. One Magnum Pass,
Mobile, AL 36618, USA
Fwu-Shan.Shieh@Minolta-QMS.com
[2]Compaq Computer Corporation, 334 South Street,
Shrewsbury, MA 01545-4112, USA
Carolyn.McCreary@compaq.com

Abstract. The stability is an essential issue for incremental drawings. To allow stable updating, means to modify graph slightly (such as adding or deleting an edge or a node) without changing the layout dramatically from previous layout. In this paper, a method for achieving stable incremental directed graph layout by using clan-based graph decomposition is described. For a given directed graph, the clan-based decomposition generates a parse tree. The parse tree, which is used for layout, is also employed in locating changes and maintaining visual stability during incremental drawing. By using the generated parse tree, each incremental update can be done very efficiently.

1 Introduction

Directed graphs are an excellent means of conveying the structure and operation of many types of systems. In order to have a meaningful and understandable hand drawn graph, much time is required to plan how the graph should be organized on the page. It is especially hard to hand draw an understandable graph containing a huge number of nodes and edges. In addition, it is difficult for a user to draw a graph when the data is generated by applications (e.g., dialogue state diagrams generated by reverse engineering [1]). In the past decades, several visualization systems have been created for static (automatic) drawings [see 3 & 11 for lists]. Static drawings are not completely satisfactory because in many situations the displayed drawings are subject to change from time to time by the user (such as manual editing, browsing large graphs, and visualizing dynamic graphs) [10]. For dynamic drawings, stable incremental updating where the placement of only a minimal number of nodes and edges are modified, is essential [10, & 12]. Currently, only a few Sugiyama-based dynamic drawing systems have been developed for acyclic directed graphs [6, 10, & 15], and general directed graphs [14]. Based on the experience gained from clan-based graph drawings [8, 9, & 13], the parse tree generated by clan-based decomposition can be used to locate updates and generate stable incremental drawings easily [12].

J. Marks (Ed.): GD 2000, LNCS 1984, pp. 384-395, 2001.
© Springer-Verlag Berlin Heidelberg 2001

2 Clan-Based Graph Drawing

Clan-based graph decomposition parses a directed acyclic graph (DAG) into a hierarchy of subgraphs. These new subgraphs generated by the decomposition are called clans and a clan is classified as one of three types: (a) **series**, (b) **parallel**, and (c) **primitive** [2, 4, and 5].

By using Clan-based graph decomposition, any digraph can be decomposed into an inclusion tree, known as the parse tree, of subgraphs (clans) whose leaves are singleton clans (graph nodes) and whose internal nodes are complex clans (series or parallel) built from their descendants. The primitive clans are decomposed into series and parallel clans by augmenting edges from all the source nodes of the primitive to the union of the children of the sources [4, 5]. After decomposition, a bounding box with computed dimension is associated with each clan and the nodes in the clan are assigned locations within the bounding box. The generated parse tree of the graph with bounding boxes attributed is used to provide geometric interpretations to the graph. To show the directed graph where the edges uniformly point downward (or upward in the case of a reverse edge), the series clans are displayed vertically and connected by inter-clan edges, and the parallel clans are displayed horizontally with no edges between them. To achieve an aesthetically pleasing layout, the nodes are centered. Figure 1 shows a graph, parse tree, and node layout. . For the details about clan-based graph drawing, please refer to [7, 8, and 9].

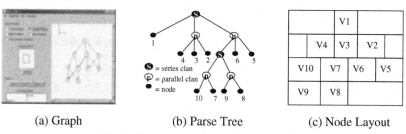

| (a) Graph | (b) Parse Tree | (c) Node Layout |

Fig. 1. Graph, Parse Tree, and Node Layout

3 Clan-Based Incremental Drawing

In order to have stable incremental drawing for clan-based graph decomposition, any layout computation for a successive drawing should be limited to a minimum area that contains updates only. Since parallel clans contain no inter-clan connections, this limited area for clan-based drawing is a series clan (called minimum series clan, MSC). After the MSC is identified, the layout algorithm is applied to the MSC only.

During the incremental drawing, the previous graph drawing and its corresponding parse tree, attributed with bounding boxes, are used to locate the MSC for the next graph and its drawing. The MSC contains all nodes affected by the updates except the added nodes, which are not in the previous drawing. The updates could be multiple node or edge insertions and deletions. For an update, the affected nodes include (a) nodes added, (b) nodes deleted, (c) nodes connected to deleted nodes, (d) nodes connected to added nodes, (e) nodes connected by added edges, (f) nodes connected by deleted nodes, and (g) user selected nodes.

The following notations are used to denote graph objects in iteration i:
(1) G_i , graph.
(2) E_i , all edges of graph G_i.
(3) V_i, all nodes of graph G_i.
(4) D_i, drawing of graph G_i.
(5) T_i, parse tree of drawing D_i with its bounding box and position attributes.
(6) E_{add}, edges added to drawing D_{i-1}.
(7) E_{del}, edges deleted from drawing D_{i-1}.
(8) N_{add}, nodes added to drawing D_{i-1}.
(9) N_{del}, nodes deleted from drawing D_{i-1}.
(10) $N_{c-add-n}$, nodes connected to added nodes N_{add}.
(11) $N_{c-del-n}$, nodes connected to deleted nodes N_{del}.
(12) $N_{c-add-e}$, nodes connected by added edges E_{add}.
(13) $N_{c-del-e}$, nodes connected by deleted edges E_{del}.
(14) N_{sel}, selected nodes.
(15) $N_{affected}$, nodes affected by update. $N_{affected} = N_{add} \cup N_{del} \cup N_{c-add-n} \cup N_{c-del-n} \cup N_{c-add-e} \cup N_{c-del-n} \cup N_{sel}$.
(16) P_{i-1}, array of clan tree pointers to leaf nodes of T_{i-1}.
The **msc** and **act** subscripts denote graph objects of the minimum series clan and the subgraph that requires layout computation, respectively.

When few changes are made from one iteration of the graph to the next, the new graph can be drawn by:
(1) identifying the MSC, C_{msc}, and its corresponding graph, G_{msc},
(2) adding/deleting nodes and edges from G_{msc} to form the affected graph, G_{act},
(3) computing the parse tree of G_{act},
(4) determining the layout of G_{act} from the parse tree, and
(5) scaling G_{act} to fit in the space occupied by G_{msc}.

Figure 2 shows a graph drawing, parse tree, and its updated graph. The affected nodes $N_{affected}$ for this update are nodes 7, 9, and 17.   From Figure 2 (b) parse tree, the MSC that contains $N_{affected}$ - N_{add} is series clan S4. For Figure 2 (a), the G_{msc} consists of nodes (7, 8, 9) and edges ((7, 8), (8, 9)). After G_{msc} is found, the clan-based layout algorithm will be applied only to sub-graph G_{act} of current graph. The sub-graph G_{act} can be identified as G_{act}.nodes = G_{msc}.nodes $\cup N_{add}$ - N_{del} and G_{act}.edges = G_{msc}.edges $\cup$ E_{add} - E_{del}. In Figure 2 (c), the G_{act} consists of nodes (7, 8, 9, 17) and edges ((7, 8), (8, 9), (7, 17), (19, 9)). After the layout algorithm is applied to G_{act}, the parse tree T_{act} and drawing D_{act} are generated (as shown on Figure 3). The size and position of G_{act}'s drawing D_{act} are attributed in parse tree T_{act}.

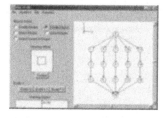

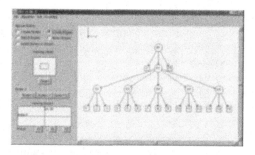

(a) Graph Drawing   (b) Parse Tree for (a)

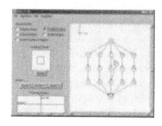

(c) Node 17 and Edges (7, 17) & (17, 9) Added to (a)

Fig. 2. Graph Drawing, Parse Tree, and Updated Graph

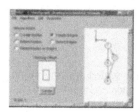   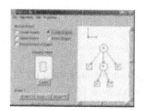   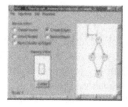

(a) Graph G_{act} of Fig. 2 (c) (b) Parse Tree of graph D_{act} (c) Drawing of Graph G_{act}

Fig. 3. Graph G_{act}, Parse Tree, and Drawing

In order to minimize the number of nodes that must be moved for stable incremental drawing, only the G_{act} is recomputed for current graph G_i, and the drawing D_{act} of G_{act} is sized to be contained in the area used by G_{msc}'s drawing D_{msc}. The size and position of G_{msc}'s drawing D_{msc} are attributed in C_{msc}. If the size of D_{act} is greater than D_{msc}, the D_{act} is scaled down. If the size of D_{act} is smaller than D_{msc}, the D_{act} is positioned in the center of the area used by D_{msc}.

The MSC C_{msc} in T_{i-1} is replaced by graph G_{act}'s parse tree T_{act}. The modified parse tree becomes the current parse tree T_i for current graph G_i. Figure 4 shows the new parse tree T_i and it's drawing D_i of Figure 2 (c)'s incremental update.

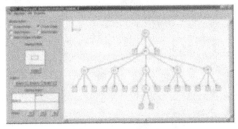

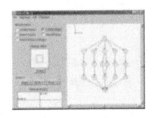

(a) Parse Tree from Fig. 2(b) with   (b) Drawing of Figure 2(c)
    Series Clan S4 Replaced

Fig. 4. Drawing for Figure 2 (c)'s Incremental Update

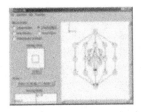

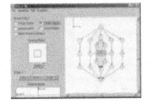

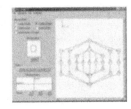

(a) Updated Graph of   (b) Drawing Shows   (c) New Drawing of (b)
    Figure 4(b)   Nodes Overlapped

Fig. 5. An Updated Graph and Drawings for Fig. 4 (b)

4 Insuring Readability

When D_{act} is scaled down to fit in the area used by D_{msc}, the scaled drawing might be unreadable. Figure 5 (b) shows this problem after an update. Figure 5 (a) is an updated graph of Figure 4 (b)'s drawing. In this updated graph, nodes (18, 19, 20, 21, 22, 23, 24) and edges ((7, 18), (18, 19), (19, 20), (20, 9), (7, 21), (21, 9), (7, 22), (22, 23), (23, 24), (24, 9)) are added. Figure 5 (b) is the drawing for Figure 5 (a). In the Figure 5 (b), the scaled down drawing D_{act} has overlapping nodes that make it a problem for the user to read the drawing. So, a maximum scale limit is needed to ensure readability. Figure 5 (c) shows new drawing of Figure 5 (a). In Figure 5 (c), the scale limit is applied. After the scale limit is reached, the D_{act} is given more space to maintain the readability. Although only the subtree rooted at MSC is replaced, the bounding box and placement attributes of the entire parse tree must be computed. For the example in Figure 5, this has the effect of moving nodes (1, 2, 3, 4, 5, and 6) to the left and nodes (10, 11, 12, 13, 14, and 15) to the right.

5 Locality of Incremental Drawing

Successive modifications to graphs often occur within a small geometric or logical neighborhood [10]. In order to build more stable drawing with fewer computations, the algorithms should take advantage of this locality property. When adjusting the previous clan tree T_{i-1} and the previous drawing D_{i-1} to make extra space for D_{act}, if the space created is the exact space needed by D_{act}, then any nodes added to the current G_{act} might cause T_i and D_i to be adjusted again. If the graph is expected to grow, the scale factor for G_{act} should be increased.

Figure 6 shows an example of extra space created for future updates. Figures 6 (b) is successive drawings of Figure 6 (a) and Figure 6 (c) is successive drawing of Figure (b). In Figures 6 (b) and 6 (c), nodes (1, 2, 3, 4, 5, 6, 10, 11, 12, 13, 14, 15) are not re-positioned because the possible extra space needed by successive drawings has been generated during Figure 6 (a) drawing layout computation.

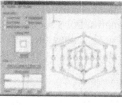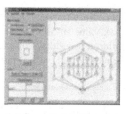

(a) New Drawing of	(b) Successive Drawing	(c) Successive Drawing
Fig. 5 (b)	of (a)	of (b)

Fig. 6. New Update Drawing of Figure 5 (c) with Nodes and Edges Added

6 Reposition Nodes Not in the Minimum Series Clan

If more space is needed, what nodes need be re-positioned or re-sized? Let $P_{msc\text{-}root}$ be the path from C_{msc} to root in previous parse tree T_{i-1}, and L_{adjust} be extra length and W_{adjust} be extra width needed by D_{act}. To make extra space, for each clan C_{path} along the path $P_{msc\text{-}root}$ needs to add W_{adjust} to C_{path}'s width and L_{adjust} to C_{path}'s length. Also, along the path $P_{msc\text{-}root}$,

1. all left siblings of C_{path} whose parent is a parallel clan need to be shifted left with $W_{adjust} / 2$ distance,
2. all right siblings of C_{path} whose parent is a parallel clan need to be shifted right with $W_{adjust} / 2$ distance, and
3. all right siblings of C_{path} whose parent is a series clan need to be shifted down with L_{adjust} distance.

Figure 7 is a parse tree with C_{msc} identified. The path $P_{msc\text{-}root}$ includes clans C_{msc}, P7, S5, P2, and S1. In this parse tree, if more space is needed :

1. node 64 and clan S4's nodes need to be shifted to the left,
2. clan S10's nodes and clan S6's nodes need to be shifted to the right, and
3. and node 65, clan P8's nodes, and P3's nodes need to be shifted down.

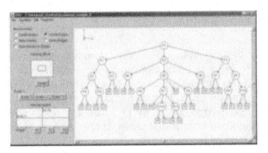

Fig. 7. Parse Tree with C_{msc} identified

7 Resizing Affected Clans

Is it necessary to resize all clans along the path $P_{msc\text{-}root}$? For a parallel clan, its length is the max value of all children's lengths. For a series clan, its width is the max value of all children's widths. For each clan C_{path} along the path $P_{msc\text{-}root}$, if both the C_{path}'s width and length do not exceed parent's max values, it is not necessary to resize parent.

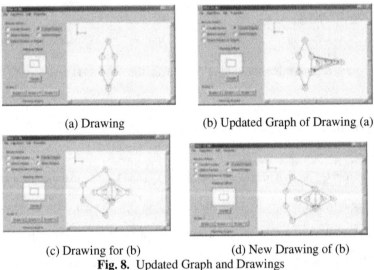

(a) Drawing   (b) Updated Graph of Drawing (a)

(c) Drawing for (b)   (d) New Drawing of (b)

Fig. 8. Updated Graph and Drawings

In some situations, it is not necessary to re-position siblings along the path $P_{msc\text{-}root}$ or resize clans which contains C_{msc} when the drawing D_i requires extra space. In Figure 8 (b), nodes (6, 7, 8, 9) and edges ((3, 6), (6, 4), (3, 7), (7, 4), (3, 8), (8, 4), (3, 9), (9, 4)) are added to Figure 8 (a) drawing. Figure 8 (c) is the drawing for Figure 8 (b). In Figure 8 (c), nodes 1 and 2 are shifted to the left to make some space for updates. However, the shift is not necessary. Since there are no nodes at the right hand side of nodes 3 and 4, it would be more reasonable for updates to grow toward the right without shifting nodes 1 and 2. Figure 8 (d) shows a different drawing of Figure 8 (b). In Figure 8 (d), nodes 1 and 2 are not moved.

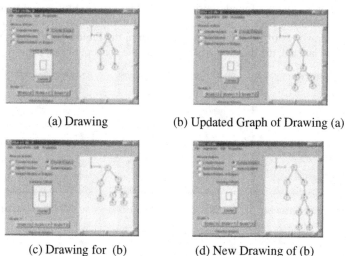

(a) Drawing   (b) Updated Graph of Drawing (a)

(c) Drawing for (b)   (d) New Drawing of (b)

Fig. 9. Updated Graph and Drawings

Figure 9 shows a case when re-sizing is not necessary. In Figure 9 (b), nodes are appended to the node 4 of Figure 9 (a). Figure 9 (c) is the drawing for Figure 9 (b). In Figure 9 (c), the sub-graph containing nodes (3, 4, 5, 6, 7, and 8) is scaled even when there are no nodes below them. In this case, the graph should be able to grow downward freely without affecting the drawing stability. Figure 9 (d) shows a different drawing of Figure 9 (b). The path $P_{msc\text{-}root}$ can be used to identify the cases where re-sizing / re-positioning is not necessary:

1. along the path $P_{msc\text{-}root}$, if there are no left siblings for a clan whose parent is a parallel clan, the drawing D_i can grow toward left without shifting other clans to the right,
2. along the path $P_{msc\text{-}root}$, if there are no right siblings for a clan whose parent is a parallel clan, the drawing D_i can grow toward right without shifting other clans to the left, and
3. along the path $P_{msc\text{-}root}$, if there are no right siblings for a clan whose parent is a series clan, the drawing D_i can grow downward without scaling down length.

8 Incremental Drawing of Cyclic Graphs

Using the depth first search to find an edge to be reversed for the cyclic graph may not be suitable for some applications [12]. Figure 10 shows a problem in incremental drawing when only one edge is drawn upward for each cycle of cyclic graphs. Figure 10 (a) shows a path of a company's system code release. In Figure 10 (b), a new path for bugs report is added. Figure 10 (c) is the drawing of Figure 10 (b). The Figure 10 (c) does not represent the two paths in the expected way. If edge labels are removed (as Figure 10 (d)), it will be even more difficult for the user to visualize the concept of two different paths. In order to improve this situation, the visual input graph nodes' position information is used. During the graph layout computation, if upward edges contributed to a cycle are found, those edges will be reversed for layout computation and then be changed back to upward after layout computations done. Figure 10 (e) shows new drawings of Figure 10 (b). As shown in the Figure 10 (f), even when the edges are not labeled, the drawing still represents the concept of two paths.

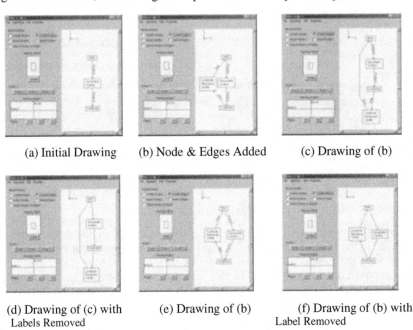

   (a) Initial Drawing   (b) Node & Edges Added   (c) Drawing of (b)

   (d) Drawing of (c) with   (e) Drawing of (b)   (f) Drawing of (b) with
       Labels Removed   Label Removed

Fig. 10. Updated Graph and Drawings of Incremental Cyclic Drawing

By using clan-based drawing for incremental drawing, the layout computation only applies to the affected sub-graph G_{act} instead of the entire graph. Sometimes when the sub-graph is extracted from the entire graph, its original role in the entire graph might be missing. The Figure 11 (b) is an update of Figure 11 (a), and Figure 11 (c) is Figure 11 (b)'s G_{act} which requires layout computation. The G_{act} is not a cyclic graph, but G_{act} is part of the cycles of Figure 11 (b) cyclic graph originally. Since G_{act} is not a cyclic graph, no edges are reversed during the layout computation. Figure 11 (d) shows the incorrect updated drawing of Figure 11 (b). To insure proper edge direction, upward edges need be checked to see if those edges are part of cycles of original graph. If an upward edge contributes to a cycle, this upward edge will be reversed during the computation and reversed again after the computation. Figure 11 (e) shows the correct drawing of Figure 11 (b).

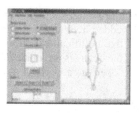 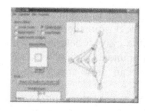 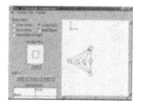

(a) Initial Drawing   (b) Nodes & Edges Added   (c) Sub-graph of (b)

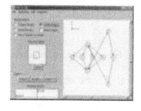 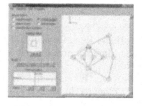

(d) Incorrect Drawing of (b)   (e) Correct Drawing of (b)

Fig. 11. Drawing and Graph with Upward Edges Added

9 Incremental Drawing Examples

Figure 12 shows a series of incremental drawings created by the clan-based drawing algorithm. The Figures 12 (b), (e), and (f), show that some upward paths were added. Those upward paths are part of cycles. After nodes (11, 12, 13) and edges ((4, 11), (11, 12), (12, 13), (13, 6)) are added to Figure 12 (f), the length of new drawing, Figure 12 (g), is not changed because the minimum affected area still has enough space for updates. Figures 13 (n) is the new drawing of 12 (m) after upward edges ((7, 18), (7, 20)) are added. The edges ((7, 18), (7, 20)) are not part of cycles, so node 7 is moved to above nodes 18 and 20.

10 Conclusion

Clan-based graph drawing using a parse tree improves the incremental drawing stability for directed graph drawings in several areas:
(1) The attributed parse tree provides an easy way to layout graphs with nodes of different sizes. A node can be spanned over more than one level in drawing if necessary. During the incremental updates, if the change is only to enlarge nodes, the new updates might be done very easily without re-positioning other nodes.
(2) The utilization of parse trees allows the incremental drawing to be stable without sacrificing aesthetic criteria and speed. By using clan-based drawing, the incremental drawing can be done very efficiently because the changed and recomputed area can be localized with the provided parse tree.

(3) No extra constraints are needed to maintain incremental drawing stability. For clan-based drawing, the node position constraints are embedded in the parse tree.
(4) By using the parse tree, it is easy to determine whether the modifications are in the interior or the exterior boundary area of a drawing. If the changes are made to exterior area, the updates can grow freely toward the open area without re-positioning other nodes.
(5) By using the parse tree to locate the minimum affected area of updates, the locality issue can be considered for the future stable updates.

References

1. J. H. Cross II and R. S. Dannelly, "Reverse Engineering Graphical Representations of X Source Code," International Journal of Software Engineering and Knowledge Engineering, Spring, 1996.
2. A. H. Deutz, A. Ehrenfeucht, G. Rozenberg, "Clans and regions in 2-structures," Theoretical Computer Science, 129, 207-262, 1994.
3. G. Di Battista, P. Eades, R. Tamassia, I. Tollis, "Algorithms for Drawing Graphs: an Annotated Bibliography", Computation Geometry: Theory and Applications, 4(5):235-282, 1994.
4. A. Ehrenfeucht and G. Rozenberg, "Theory of 2-Structures, Part I: Clans, Basic Subclasses, and Morphisms," Theoretical Computer Science, Vol. 70, 277-303, 1990.
5. A. Ehrenfeucht and G. Rozenberg, "Theory of 2-Structures, Part II: Representation Through Labeled Tree Families," Theoretical Computer Science, Vol. 70, 305-342, 1990.
6. M. Frohlich,"Incremental Graphout in Visualization System – daVinci," PhD thesis, Department of Computer Science, The University of Bremen, Germany, November 1997.
7. C. M. McCreary, R. O. Chapman, and F. S. Shieh, "Using Graph Paring for Automatic Graph Drawing", IEEE Trans. on Systems Man, and Cybernetics -- Part A: Systems and Humans, Vol. 28, No. 5, 545-561, 1998.
8. C. L. McCreary and A. Reed, "A Graph Parsing Algorithm and Implementation," Tech. Rpt. TR-93-04, Dept. of Comp. Sci and Eng., Auburn U. 1993.
9. C. McCreary, F. S. Shieh, and H. Gill, "CG: a Graph Drawing System Using Graph-Grammar Parsing," Lecture Notes in Computer Science, Vol. 894, 270-273, Springer-Verlag, 1995.
10. S. C. North, "Incremental Layout in DynaDAG," Lecture Notes in Computer Science, Vol. 1027, 409 - 418, Springer-Verlag, 1996.
11. G. Sander, "Graph Drawing Tools and Related Work,"
     http://www.cs.uni-sb.de/RW/users/sander/html/gstools.html
12. F. S. Shieh, "Stability and Topology of Graph Drawing," Auburn University, Ph.D. dissertation, 2000.
13. F. S. Shieh, and C. L. McCreary, "Directed Graphs Drawing by Clan-based Decomposition," Lecture Notes in Computer Science, Vol. 1027, 472 - 482, Springer-Verlag, 1996.
14. Tom Sawyer, "Graph Toolkit". http://www.tomsawyer.com.
15. K. Sugiyama, S. Tagawa and M. Toda, "Methods for Understanding of Hierarchical System Structures," IEEE Trans. on Sys. Man, and Cyb., SMC-11, 109-125, 1981.

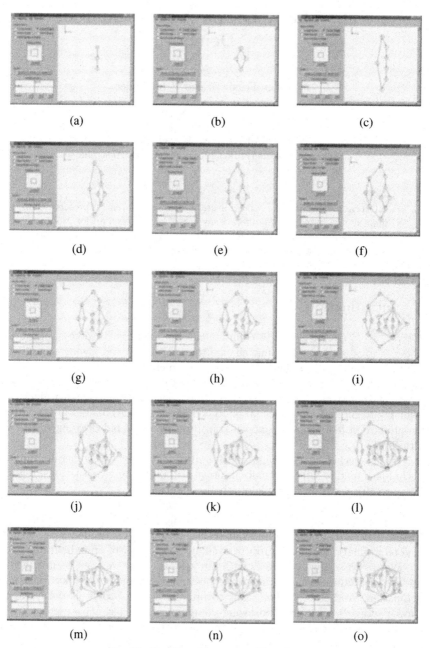

Fig. 12. Clan-Based Incremental Drawings

The Marey Graph Animation Tool Demo

Carsten Friedrich and Peter Eades

The University of Newcastle
University Drive
Callaghan, NSW 2308
{friedric,eades}@cs.newcastle.edu.au Australia

Abstract Enabling the user of a graph drawing system to preserve the mental map between two different layouts of a graph is a major problem. In this paper we present Marey, a system that can smoothly transform one drawing of a graph into another without any restrictions to the class of graphs or type of layout algorithm.[1]

1 Introduction

Graphs are a common way to communicate information. In many applications these graphs are not static but change their structure and layout according to user and application actions. Preserving the mental map during these changes has been identified to be crucial for the usability of a system [2]. There are two possible approaches to this problem. Either develop graph drawing algorithms that try to minimize changes [3] or to communicate the changes in the form of an animation [8,6,7], i.e. a smooth transition from the old drawing to the new drawing.

While specialized algorithms work quite well in practice, general animation techniques tend to fail to actually improve usability in many situations. Figure 1 on page 397 shows an example of a bad animation[2].

In this paper we introduce Marey, a system which implements a variety of different animation techniques which can be used to display the changes in the graph structure or the graph layout. Furthermore we try to identify criteria for a good animation and measures to evaluate the quality of an animation.

We try to achieve two goals with our system. First we want to use the tool in further studies and experiments to better understand the properties of good animation. And secondly, we want to provide a set of tools in the form of a Java package to the graph drawing community which can be used to create applications using graph animations not restricted to specific layout algorithms or classes of graphs.

---

[1] This work was supported by DSTO Australia

[2] All examples, in the form of mpeg videos, and a free mpeg player for MS Windows NT as well as references to free players for Unix are available from http://www.cs.newcastle.edu.au/~friedric/gd00/

J. Marks (Ed.): GD 2000, LNCS 1984, pp. 396–406, 2001.

Figure 1. Example of a bad animation. The nodes from the drawing on the left are moved to their new position in the drawing on the right using a linear interpolation. The drawing in the middle is a snapshot of the animation where nodes lie very close to each other. Individual node movements are difficult to follow at this stage. The mpeg file is `http://www.cs.newcastle.edu.au/~friedric/gd00/a.mpg`

2   Model for a Good Animation

An animation is a sequence of images. This sequence is characterized by subtle but highly structured changes between consecutive images over space and over time. The changes are perceived as movement of the corresponding objects in the image by the human brain. A detailed analysis of these mechanisms is beyond the scope of this paper. For an introduction to human perception of moving pictures see e.g. [1].

In the case of a graph animation, the images are drawings of graphs. The changes in the drawings are changes in the positions of the nodes and edges.

The animation should help the user to maintain the mental map of a changing graph. Major changes to the drawing of a graph usually occur when the user applies a layout algorithm which provides a different view of the graph or when the structure of the graph changes in a way which makes it necessary to recompute the layout. Examples for structural changes in graphs are collapsing or expanding sub graphs in clustered graphs, navigation in infinite graphs[3] or graphs such as graph A of the 1999 Graph Drawing Contest [4] which represents the changes in the cast of a soap opera.

2.1   General Goals

The following general goals have been identified for a good animation between two graph layouts.

1. Preserve the mental map
2. Communicate the structural changes in the graph

Apart from preserving the mental map between two drawings, in the case where the new layout was triggered by changes to the graph structure it is important to communicate that the graph changed and what these changes were to the user.

---

[3] E.g. www-based graphs

2.2   Criteria for a Good Animation

The following aesthetic criteria have been identified to characterize a good animation.

The movements of nodes and edges should be easy to follow.

The movements of the graph should be structured. The more structure the user can identify in the movements, the easier it is to preserve the mental map of the graph. E.g. if nodes of a sub-graphs do not change their positions in relation to each other, the quality of the animation increases significantly if these nodes move in a uniform way.

The transition from source to destination should be smooth. The movements should be performed in small steps and adequately fast to help the human brain to perceive and interpret the movement.

Displaying non-existing structures should be avoided. An often neglected problem in graph drawings is the case where the drawing displays some structure which does not exist in the graph [12]. Figure 2 on page 398 shows an example for two layouts of the same graph, where the second layout could lead the user wrongly to assume that the graph is a simple path. Similar problems can occur easily during an animation, as the human brain tends to be quite imaginative when it tries to interpret moving images [11].

http://www.cs.newcastle.edu.au/~friedric/gd00/d.mpg shows an example for an animation which wrongly suggests the transformation to be the rotation of a three-dimensional graph.

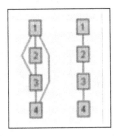

Figure 2. Example of a misleading layout

The individual drawings should satisfy aesthetic criteria. such as minimizing edge crossings, maximizing the smallest angle, etc; see [2] for details.

Some of these criteria are NP hard to optimize, e.g. crossing minimization, and often not all criteria can be achieved in one solution. E.g. it might be necessary to move nodes along a non-optimal path to avoid edge crossings, or it might be preferable to accept some edge crossings instead of watching the graph untangle in a complex and confusing way.

2.3   Measures for a Good Animation

To be able to actually evaluate how well an animation matches our criteria we derived the following possible measures.

Minimize temporary edge crossings. Edge crossings in a graph drawing generally reduce readability. It seems that this is also valid for animations. If the animation avoids introducing unnecessary edge crossings on the way then it is easier for the user to follow the movements.

Maintain a minimal distance between nodes which do not move uniformly. If nodes lie close to each other, it is more difficult to follow their individual movements than if they are further apart. Of course, if two nodes lie next to each other in the source drawing and in the target drawing, it would be better to move them uniformly close to each other to their destination than to separate them first.

Maximize symmetries in movement. Symmetry in a drawing helps the user to understand the structure of a graph. In an animation symmetry of movement makes it easier to understand the structure of the movement. A formal measure for symmetric node movement can be derived by extending the model in [9].

Minimize the length of the path of a node. A node which moves on a straight line between its source and destination clearly moves a minimal distance. However some criteria may prevent straight line movement. In such cases we require a minimum distance movement to help the user to follow and anticipate the node movement.

Provide smooth transitions and adequate speed. Smooth transitions and an adequate speed are obvious criteria for a good animation. If the transition steps are to big the user is no longer able to perceive a movement. If the speed is too slow the user will become impatient and stop following the animation. If it is too fast the user will not be able to keep track of the nodes and comprehend the movements.

Formal metrics can be derived from the above. E.g. it is possible to count the number of temporary edges or compute the length of the node paths.

3   Architecture of Marey

As our tool was implemented using Java, the architecture of the system was developed following object oriented design patterns.

The graph itself, the layout algorithms, the graph modifying algorithms, and the animation engine are implemented as separate modules. They exchange information and trigger actions using a defined API.

In a typical use case the user invokes some action which results in major changes in the graph drawing, e.g. applying a layout algorithm or changing the graph structure followed by a layout algorithm. Figure 3 on page 400 shows the data-flow for these cases.

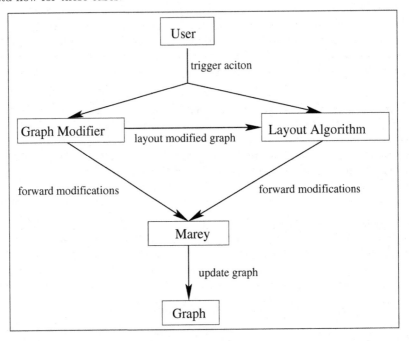

Figure 3. Architecture

Note that there is usually no need for the user to directly access the animation module. In our prototype the user is able to access the Marey module to take snapshots of graphs and generate an animation between them. Figure 4 on page 401 shows the control panel of the Marey module.

A special Java class is used to store the changes which should be animated[4].

Modules can either provide instances of that class to the animation module or tell the animation module to take a snapshot of the current state of the graph and compute the differences to the last snapshot.

---

[4] Currently only node movements and visibility of nodes and edges can be animated. The system is easily extensible and more features are currently being developed

Figure 4. Control Panel

3.1   The Animation Process

When the animation module has enough information in the form of snapshots or change-objects the animation itself can be invoked. The animation between two states consists of six steps.

1. Hide vanishing nodes and edges
2. Adjust absolute positions (optional).
3. Rotate graph (optional).
4. Scale graph (optional).
5. Move nodes to their final position
6. Show newly visible elements

1. Hide vanishing nodes. It is difficult for an animation engine to animate the removal of graph elements, as they are usually no longer existent in the graph when the animation is generated. Our proposed work around for the modifying module is to only hide the elements, generate the animation and after the animation remove them from the graph. As these nodes do not have a destination, hiding happens at the beginning of the animation process.

They can simply be hidden after the first step, or slowly fade out.

The basic idea behind the next three steps is to minimize unstructured movement during the animation. We try to break up the animation into a uniform transformation, a uniform scaling, a uniform rotation to all elements, and a final individual movement for each node. The three uniform steps can always be easily performed without destroying the mental map and thereby minimize the movements in the more critical individual movements.

2. Adjust absolute positions. To minimize node movements during the animation the layouts are adjusted using a uniform translation of all nodes to reduce distances between old and new positions.

Either the barycenter or the center of the bounding boxes of the graphs is computed. The layouts are then transformed to overlap the two centers.

There are two possible approaches to this step. In most cases the absolute position of a graph is irrelevant. In this case time can be saved by just moving the new layout towards the old layout and not animate this operation at all.

Alternatively, as there are cases where the absolute position of a graph actually matters, the old layout can be moved towards the new layout in an animation.

Furthermore, in the case where only a few nodes move between two layouts, it might be preferable not to adjust the absolute positions at all as this would cause all nodes to move during the animation. Therefore this step is optional.

3. Rotate graph. We compute the average rotation of all nodes around the barycenter or the center of the bounding box of the graph, depending on the last step. We then rotate the graph at this angle.

The time needed to display the rotation may be relatively long in a simple animation. The extra time can out-weight the improvement achieved by the rotation in this case. This step is therefore optional.

If the orientation of the target drawing does not matter the target drawing can be rotated to match the orientation of the source drawing without animation.

4. Scaling. There are different approaches to compute a scaling. The graph can be scaled so that the bounding boxes of the two layouts have the same dimensions or it could be scaled by the average, minimal or maximal scaling factor over all nodes.

Again, there are cases where scaling does not improve the animation and this step is therefore optional.

If the size of the target drawing does not matter, the target drawing can be resized to match the source drawing without animation.

5. Move nodes to their final positions. For the final movements several algorithms are available. The easiest approach is to move the nodes the remaining way to their new positions on a linear path. Alternatively these movements can be restricted to certain general directions at a time or broken up in uniform movements, e.g. first move to the x-positions and then to the y-positions. The advantage of these approaches is that the movements can be computed very fast. The disadvantage is that none of our criteria for a good animation is enforced.

However more sophisticated approaches are more appropriate in some cases. We have developed an adaptation of the force directed layout approach described in [5] to move the nodes to their final positions. The repulsive forces are similar to the static version, whereas instead of attracting edges, nodes are attracted to their destination. This approach provides a minimal distance between nodes at all times in most cases and thereby increase traceability. Problems with this approach are to update the forces fast enough to be able to provide a fast and smooth animation and secondly to guarantee an efficient movement of the nodes to their final destination.

A modification of [3] might be able to avoid new edge crossings at the same time.

Other approaches are imaginable, and are currently being developed, which try to minimize our measures for good animation directly.

Show newly visible elements. The last step adds graph elements to the drawing which did not exist or were invisible at the start of the animation. As for the first step this can be done within one step or using a slower fade-in.

4   Examples

The following examples show typical use cases of the animation module. The black and white printed pictures in this section try to capture the animation process. Of course this is only possible to a certain degree. All examples are available as mpeg videos on the following web site:

http://www.cs.newcastle.edu.au/~friedric/gd00

Note that some freely available mpeg display programs do not support the features necessary to play the movies. References to free and working mpeg-players for Windows NT and Unix are provided on this page.

4.1   Applying a New Layout

The first example shows the animation between two force directed layouts. The structure of the drawing does not change much apart from a rotation of the graph of about 180 degree. A direct movement of the nodes from the source to the target would result in a confusing animation as shown in figure 1. Figure 5 shows how Marey breaks up the movement into a rotation and a final movement of the nodes.

The second example shows the transition from an hierarchical type layout to an force directed layout of a graph. Snapshots of the animation are shown in figure 6.

4.2   Sub-graph Expansion

The example in figure 7 shows a clustered graph. The center node, which represents a group of nodes is expanded and a new layout is computed using a force directed algorithm. The animation shows the change from the old layout to the new layout.

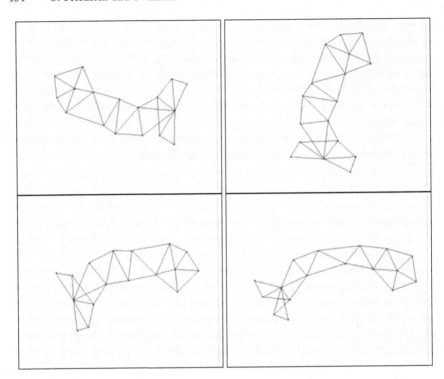

Figure 5. The drawing in the upper left corner is the initial drawing, the drawing in the lower right corner is the target drawing. The animation is broken into a rotation of the graph and a subsequent movement of the node to their final position. The mpeg file is http://www.cs.newcastle.edu.au/~friedric/gd00/b.mpg

5   Status and Future Plans

The system currently implements a basic set of animation techniques and further algorithms are being developed. The tool is scheduled to be released as a freely available Java package in the second half of 2000. A prototype is currently integrated into the InVision framework[10].

To gain a deeper understanding of the criteria for a good animation user experiments are planned for the near future. We hope the results of these experiments will enable us to develop new and better animation techniques and algorithms to be integrated into our tool.

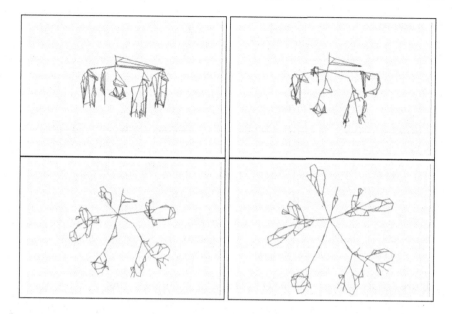

Figure 6. Morphing between a hierarchical layout and a force directed layout using a linear interpolated path for the nodes. The mpeg file is http://www.cs.newcastle.edu.au/~friedric/gd00/c.mpg

References

1. Edward Adelson. Mechanisms for motion perception. *Optics and Photonics News*, August:24–30, 1991.
2. Giuseppe Di Battista, Peter Eades, Roberto Tamassia, and Ioannis G. Tollis. *Graph drawing: algorithms for the visualization of graphs*. Prentice-Hall Inc., 1999.
3. F. Bertault. A force-directed algorithm that preserves edge-crossing properties. *Information Processing Letters*, 74(1–2):7–13, 2000.
4. Franz J. Brandenburg, Michael Jünger, Joe Marks, Petra Mutzel, and Falk Schreiber. Graph-drawing contest report. In *Proc. of the 7th Internat. Symposium on Graph Drawing (GD'99)*, pages 400–409, 1999.
5. Frick, Ludwig, and Mehldau. A fast adaptive layout algorithm for undirected graphs. In *Proc. of the DIMACS International Workshop on Graph Drawing (GD'94)*, 1994.
6. C. Friedrich. The ffGraph library. Technical Report 9520, Universität Passau, Dezember 1995.
7. http://www.mpi sb.mpg.de/AGD/. *The AGD-Library User Manual Version 1.1.2*. Max-Planck-Institut für Informatik.
8. Mao Lin Huang and Peter Eades. A fully animated interactive system for clustering and navigating huge graphs. In Sue H. Whitesides, editor, *Proc. of the 6th Internat. Symposium on Graph Drawing (GD'98)*, pages 374–383, 1998.
9. J. Manning. *Geometric Symmetry in Graphs*. PhD thesis, Purdue University, Department of Computer Sciences, 1990.

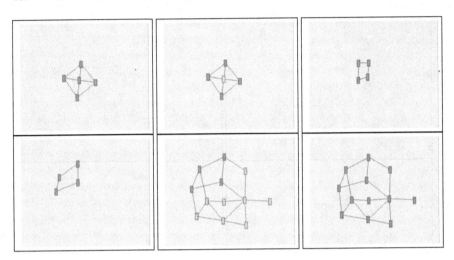

Figure 7. Sub-graph expansion. The node in the middle of the first image is expanded to a sub-graph. The sequence shows how the node is faded out, how the remaining nodes are moved to their new position, and how the new nodes are faded in. The mpeg file is `http://www.cs.newcastle.edu.au/~friedric/gd00/expand.mpg`

10. T.R. Pattison, R.J. Vernik, D.P.J. Goodburn, and M.P. Phillips. Rapid assembly and deployment of domain visualisation solutions. In *Submitted to IEEE Visualisation 2000*, 2000.
11. Robert Sekuler and Randolph Blake. *Perception*. McGraw-Hill Publishing company, 1990.
12. Richard J. Webber. *Finding the Best Viewpoints for three-dimensional graph drawings*. PhD thesis, University of Newcastle (Australia), 1998.

Graph Data Format Workshop Report

Ulrik Brandes[1], M. Scott Marshall[2], and Stephen C. North[3]

[1] University of Konstanz, Department of Computer & Information Science
Box D 188, 78457 Konstanz, Germany. Ulrik.Brandes@uni-konstanz.de
[2] Centrum voor Wiskunde en Informatica, P.O. Box 94079, 1090 GB Amsterdam,
The Netherlands. scott@cwi.nl
[3] AT&T Research, 180 Park Ave., Bldg. 103, Florham Park, NJ 07932-0971, U.S.A.
north@research.att.com

Abstract. Prompted by the increasing demand for a standard exchange format for graph data, an informal workshop was held in conjunction with Graph Drawing 2000. The participants identified requirements for such a standard and formed a group to work out a proposal. The current status of this effort is publicly available at
http://www.graphdrawing.org/data/format/.

1 Introduction

Graph drawing tools need to store and exchange graph data. Despite several earlier attempts to define a standard, no agreed-upon format is widely accepted, and indeed, many packages support only their own custom format.

Motivated by the goals of tool interoperability, access to benchmark data sets, and, most importantly, data exchange over the Web, the Steering Committee of the Graph Drawing Symposium started a new initiative. In a concerted effort, a standard format will be defined and proposed to developers of graph drawing and related software.

As a first step towards this end, an informal workshop was announced on the GD 2000 Web page[1] and held the day before the start of the symposium. The workshop featured two invited presentations and plenty of discussion.

Date and location: 20 September 2000, 14:00—17:30, Colonial Williamsburg
Local organization: Joe Marks and Kathy Ryall
Minutes: Helen Purchase
Chair: Ulrik Brandes

14:00–14:20 Stephen North: *Introduction to XML*
14:20–14:40 Scott Marshall: *Survey of Graph Data Formats*
15:00–17:30 Discussion blocks on
- general requirements
- specific goals
- future activities

---

[1] http://www.cs.virginia.edu/~gd2000/

J. Marks (Ed.): GD 2000, LNCS 1984, pp. 407–409, 2001.

The widespread interest in a standard was underscored by the unexpectedly large number of participants. With several others noting that they were, unfortunately, unable to attend, the following persons were present at the meeting:

- Ulrik Brandes (University of Konstanz)
- Stina Bridgeman (Brown University)
- Giuseppe Di Battista (University of Rome III)
- Peter Eades (University of Sydney)
- Ashim Garg (State Universty of New York)
- Carsten Gutwenger (Max Planck Institute for Computer Science)
- Michael Himsolt (DaimlerChrysler Research)
- Seokhee Hong (University of Sydney)
- Michael Jünger (University of Cologne)
- Michael Kaufmann (University of Tübingen)
- Sebastian Leipert (University of Cologne)
- Giuseppe Liotta (University of Perugia)
- Joe Marks (Mitsubishi Electric Research)
- Scott Marshall (CWI Amsterdam)
- Petra Mutzel (Vienna University of Technology)
- Stephen North (AT&T Research)
- Maurizio Patrignani (University of Rome III)
- Helen Purchase (University of Queensland)
- Kathy Ryall (University of Virginia)
- Galina Shubina (Brown University)
- Susan Sim (University of Toronto)
- Roberto Tamassia (Brown University)
- Luca Vismara (Brown University)
- Vance Waddle (IBM Research)

2   Summary of Discussion

The discussion was held in an open and constructive atmosphere. None of the topics required vote-taking, since each was discussed until either there was no noticeable dissent or a topic was identified as requiring more detailed investigation.

In a first round of discussion, the scope of the project was delineated by gathering potentially relevant features of a common exchange format. Among such requirements were the support of hypergraphs, clustering, hierachical graphs, graph properties and corresponding certificates, dynamic graphs, 3D graphics, animation, and even layout algorithms as part of the description of a graph drawing. In addition, issues such as scalability and human readability were addressed briefly.

Given this vast array of options, the unanimous decision was to focus on a simple format that should be sufficient for the most common cases, yet extensible to incorporate more advanced features. However, it became obvious that even for a simple format, several difficult decisions need to be made. At this point,

Susan Sim was asked to report on the current state of these decisions in a similiar project called GXL,[2] aimed at defining an exchange format for graphs arising in software reengineering and graph transformation. Subsequently, several of these issues were discussed in some depth, for instance, wether a file should contain unique identifiers for nodes (yes) and edges (optionally), or whether a graph should be allowed to have both directed and undirected edges (yes), or parallel edges (yes) or loops (yes).

While opinions on some of these details differ from the path currently taken in GXL, it was agreed that the proposal should be defined in cooperation with the GXL consortium and other relevant groups.

3   Results

The consensus among the participants was that a graph data format should have a layered architecture that conceptually separates the following four aspects (in order of increasing level of detail):

1. structure (vertices, edges, incidences, attributes)
2. topology (ordering of edges, crossings, etc.)
3. shape (bends, feature shapes, etc.)
4. geometry (positions, feature sizes, etc.)
5. rendering (colors, styles, icons, textures, etc.)

The number of layers of the format need not necessarily match this list, but it should be possible to omit higher layers from a graph description. The design should be open to allow formats for applications other than graph drawing to specify other characteristics on top of the lower layers. In particular, the goal is to collaborate with groups working on a data exchange format for software reengineering and graph transformations.[3]

While extensions to incorporate layout algorithms, dynamic graphs, clustering, and so on may not be part of the initial proposal, the need for such extensions should be taken into account.

As a result of the discussion, a working group was formed, and several others have joined since, or expressed interest in contributing. The group will focus on defining the layer structure with extension points. It is understood that a draft proposal will be made available for public review before the 2001 Symposium on Graph Drawing (GD 2001), and that the proposal will be considered for publication in the GD 2001 proceedings. The current status of the project is maintained at http://www.graphdrawing.org/data/format/.

---

[2] Graph Exchange Language, see http://www.gupro.de/GXL/.

[3] See http://www.cs.toronto.edu/~simsuz/wosef/ for more information.

Graph-Drawing Contest Report

Franz Brandenburg[1], Ulrik Brandes[2],
Michael Himsolt[3], and Marcus Raitner[1]

[1] Universität Passau
Passau, Germany
brandenb@fmi.uni-passau.de
[2] Universität Konstanz
Konstanz, Germany
[3] DaimlerChrysler Forschungszentrum Ulm
Ulm, Germany

Abstract. This report describes the Seventh Annual Graph Drawing Contest, held in conjunction with the 2000 Graph Drawing Symposium in Williamsburg, Virginia. The purpose of the contest is to monitor and challenge the current state of the art in graph-drawing technology [3,4, 6,7,5,2].

1 Introduction

Text descriptions of the four categories for the 2000 contest were available via the World Wide Web (WWW) [10]. Eight separate submissions were received, containing 33 different graph drawings, and one live demonstration. Moreover there were four spontaneous mobiles of space clusters. The winners for the Categories were selected by the contest organizers and Joe Marks. Conflicts of interest were avoided on an honor basis. The winning entries are described below.

2 Winning Submissions

2.1 Category A

The graph given for Category A has been generated with DaimlerChrysler's C++ Analyzer. The C++ Analyzer is a tool for static analysis of C and C++ software. It supports visualization of software with graphs, browsing and cross referencing as well as computation of metrics.

This particular graph is a real world example and visualizes the class relationships in the C++ Analyzer itself. The graph consists of a large number of individual components and contains several high degree nodes. The challenge for this particular contest entry was not to visualize any interesting graph properties (there probably aren't any), but to find a nice and comprehensible layout for a rather large graph.

Among the submitted entries the winner Nikola S. Nikolov from the University of Limerick gave the best analysis and display of the underlying structure.

J. Marks (Ed.): GD 2000, LNCS 1984, pp. 410–418, 2001.

His drawings were made with an ILP-based system developed by the Graph Drawing research group at the University of Limerick

The first step was to separate the 26 connected components of the graph. Then each component was drawn separately by using the ILP-based algorithms for layering and crossing minimization. All the drawings have minimum number of dummy nodes for the layering algorithm (subject to the given input parameters) and minimum number of edge crossings for the resulting layering. The final drawings of the five big components were chosen among alternative solutions made with different input parameters to the layering algorithm. It was tried to keep the dimensions of the drawing within reasonable bounds and to achieve as few edge crossings as possible. Manual editing was needed to tune the position of the nodes and to tidy the labels.

In the two biggest components the arrows of the edges were removed to achieve a clearer drawing. All the edges without arrows point downwards.

The nodes of component 2 are colored in three colors: white, yellow (grey in the print), blue (black in the print). Each yellow node is connected by an edge to the blue node with label "SCPos". These edges were removed to achieve a better drawing. All these edges have a yellow source node and the blue node "SCPos" as a destination node. Figure 1 shows this component.

2.2   Category B

In the analysis of social networks, graph drawing is increasingly recognized as a tool to effectively support visual exploration and communication of findings. The graphs given for Category B addressed one of the major challenges in this area, namely the comparison of networks of similar type.

The data for Category B were provided by Maryann M. Durland, an independent business consultant. Within a sizeable corporation, two teams working in similar projects were asked how frequent they would be in contact with other members of their team, with their clients, and with domain experts external to their team. The data thus consists of two directed graphs, each with three types of vertices (team members, clients, external experts) and integer edge weights ranging from 1 to 4 (indicating quarterly, monthly, biweekly, and weekly contact, respectively).

As in many graph drawing contests before, the winning entry was submitted by Vladimir Batagelj and Andrej Mrvar from the University of Ljubljana, Slovenia. Figure 2 shows their circular layouts of the two networks, where team members, clients, and experts are located on the inner, middle, and outer circle, respectively. Essentially, the circular ordering is determined by the mutual overlap of pairwise vertex neighborhoods, namely by applying a heuristic for the Traveling Salesperson Problem to a complete graph with edge weights that quantify neighborhood similarity. Edge weights are depicted by different gray scales.

These images clearly show how both teams divide into a client/expert interface and a group working internally. However, the teams also differ quite visibly in both their internal and external communication. The drawings thus readily enable comparison and interpretation of the way these teams have organized their contacts.

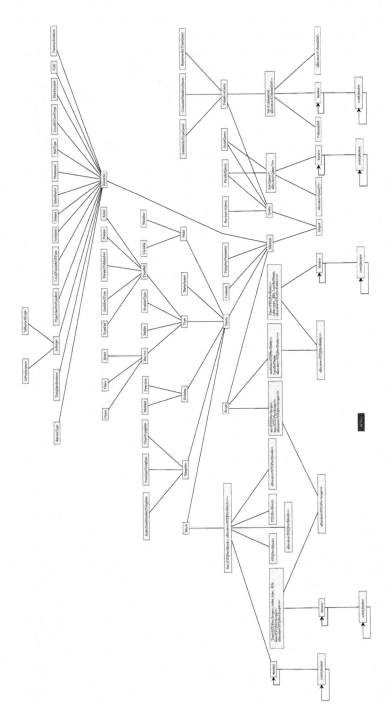

Fig. 1. Winning entry for Category A (original in color)

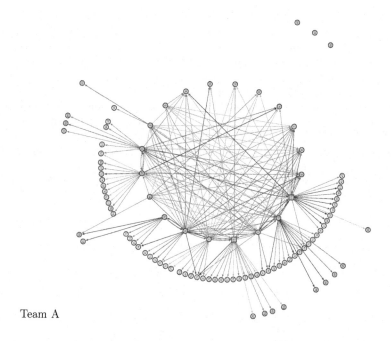

Team A

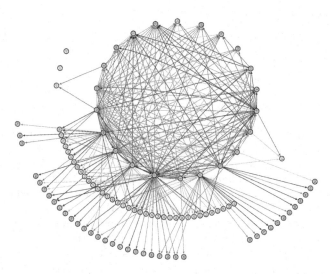

Team B

Fig. 2. Winning entry for Category B (original in color)

2.3   Category C

The data for the Category C graph were provided by John Carlis of the University of Minnesota and John Hanna of US West in Minneapolis. The graph describes a real in-use data-model, although all the labels were changed to Xs to protect the intellectual investment.

A data model is a technology-independent statement of the kinds of data to be remembered by a database management system. For this contest we have simplified the drawing to include just entities, attributes, and binary relationships. Graphically, each entity appears as a box (a rectangle), each attribute appears as text within an entity's box, and each relationship appears as a line between two (not necessarily distinct) entities. The quality of the graph significantly impacts the design process. A poorly drawn graph means an unreadable model.

Unfortunately there were no entries for graph C and it is kept for future competitions.

2.4   Category D

Category D is the free or artistic category. There is no explicit challenge graph but a framework to combine arts and graphs. The entries to category D were divided into three sub-categories

D1. Jan Adamec from Charles University, Prague took a composition of Miró and converted the painting into a graph by putting vertices at intersection points and adding some vertices and edges.

Each of its nine components was drawn separately using a modified spring algorithm. No further manual editing (except of adding colors) was done.

D2. Christian A. Duncan, Pawel Gajer, Michael T. Goodrich and Stephen G. Kobourov submitted a stereo-graphic hologram of the graph corresponding to the Sierpinski pyramid of order 7, see Fig. 4 for a 2-D drawing of their 3-D model.

The Sierpinski pyramid is a classic fractal [11]. Whereas traditionally the image is defined with fixed vertices and edges, they made theirs a fractal graph with no specific embedding. They used a graph embedder to embed the representation of the graph producing interesting and beautiful results.

The Sierpinski pyramid graph is created by a recursive procedure, parametrized on the order of the recursion. As in the 2-D case, at each iteration, every pyramid is divided into five congruent smaller pyramids with the central pyramid removed. In a Sierpinski pyramid of order k the number of vertices is $|V_k| = \frac{4^k}{2} + 2$ and the number of edges is $|E_k| = 6(|V| - 4) + 12 = 3 \times 4k$. in their example the Sierpinski pyramid of order 7 has 8,194 vertices and 49,152 edges.

Given the parameter k the adjacency matrix of a graph which corresponds to the Sierpinski pyramid of order k was generated. The graph is then drawn using the GRIP system [9] without any modification to the resulting drawing. The drawing method of the GRIP system uses a multi-dimensional algorithm

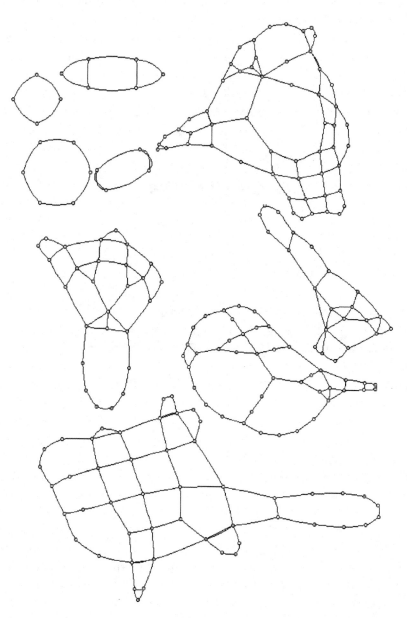

Fig. 3. Jan Adamec's version of Miró's painting (original in color)

for drawing large graphs [8]. The algorithm takes advantage of the symmetric nature of the Sierpinski pyramid to quickly produce a final drawing. Altogether, the generation of the graph, the preprocessing, and final 3-D drawing took 22 seconds on a 550Mhz Pentium II with 128MB of RAM.

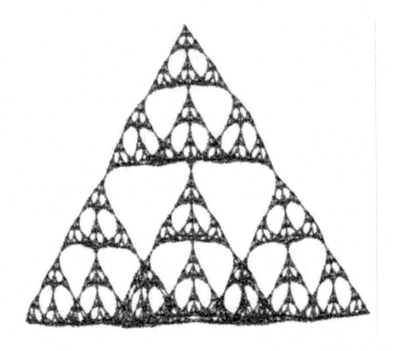

Fig. 4. 2-D drawing of the 3-D hologram of the Sierpinski pyramid of order 7.

The hologram of the drawing was created using the stereo-graphic image method [1]. The hologram captures the 3-D nature of the drawing to produce an effect similar to that of rotation of the original 3D model in OpenGL. The hologram itself was presented at the conference and is exhibited at the University of Arizona.

D3. was Joe Marks' surprise for the participants of the symposium. They found a space cluster in their GD'2000 bag. A space cluster consists of 72 elastic rods and 24 plastic intersections. The participants were encouraged to construct three dimensional graphs. There were sharp time restrictions: three breaks and one night session. Four teams submitted their assemblies, which were aesthetically nice 3D graphs.

The winner is "The Rose" by Robby Schönfeld from the University of Halle, Germany and Nikola S. Nikolov, from the University of Limerick, Ireland. Figure 5 shows the photo. This is the largest finite graph which is both maximal planar and regular.

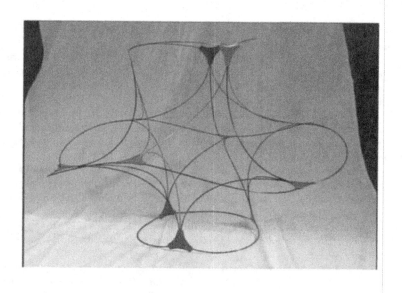

Fig. 5. The "Rose"

3   Observations and Conclusions

The organization of this year's graph drawing competition had been shifted from its initiator Joe Marks to Franz Brandenburg. Confronted with the three tasks of collecting challenge graphs, raising money from sponsors and finally ranking the entries, the first one was the hardest.

As in the past years, the graph drawing competition had a scientific and a fun category. Several times the posted challenge graphs have served to initiate research or give a push in a particular direction. Winning entries are referenced in several papers. This year, the data were taken from real life problems. Using well-founded approaches the winners made the underlying meaning of the data visible.

Unfortunately, the number of contributions to the graph drawing competitions has reached a low point. We'll try to attract more members inside and outside the graph drawing community to contribute to the graph drawing competition.

Acknowledgements. Sponsorship for this contest was provided by AT&T Research, Florham Park, Daimler Chrysler Forschungszentrum, Ulm, and Tom Sawyer Software, Berkeley.

References

1. S. A. Benton. Survey of display hologram types. In *XXSPIE Industrial and Commercial Applications ofHolography*, 1983.
2. F. J. Brandenburg, M. Jünger, J. Marks, P. Mutzel, and F. Schreiber. Graph-drawing contest report. In J. Kratochvìl, editor, *Lecture Notes in Computer Science: 1731 (Proceedings of the Symposium on Graph Drawing GD '99)*, pages 400–409, Berlin, September 1999. Springer.
3. P. Eades and J. Marks. Graph-drawing contest report. In R. Tamassia and I. G. Tollis, editors, *Lecture Notes in Computer Science: 894 (Proceedings of the DIMACS International Workshop on Graph Drawing '94)*, pages 143–146, Berlin, October 1994. Springer.
4. P. Eades and J. Marks. Graph-drawing contest report. In F. J. Brandenburg, editor, *Lecture Notes in Computer Science: 1027 (Proceedings of the Symposium on Graph Drawing GD '95)*, pages 224–233, Berlin, September 1995. Springer.
5. P. Eades, J. Marks, P. Mutzel, and S. North. Graph-drawing contest report. In S. H. Whitesides, editor, *Lecture Notes in Computer Science: 1547 (Proceedings of the Symposium on Graph Drawing GD '98)*, pages 423–435, Berlin, August 1998. Springer.
6. P. Eades, J. Marks, and S. North. Graph-drawing contest report. In S. North, editor, *Lecture Notes in Computer Science: 1190 (Proceedings of the Symposium on Graph Drawing GD '96)*, pages 129–138, Berlin, September 1996. Springer.
7. P. Eades, J. Marks, and S. North. Graph-drawing contest report. In G. DiBattista, editor, *Lecture Notes in Computer Science: 1353 (Proceedings of the Symposium on Graph Drawing GD '97)*, pages 438–445, Berlin, September 1997. Springer.
8. P. Gajer, M. T. Goodrich, and S. G. Kobourov. A fast multi-dimensional algorithm for drawing large graphs. In *To appear in Proceedings of the 8th Symposium on Graph Drawing*, 2000.
9. P. Gajer and S. G. Kobourov. Grip: Graph drawing with intelligent placement. In *To appear in Proceedings of the 8th Symposium on Graph Drawing*, 2000.
10. http://www.infosun.fmi.uni-passau.de/GD2000/.
11. B. B. Mandelbrot. *The Fractal Geometry of Nature*. W.H. Freeman and Co, New York, rev. 1983.

Author Index

Lecture Notes in Computer Science

For information about Vols. 1–1904
please contact your bookseller or Springer-Verlag

Vol. 1905: H. Scholten, M.J. van Sinderen (Eds.), Interactive Distributed Multimedia Systems and Telecommunication Services. Proceedings, 2000. XI, 306 pages. 2000.

Vol. 1906: A. Porto, G.-C. Roman (Eds.), Coordination Languages and Models. Proceedings, 2000. IX, 353 pages. 2000.

Vol. 1907: H. Debar, L. Mé, S.F. Wu (Eds.), Recent Advances in Intrusion Detection. Proceedings, 2000. X, 227 pages. 2000.

Vol. 1908: J. Dongarra, P. Kacsuk, N. Podhorszki (Eds.), Recent Advances in Parallel Virtual Machine and Message Passing Interface. Proceedings, 2000. XV, 364 pages. 2000.

Vol. 1909: T. Yakhno (Ed.), Advances in Information Systems. Proceedings, 2000. XVI, 460 pages. 2000.

Vol. 1910: D.A. Zighed, J. Komorowski, J. Żytkow (Eds.), Principles of Data Mining and Knowledge Discovery. Proceedings, 2000. XV, 701 pages. 2000. (Subseries LNAI).

Vol. 1911: D.G. Feitelson, L. Rudolph (Eds.), Job Scheduling Strategies for Parallel Processing. VII, 209 pages. 2000.

Vol. 1912: Y. Gurevich, P.W. Kutter, M. Odersky, L. Thiele (Eds.), Abstract State Machines. Proceedings, 2000. X, 381 pages. 2000.

Vol. 1913: K. Jansen, S. Khuller (Eds.), Approximation Algorithms for Combinatorial Optimization. Proceedings, 2000. IX, 275 pages. 2000.

Vol. 1914: M. Herlihy (Ed.), Distributed Computing. Proceedings, 2000. VIII, 389 pages. 2000.

Vol. 1915: S. Dwarkadas (Ed.), Languages, Compilers, and Run-Time Systems for Scalable Computers. Proceedings, 2000. VIII, 301 pages. 2000.

Vol. 1916: F. Dignum, M. Greaves (Eds.), Issues in Agent Communication. X, 351 pages. 2000. (Subseries LNAI).

Vol. 1917: M. Schoenauer, K. Deb, G. Rudolph, X. Yao, E. Lutton, J.J. Merelo, H.-P. Schwefel (Eds.), Parallel Problem Solving from Nature – PPSN VI. Proceedings, 2000. XXI, 914 pages. 2000.

Vol. 1918: D. Soudris, P. Pirsch, E. Barke (Eds.), Integrated Circuit Design. Proceedings, 2000. XII, 338 pages. 2000.

Vol. 1919: M. Ojeda-Aciego, I.P. de Guzman, G. Brewka, L. Moniz Pereira (Eds.), Logics in Artificial Intelligence. Proceedings, 2000. XI, 407 pages. 2000. (Subseries LNAI).

Vol. 1920: A.H.F. Laender, S.W. Liddle, V.C. Storey (Eds.), Conceptual Modeling – ER 2000. Proceedings, 2000. XV, 588 pages. 2000.

Vol. 1921: S.W. Liddle, H.C. Mayr, B. Thalheim (Eds.), Conceptual Modeling for E-Business and the Web. Proceedings, 2000. X, 179 pages. 2000.

Vol. 1922: J. Crowcroft, J. Roberts, M.I. Smirnov (Eds.), Quality of Future Internet Services. Proceedings, 2000. XI, 368 pages. 2000.

Vol. 1923: J. Borbinha, T. Baker (Eds.), Research and Advanced Technology for Digital Libraries. Proceedings, 2000. XVII, 513 pages. 2000.

Vol. 1924: W. Taha (Ed.), Semantics, Applications, and Implementation of Program Generation. Proceedings, 2000. VIII, 231 pages. 2000.

Vol. 1925: J. Cussens, S. Džeroski (Eds.), Learning Language in Logic. X, 301 pages 2000. (Subseries LNAI).

Vol. 1926: M. Joseph (Ed.), Formal Techniques in Real-Time and Fault-Tolerant Systems. Proceedings, 2000. X, 305 pages. 2000.

Vol. 1927: P. Thomas, H.W. Gellersen, (Eds.), Handheld and Ubiquitous Computing. Proceedings, 2000. X, 249 pages. 2000.

Vol. 1928: U. Brandes, D. Wagner (Eds.), Graph-Theoretic Concepts in Computer Science. Proceedings, 2000. X, 315 pages. 2000.

Vol. 1929: R. Laurini (Ed.), Advances in Visual Information Systems. Proceedings, 2000. XII, 542 pages. 2000.

Vol. 1931: E. Horlait (Ed.), Mobile Agents for Telecommunication Applications. Proceedings, 2000. IX, 271 pages. 2000.

Vol. 1658: J. Baumann, Mobile Agents: Control Algorithms. XIX, 161 pages. 2000.

Vol. 1756: G. Ruhe, F. Bomarius (Eds.), Learning Software Organization. Proceedings, 1999. VIII, 226 pages. 2000.

Vol. 1766: M. Jazayeri, R.G.K. Loos, D.R. Musser (Eds.), Generic Programming. Proceedings, 1998. X, 269 pages. 2000.

Vol. 1791: D. Fensel, Problem-Solving Methods. XII, 153 pages. 2000. (Subseries LNAI).

Vol. 1799: K. Czarnecki, U.W. Eisenecker, Generative and Component-Based Software Engineering. Proceedings, 1999. VIII, 225 pages. 2000.

Vol. 1812: J. Wyatt, J. Demiris (Eds.), Advances in Robot Learning. Proceedings, 1999. VII, 165 pages. 2000. (Subseries LNAI).

Vol. 1932: Z.W. Raś, S. Ohsuga (Eds.), Foundations of Intelligent Systems. Proceedings, 2000. XII, 646 pages. (Subseries LNAI).

Vol. 1933: R.W. Brause, E. Hanisch (Eds.), Medical Data Analysis. Proceedings, 2000. XI, 316 pages. 2000.

Vol. 1934: J.S. White (Ed.), Envisioning Machine Translation in the Information Future. Proceedings, 2000. XV, 254 pages. 2000. (Subseries LNAI).

Vol. 1935: S.L. Delp, A.M. DiGioia, B. Jaramaz (Eds.), Medical Image Computing and Computer-Assisted Intervention – MICCAI 2000. Proceedings, 2000. XXV, 1250 pages. 2000.

Vol. 1937: R. Dieng, O. Corby (Eds.), Knowledge Engineering and Knowledge Management. Proceedings, 2000. XIII, 457 pages. 2000. (Subseries LNAI).

Vol. 1938: S. Rao, K.I. Sletta (Eds.), Next Generation Networks. Proceedings, 2000. XI, 392 pages. 2000.

Vol. 1939: A. Evans, S. Kent, B. Selic (Eds.), «UML» – The Unified Modeling Language. Proceedings, 2000. XIV, 572 pages. 2000.

Vol. 1940: M. Valero, K. Joe, M. Kitsuregawa, H. Tanaka (Eds.), High Performance Computing. Proceedings, 2000. XV, 595 pages. 2000.

Vol. 1941: A.K. Chhabra, D. Dori (Eds.), Graphics Recognition. Proceedings, 1999. XI, 346 pages. 2000.

Vol. 1942: H. Yasuda (Ed.), Active Networks. Proceedings, 2000. XI, 424 pages. 2000.

Vol. 1943: F. Koornneef, M. van der Meulen (Eds.), Computer Safety, Reliability and Security. Proceedings, 2000. X, 432 pages. 2000.

Vol. 1945: W. Grieskamp, T. Santen, B. Stoddart (Eds.), Integrated Formal Methods. Proceedings, 2000. X, 441 pages. 2000.

Vol. 1948: T. Tan, Y. Shi, W. Gao (Eds.), Advances in Multimodal Interfaces – ICMI 2000. Proceedings, 2000. XVI, 678 pages. 2000.

Vol. 1949: R. Connor, A. Mendelzon (Eds.), Research Issues in Structured and Semistructured Database Programming. Proceedings, 1999. XII, 325 pages. 2000.

Vol. 1950: D. van Melkebeek, Randomness and Completeness in Computational Complexity. XV, 196 pages. 2000.

Vol. 1951: F. van der Linden (Ed.), Software Architectures for Product Families. Proceedings, 2000. VIII, 255 pages. 2000.

Vol. 1952: M.C. Monard, J. Simão Sichman (Eds.), Advances in Artificial Intelligence. Proceedings, 2000. XV, 498 pages. 2000. (Subseries LNAI).

Vol. 1953: G. Borgefors, I. Nyström, G. Sanniti di Baja (Eds.), Discrete Geometry for Computer Imagery. Proceedings, 2000. XI, 544 pages. 2000.

Vol. 1954: W.A. Hunt, Jr., S.D. Johnson (Eds.), Formal Methods in Computer-Aided Design. Proceedings, 2000. XI, 539 pages. 2000.

Vol. 1955: M. Parigot, A. Voronkov (Eds.), Logic for Programming and Automated Reasoning. Proceedings, 2000. XIII, 487 pages. 2000. (Subseries LNAI).

Vol. 1956: T. Coquand, P. Dybjer, B. Nordström, J. Smith (Eds.), Types for Proofs and Programs. Proceedings, 1999. VII, 195 pages. 2000.

Vol. 1960: A. Ambler, S.B. Calo, G. Kar (Eds.), Services Management in Intelligent Networks. Proceedings, 2000. X, 259 pages. 2000.

Vol. 1961: J. He, M. Sato (Eds.), Advances in Computing Science – ASIAN 2000. Proceedings, 2000. X, 299 pages. 2000.

Vol. 1963: V. Hlaváč, K.G. Jeffery, J. Wiedermann (Eds.), SOFSEM 2000: Theory and Practice of Informatics. Proceedings, 2000. XI, 460 pages. 2000.

Vol. 1964: J. Malenfant, S. Moisan, A. Moreira (Eds.), Object-Oriented Technology. Proceedings, 2000. XI, 309 pages. 2000.

Vol. 1965: Ç. K. Koç, C. Paar (Eds.), Cryptographic Hardware and Embedded Systems – CHES 2000. Proceedings, 2000. XI, 355 pages. 2000.

Vol. 1966: S. Bhalla (Ed.), Databases in Networked Information Systems. Proceedings, 2000. VIII, 247 pages. 2000.

Vol. 1967: S. Arikawa, S. Morishita (Eds.), Discovery Science. Proceedings, 2000. XII, 332 pages. 2000. (Subseries LNAI).

Vol. 1968: H. Arimura, S. Jain, A. Sharma (Eds.), Algorithmic Learning Theory. Proceedings, 2000. XI, 335 pages. 2000. (Subseries LNAI).

Vol. 1969: D.T. Lee, S.-H. Teng (Eds.), Algorithms and Computation. Proceedings, 2000. XIV, 578 pages. 2000.

Vol. 1970: M. Valero, V.K. Prasanna, S. Vajapeyam (Eds.), High Performance Computing – HiPC 2000. Proceedings, 2000. XVIII, 568 pages. 2000.

Vol. 1971: R. Buyya, M. Baker (Eds.), Grid Computing – GRID 2000. Proceedings, 2000. XIV, 229 pages. 2000.

Vol. 1972: A. Omicini, R. Tolksdorf, F. Zambonelli (Eds.), Engineering Societies in the Agents World. Proceedings, 2000. IX, 143 pages. 2000. (Subseries LNAI).

Vol. 1973: J. Van den Bussche, V. Vianu (Eds.), Database Theory – ICDT 2001. Proceedings, 2001. X, 451 pages. 2001.

Vol. 1974: S. Kapoor, S. Prasad (Eds.), FST TCS 2000: Foundations of Software Technology and Theoretical Computer Science. Proceedings, 2000. XIII, 532 pages. 2000.

Vol. 1975: J. Pieprzyk, E. Okamoto, J. Seberry (Eds.), Information Security. Proceedings, 2000. X, 323 pages. 2000.

Vol. 1976: T. Okamoto (Ed.), Advances in Cryptology – ASIACRYPT 2000. Proceedings, 2000. XII, 630 pages. 2000.

Vol. 1977: B. Roy, E. Okamoto (Eds.), Progress in Cryptology – INDOCRYPT 2000. Proceedings, 2000. X, 295 pages. 2000.

Vol. 1979: S. Moss, P. Davidsson (Eds.), Multi-Agent-Based Simulation. Proceedings, 2000. VIII, 267 pages. 2001. (Subseries LNAI).

Vol. 1983: K.S. Leung, L.-W. Chan, H. Meng (Eds.), Intelligent Data Engineering and Automated Learning – IDEAL 2000. Proceedings, 2000. XVI, 573 pages. 2000.

Vol. 1984: J. Marks (Ed.), Graph Drawing. Proceedings, 2001. XII, 419 pages. 2001.

Vol. 1987: K.-L. Tan, M.J. Franklin, J. C.-S. Lui (Eds.), Mobile Data Management. Proceedings, 2001. XIII, 289 pages. 2001.

Vol. 1989: M. Ajmone Marsan, A. Bianco (Eds.), Quality of Service in Multiservice IP Networks. Proceedings, 2001. XII, 440 pages. 2001.